Laser Control of Atoms and Molecules

*In bright memory of my mother,
Anna Vasil'evna Letokhova*

Laser Control of Atoms and Molecules

Vladilen Letokhov
Institute of Spectroscopy
Russian Academy of Sciences

OXFORD
UNIVERSITY PRESS

Great Clarendon Street, Oxford OX2 6DP

Oxford University Press is a department of the University of Oxford.
It furthers the University's objective of excellence in research, scholarship,
and education by publishing worldwide in

Oxford New York

Auckland Cape Town Dar es Salaam Hong Kong Karachi
Kuala Lumpur Madrid Melbourne Mexico City Nairobi
New Delhi Shanghai Taipei Toronto

With offices in

Argentina Austria Brazil Chile Czech Republic France Greece
Guatemala Hungary Italy Japan Poland Portugal Singapore
South Korea Switzerland Thailand Turkey Ukraine Vietnam

Oxford is a registered trade mark of Oxford University Press
in the UK and in certain other countries

Published in the United States
by Oxford University Press Inc., New York

© Oxford University Press 2007

The moral rights of the author have been asserted
Database right Oxford University Press (maker)

First published 2007

All rights reserved. No part of this publication may be reproduced,
stored in a retrieval system, or transmitted, in any form or by any means,
without the prior permission in writing of Oxford University Press,
or as expressly permitted by law, or under terms agreed with the appropriate
reprographics rights organization. Enquiries concerning reproduction
outside the scope of the above should be sent to the Rights Department,
Oxford University Press, at the address above

You must not circulate this book in any other binding or cover
and you must impose the same condition on any acquirer

British Library Cataloguing in Publication Data

Data available

Library of Congress Cataloging in Publication Data

Letokhov, V. S.
  Laser control of atoms and molecules / Vladilen Letokhov.
    p. cm.
  ISBN-13: 978-0-19-852816-6 (alk. paper)
  1. Quantum optics.   2. Laser cooling.   3. Laser beams.
    4. Laser spectroscopy.   I. Title.
  QC446.2.L48 2007
  535.8'4—dc22   2006032598

Typeset by Newgen Imaging Systems (P) Ltd., Chennai, India
Printed in Great Britain
on acid-free paper by Biddles Ltd., www.biddles.co.uk

ISBN 978-0-19-852816-6

10 9 8 7 6 5 4 3 2 1

Preface

This book has a very personal shade to me. I started my research activities as a diploma student, first at the Optics Laboratory (1960) and then at the Radiophysics Laboratory (1963) of the P. N. Lebedev Physics Institute of the USSR Academy of Sciences, under the direction of the founder of that laboratory, Professor Nikolai G. Basov. I was more than highly impressed by the possibility of controlling the emission of light by a substance in a nonequilibrium state by stimulated emission of radiation in a laser. It was almost magic when Professor Basov would set about discussing any laser ideas, starting with the very fundamentals—the number of photons in a single quantum state of a system. Although I was not a novice in optics (during my student years I had already written some published papers on the statistical processing of optical images and made a number of inventions in the field of aerospace optics that were successfully realized), the capabilities of coherent light seemed fantastic to me. Whatever problem one might touch upon, everywhere there was scope for new suggestions and ideas. Against this background, the important problems associated with the development of ultrahigh-power lasers that were central to the P. N. Lebedev Physics Institute in the sixties and seventies of the past century seemed premature to me, and I started to leave them, dreaming of ways to use laser light at the atomic–molecular level. I have wholly devoted myself to these problems since 1970, when I joined the newly established Institute of Spectroscopy in Troitsk, near Moscow, and began to form a team of like-minded persons, mainly students, graduates, and postgraduates of the Moscow Physical-Technical Institute. They believed in my ideas, and I recall that critical experiments were being conducted day and night. It is this team of like-minded persons that I am largely indebted to for the advances made in the development of this avenue of scientific exploration, despite the rather primitive experimental resources at their disposal, much inferior to those of Western laboratories.

Since then I have devoted almost 40 years of my life to seemingly very different problems, but to me almost all of them can be lumped together under the general term "laser control of atoms and molecules." This implies the laser selection of atomic/molecular velocities for the purposes of Doppler-free spectroscopy, laser control of the position and velocity of atoms, that is, the laser trapping and cooling of atoms, and laser control of atomic/molecular processes (ionization and dissociation) with a view to detecting single atoms and molecules and, particularly, separating isotopes and nuclear isomers. Over the course of the past years the principal problems posed have been successfully solved, and many of them have evolved remarkably in subsequent investigations by the international community. For example, the solution of the problem of the laser cooling and trapping of atoms has given birth to a new field in the physics of ultracold matter, namely, quantum atomic and molecular gases. Laser noncoherent control of unimolecular processes has found an interesting extension in the field of laser coherent control of molecules. The concept of laser control of position has been successfully demonstrated with microparticles (optical tweezers), concurrently with investigations into atomic control. The laser photoionization of molecules on surfaces has led to the development of novel techniques of laser-assisted mass spectrometry of macromolecules, and so on.

At some time I felt the need to give readers my view of the concept as a whole, even though many monographs, including those written with my participation, have already been devoted to its various parts. Naturally, in a single and not very large book, each of the problems can only be treated rather fragmentarily, without many important details. This is an unavoidable shortcoming of the book, but at the same time, it is its potential advantage for those readers who wish to take a look at a problem as a whole first and then turn their attention to the details.

Any book, especially one written in the wake of fresh events by one of their participants, who is bound to be subjective, has a short lifetime. Quoting Freeman Dyson, my idol of wide and independent outlook on the surrounding world, a book that comes out at an instant t, containing the entire information acquired for a time $(t-T)$, gets out of date at an instant $(t+T)$. This standpoint seems optimistic to me, for it fails to take into account the exponential rate of development of science in any given area and the subsequent rapid change of the line of inquiry on a timescale τ_0. If $T \gg \tau_0$, the book will get out of date even faster, so that soon it might become hardly of interest to anybody. And considering the present-day development of e-information, books of a synthetic character can be of interest as an itinerary of sorts for the subsequent use of a vast amount of e-information. All this looks as if I'm defending myself for my efforts and time spent in writing this book. But having once read the book *Summing up* by Somerset Maugham, I think that my efforts have not been useless, at least to me, personally.

This book would probably have never been published but for the encouragement of my wife, Professor Tiina Karu, who regularly turned me back to it when I had lost all interest in the face of the mounting torrent of very interesting investigations that I could not wholly take part in for historical reasons since 1990.

The idea of the book was conceived in 2002 during one of my regular visits to one of my favorite places in the USA—the Optical Science Center of the University of Arizona in Tucson—by invitation of my friend, Professor Pierre Meystre, to whom I'm very grateful.

My kind friend Professor Colin Webb from Oxford University treated the idea of the book with great interest and convinced Oxford University Press of the feasibility of this project. Were it not for the encouragement and patience of Dr. Sönke Adlung from OUP, this project would not have been completed.

I would like to thank Mrs. Ol'ga Tatyanchenko for her highly qualified and fast processing of my manuscript, Mrs. Alla Makarova and Mrs. Elena Nikolaeva, librarians from the Institute of Spectroscopy, for their invariable assistance in my search for the necessary papers and books, Mr. Sergei Kittell for rendering this manuscript into English, and Mr. Alexander Kornushkin for graphic work.

Finally, I would like to thank Dr. Sergei Aseyev, Professor Victor Balykin, Professor Alexander Makarov, Professor Vladimir Minogin, and Professor Evgeny Ryabov, my former graduate students, for their help in preparing and editing many of the chapters of this book.

Troitsk, September, 2006.

Contents

1	**Introduction**	1
	1.1 Ways to control the emission of light	1
	1.2 From the control of light to the control of atoms and molecules	7
	1.3 On the aims of this book	10
2	**Elementary radiative processes**	12
	2.1 Spontaneous emission	12
	2.2 Stimulated absorption and emission	15
	2.3 Recoil effect and Doppler effect	18
	2.4 Resonant excitation of a two-level system free from relaxation	22
	2.5 Resonant excitation of a two-level system with relaxations	26
	2.6 Radiation-scattering processes	33
3	**Laser velocity-selective excitation**	35
	3.1 Doppler broadening of optical spectral lines	35
	3.2 Homogeneous broadening mechanisms	38
	3.3 Doppler-free saturation spectroscopy	40
	3.4 Ultrahigh spectral resolution	49
4	**Optical orientation of atoms and nuclei**	54
	4.1 Optical orientation of atoms	54
	4.2 Radio-frequency spectroscopy of optically oriented atoms	58
	4.3 Spin-exchange optical pumping	61
	4.4 Coherent effects and optically oriented atoms	62
	4.5 Applications of optically pumped atoms	64
5	**Laser cooling of atoms**	68
	5.1 Introduction. History of ideas	69
	5.2 Laser radiation force on a two-level atom	72
	5.3 Quantum fluctuation effects. Temperature limits of laser cooling	76
	5.4 Doppler cooling	77
	5.5 Laser polarization gradient cooling below the Doppler limit	83
	5.6 Cooling below the recoil limit	87

6	**Laser trapping of atoms**	92
	6.1 Optical trapping	92
	6.2 Magnetic trapping	100
	6.3 Magnetooptical trapping	103
	6.4 Gravitooptical and near-field traps	106
	6.5 Optical trapping of cold atoms—new tools for atomic physics	109
7	**Atom optics**	113
	7.1 Introduction. Matter waves	113
	7.2 Reflection of atoms by light	114
	7.3 Laser focusing of an atomic beam	120
	7.4 Diffraction of atoms	127
	7.5 Atom interferometry	130
	7.6 Atomic holography	135
	7.7 Towards atom nanooptics	135
8	**From laser-cooled and trapped atoms to atomic and molecular quantum gases**	138
	8.1 Introduction	139
	8.2 Bose–Einstein condensation of atomic gases	141
	8.3 Fermi-degenerate quantum atomic gases	148
	8.4 Formation of ultracold molecules	150
	8.5 Molecular quantum gases	155
9	**Laser photoselective ionization of atoms**	158
	9.1 Introduction	158
	9.2 Resonance excitation and ionization of atoms	159
	9.3 Photoionization detection of rare atoms and radioactive isotopes	168
	9.4 Laser photoionization separation of isotopes, isobars, and nuclear isomers	175
10	**Multiphoton ionization of molecules**	182
	10.1 Photoselective resonance ionization of molecules	183
	10.2 Resonance-enhanced multiphoton ionization (REMPI) of molecules	185
	10.3 Laser desorption/ionization of biomolecules	189
11	**Photoselective laser control of molecules via molecular vibrations**	198
	11.1 Vibrationally mediated photodissociation of molecules via excited electronic states	199
	11.2 Basics of IR multiple-photon excitation/dissociation of polyatomic molecules in the ground state	201
	11.3 Characteristics of the IR MPE/D of polyatomic molecules	208
	11.4 Intermolecular selectivity of IR MPE/D for laser isotope separation	218
	11.5 Prospects for mode-selective MPE/D by IR femtosecond pulses	221

12	**Coherent laser control of molecules**	224
	12.1 Introduction to coherent optimal control	225
	12.2 Coherent control using wave packets	226
	12.3 Coherent control using quantum interference	229
	12.4 Optimal feedback control	230
	12.5 Coherent optimal control by tailored strong-field laser pulses	232
	12.6 Coherent control of large molecules in liquids	234
	12.7 Perspectives	235
13	**Related topics: laser control of microparticles and free electrons**	238
	13.1 Laser trapping of microparticles	238
	13.2 Laser control of free-electron motion	244
14	**Concluding comments**	251
	References	273
	Index	303

1
Introduction

1.1 Ways to control the emission of light

Light controls the key processes in nature. Light from the Sun provides our planet with energy and thus controls the evolution of its biosphere. The periodic absence of light stimulated the development of a wide variety of artificial light sources, starting with the candle, proceeding through the present-day electric light bulb, and ending with light-emitting diodes. Light from these sources originates from the heating of matter, the elementary act of light birth being the emission of a photon by an excited particle.

In 1900, Max Planck introduced into physics the concept of a quantum—an indivisible portion of energy that can be absorbed or given up in the process of emission. In 1905, Albert Einstein introduced the notion of the quantum of light, or photon, as a real particle of the electromagnetic field. Thereafter, Niels Bohr analyzed the photon emission process and arrived at the following fundamental conclusion: an atom is characterized by a set of energy levels, i.e. certain values of its total energy E_1, E_2, E_3, and so on. Each of these levels corresponds to a stationary atomic state in which the atom does not emit; emission only takes place when the atom jumps from one state into another; the emission frequency is related to the difference in energy between the initial and final energy levels by $\omega = (E_2 - E_1)/\hbar$, where $\hbar = h/2\pi$ is the reduced Planck's constant.

The next important step forward was taken by Einstein in 1916 in his famous work entitled "The emission and absorption of radiation by the quantum theory" (Einstein 1916), wherein he introduced two types of quantum transition of an atom or molecule between discrete quantum states, namely:

1. A *spontaneous transition* of an excited atom to a state with a lower energy, accompanied by the emission of a quantum of light. As Einstein put it, "This transition takes place without any external influence. One can hardly imagine it to be like anything else but radioactive decay." Spontaneous emission from an atom in free space is an uncontrollable process, and it is only now, when Dirac's laws of quantum electrodynamics have made it possible to describe the processes of light absorption and emission in more rigorous terms, that a search has been launched for methods to control the process of spontaneous emission of light, based on a radical modification of the surroundings of the emitting particle by replacing the spatial modes of free space with modes in organized microstructures.

2. An *induced (stimulated) transition* of an atom between energy states. This type of transition is due to radiation acting on the atom, and its probability is proportional

to the radiation intensity. If an atom that is induced to make a transition is in an excited state, it then gives an additional light quantum to the radiation, and, vice versa, if it is not initially excited, the atom then absorbs a similar quantum from the radiation.

Einstein introduced the concept of spontaneous and stimulated transitions on the basis of his astonishing intuition: the only proof was the fact that these notions helped him to easily derive Planck's formula for the thermal radiation spectrum. In 1925–1926, *quantum mechanics* was developed by the efforts of a brilliant international community of physicists, who described the quantum interaction between light, and atoms and molecules. Dirac constructed the *quantum radiation theory* using the Einsteinian idea of light quanta. The Dirac radiation theory gave a rigorous proof of Einstein's hypothesis about the spontaneous and stimulated emission of radiation, and predicted the main properties of these phenomena. A photon whose emission was induced by another photon was demonstrated to have absolutely the same characteristics as the initial photon, that is, the same propagation direction, the same energy or radiation frequency, and the same polarization. It is precisely on this *identity of photons* of stimulated radiation that the whole of quantum electronics is based, including lasers—coherent light sources. And the term "coherence" itself means that the elementary quantum emitters—atoms and molecules—emit quanta with identical characteristics.

But the phenomenon of stimulated emission of radiation cannot by itself help one either to control the emission process or, more specifically, to obtain powerful light emission. Indeed, as postulated by Einstein, there exist two types of stimulated transition: one accompanied by the emission of photons and another involving the absorption of photons. Under ordinary, equilibrium conditions, the number of atoms residing at any excited energy level is always smaller than that at any lower energy level. For that reason, the number of stimulated transitions attended by the absorption of photons always prevails in matter over the number of such transitions involving the emission of photons. This manifests itself in an obvious fashion: all bodies around us absorb radiation at least to some extent. By 1930, ideas were conceived that this situation might change cardinally if a *nonequilibrium distribution* of atoms among the excited states was produced in some way, so that the number of atoms in one of the excited states was greater than that in a lower quantum state (see the chronology of ideas in Lamb 1975). Such a nonequilibrium distribution of atoms among the available energy levels is now referred to as a *population inversion* of the levels.

The idea of the possibility of *light amplification* by the stimulated emission of radiation in a medium with an inverted population had opened a little the way to the production of flows of identical photons, but subject to one condition: a flow of equally identical photons had to be fed to the input of the amplifying medium. Otherwise, all the imperfections of the light beam at the input of the amplifying medium would be repeated in an amplified form at its output. All common light sources emit chaotic radiation, from which it is impossible to isolate a flow of identical photons of any intensity for their subsequent amplification. It took one more decisive idea, namely, to place the amplifying medium in a *resonant cavity* wherein the photon loss factor was lower than the gain factor, that is, to establish a *positive resonance feedback*. One then could start the amplification process with a single photon or a few spontaneous photons that would give birth, while repeatedly passing through the amplifying medium, to a great many photons just like themselves. This chain process of undamped multiplication of identical photons is usually called *generation*, and such a device is known as a *maser* in the case of microwaves, and an *optical quantum generator* or *laser* in the case of light.

The idea of the coherent light quantum generator came to optics from radio. The discovery of the laser occurred through a synthesis of key ideas of optics and radio. Figure 1.1 illustrates the sequence of events in the development of concepts, ideas, and experiments that led to the advent of quantum electronics. One can clearly see the potential possibility, not implemented in reality, of discovering the laser on the way to the development of the ideas of light amplification in a medium with an inverted population. The experimental verification by Heinrich Hertz of the existence of the electromagnetic waves predicted by James Maxwell marked the beginning of the intense mastering of the radio-frequency band. The development of radio engineering followed, raising the power output of radio generators, improving the sensitivity of radio receivers, and gradually shortening the wavelength of which radio waves could be exploited. The development of radar led to the mastering of the microwave band.

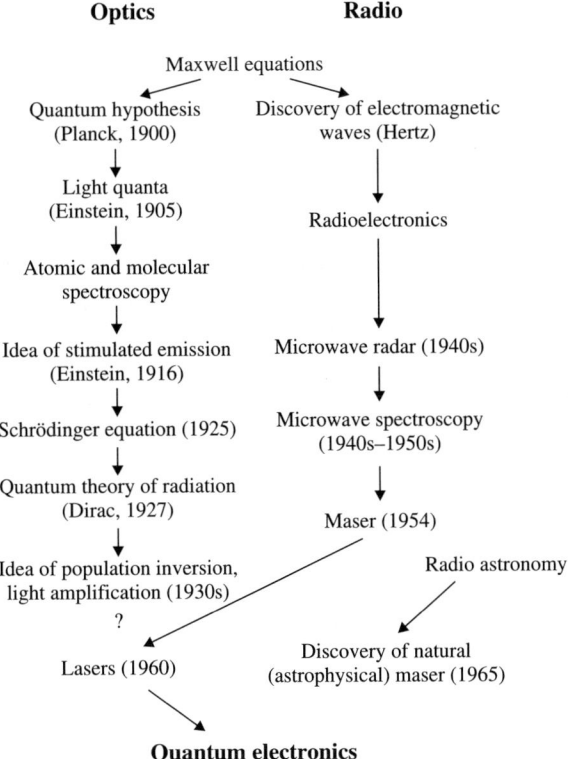

Fig. 1.1 Sequence of discoveries in optics and radio which led to the advent of quantum electronics and the control of the stimulated emission of light.

Microwave oscillators allowed the start of systematic research on the absorption spectra of molecules in the microwave band, and *microwave spectroscopy* made its appearance. It was perhaps here that the first synthesis took place of the concepts of quantum mechanics, which clearly explained microwave molecular spectra, and the concepts of experimental radio engineering.

Microwave spectroscopic studies were similar to investigations into the optical spectra of atoms and molecules, but the researchers now had at their disposal radiation sources of a much higher spectral brightness than in optics. Since spontaneous transitions in the microwave band have a very low probability, a major part is played by stimulated transitions in the interaction between radio waves and molecules. Along with the upward (absorbing) stimulated transitions, downward (radiative) stimulated transitions in excited molecules were also observed. This was the second fruitful synthesis of the predictions of quantum mechanics and the experimental capabilities of radio engineering. Using a nonuniform electric field in experiments with molecular beams, one could sort molecules in different quantum states. On the basis of such experiments, C. H. Townes and coworkers in the USA, and N. G. Basov and A. M. Prokhorov in the USSR suggested and implemented the idea of the *molecular generator*, or *maser* (Gordon *et al.* 1954; Basov and Prokhorov 1954). This new, in principle, type of electromagnetic generator relied for its operation on the preparation of "active" (excited) molecules, and the stimulated emission of energy by them in a resonant cavity sustaining undamped electromagnetic oscillations. But the most impressive achievements in quantum electronics still lay ahead, on the way to extending the maser principle to the optical band (Schawlow and Townes 1958). And soon the first *optical masers*, or *lasers*, were created (Maiman 1960).

The stimulated emission of radiation by excited particles in a resonant cavity radically changed the situation, having made it possible to *control light in space, time, frequency, and even phase*. An excited atom (or molecule) in such a cavity emits its excitation energy in the form of both spontaneous radiation into a solid angle of 4π steradians, that is, into many modes of space, and stimulated radiation emitted into a small number of controlled cavity modes in the form of a directed monochromatic beam (coherent in time and space), with an energy (or intensity) comparable to or exceeding the energy (or intensity) of the spontaneous radiation emitted into all modes. A vast arsenal of methods has been developed to *control the stimulated-emission channel*. Stimulated radiation has high brightness and low divergence, controllable over a wide wavelength range.

In essence, the unique characteristics of laser radiation brought about the one of most important instrumental revolutions of the 20th century, namely, the *laser instrumental revolution*. We have witnessed how a "conceptual scientific revolution" (Kuhn 1996)—the quantum theory—gave birth to a series of "instrumental revolutions" (Dyson 1998), including the laser, which largely determined the development of science and technology at the close of the 20th century and will quite likely become the "dynamo" of the 21st century.

To justify this far-reaching claim, one can cite as an example two avenues of scientific exploration in physics that owe their advent and progress exactly to the use of laser light. One is the discovery and development of methods for the laser cooling and trapping of atoms, which has given birth to *laser ultracold-matter physics*.

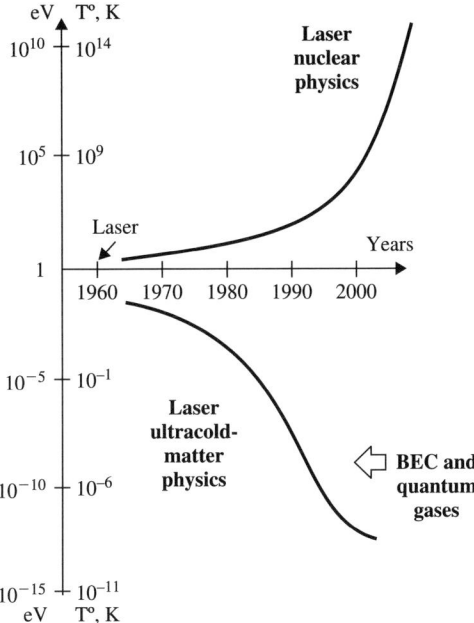

Fig. 1.2 Two examples of the development of new physical trends based on the use of laser light.

The fundamentals of this field are one of the subjects of this book. The other striking example is the development of the physics of ultraintense electromagnetic fields, thanks to the development of methods for generating and amplifying femtosecond laser pulses (Mourou et al. 1998). This field embraces a wide circle of problems, ranging from laser table-top nuclear-physics experiments to the development in the future of compact electron accelerators. Both these lines of physical research cover a remarkably wide range (almost twenty orders of magnitude) of particle energies (Fig. 1.2). Numerous other examples of the effective use of laser light are considered in this book also.

The successful development of methods to generate coherent light based on stimulated emission by matter in a thermodynamically nonequilibrium state resulted in a deeper insight being gained into the quantum theory of light and led to the rapid progress of quantum optics. Specifically, the possibility was recognized of *controlling the spontaneous emission of radiation*. This matter was first considered for the case of radio frequencies in connection with the problem of relaxation in nuclear magnetic resonance (Purcell 1946). The spontaneous emission probability was found to depend on the number of oscillation radiators (modes) per unit volume of the radiator per unit frequency range (the Purcell effect).

For free space, the density $\rho(\omega)$ of modes at the frequency $\omega = 2\pi\nu$ is defined by the Rayleigh–Jeans formula (London 1973):

$$\rho(\omega) = \frac{1}{\pi^2} \frac{\omega^2}{c^3} \tag{1.1}$$

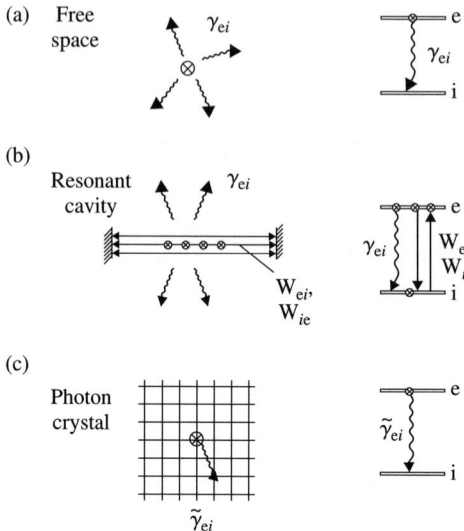

Fig. 1.3 Spontaneous and stimulated emission of light: (a) uncontrollable spontaneous radiation from an atom in free space; (b) controllable stimulated radiation into a limited number of resonant modes of an open cavity; (c) controllable spontaneous radiation from an atom in a structured environment (photonic crystal).

and so the spontaneous emission probability is governed by the number of modes within the resonance line width of the emitter (see Section 2.1 and Fig. 1.3(a)). The presence of material boundaries in the vicinity of the radiator changes the density of the field modes. By isolating a limited number of high-quality (high-Q) modes with the aid of an open resonant cavity, for example, one can obtain stimulated emission of coherent radiation, starting with spontaneous emission (provided that there exists population inversion), only into those selected modes (Fig. 1.3(b)), the probability of spontaneous emission of radiation into the rest of the huge number of free-space modes remaining unchanged. By using a high-Q microcavity, one can radically reduce the number of modes for spontaneous emission and observe that the inhibition of the spontaneous emission rates for one range of transition frequencies tends to be accompanied by an enhancement of the rates at other frequencies (Bykov 1975; Goy *et al.* 1983; Hulet *et al.* 1985). The modification of the spontaneous emission rate caused by an atomic environment is constrained by a sum rule expressed as an integral over the transition frequency (Barnett and Loudon 1996).

All of the above works pointed to the possibility of controlling the spontaneous emission rate by the proper design of the atomic environment. Yablonovich (1987) proposed that photonic-bandgap crystals could be used for controlling the propagation and emission of light (Fig. 1.3(c)). A unique advantage of photonic crystals over resonant cavities is the very wide frequency bandwidth within which spontaneous emission can be controlled. The experimental demonstration of the control of spontaneous emission from quantum dots in photonic crystals (Lodahl *et al.* 2004) has offered strong possibilities for controlling spontaneous emission on a practical scale.

Thus, within half a century after the discovery of quantum electronics, it has become possible to control, both in principle and in practice, both of the fundamental light emission processes, namely, stimulated and spontaneous emission of radiation.

1.2 From the control of light to the control of atoms and molecules

The unique and highly controllable characteristics of laser light (power from a nanowatts to a few petawatts, durations from a microsecond to a few femtoseconds, and wavelengths from the soft X-ray region to the submillimeter region) have radically extended the possibilities of exerting influence on matter, making possible the *laser control of atoms and molecules*. This is a fairly wide term, embracing numerous problems ranging from nonlinear laser spectroscopy to laser relativistic nonlinear optics. In my more narrow sense this term includes the velocity control of atoms and molecules, the control of the orientation and position of atoms, and the control of atomic and unimolecular processes, such as photoselective photoionization and photodissociation.

The laser control of the velocity distribution of atoms or molecules *at particular quantum levels* that emerged in the course of development of saturation spectroscopy free of Doppler broadening (Lamb 1964) is fairly close to the ideas considered in this book. I myself started to work on the problem of laser elimination of Doppler broadening as far back as 1965 and gradually progressed to ideas of laser confinement of atomic motion within a volume of about λ^3. Therefore, I have decided to include a brief description of the ideas of *laser velocity-selective control* of atoms and molecules.

The first ideas about the control of atoms by light (for the case of the orientation of atoms) were born even before the advent of the laser. I have in mind the fundamental work by Kastler on the optical control of the orientation of atoms produced by optical pumping (Kastler 1950). This work had an effect not only on the development of the laser, but also on the use of the laser light for the cooling of atoms. I have therefore considered it necessary to treat the optical orientation of atoms.

In the late 1960s, there was born the idea of laser control of the *position of atoms* using the gradient force in a spatially inhomogeneous light field, specifically in a standing light wave (Letokhov 1968), wherein lay the roots of the future work on the reflecting, guiding, and focusing of atoms by light, that is, the elements of atom optics with laser light. The kinetic energy of atoms under ordinary conditions is too high for this idea to be directly realized. The situation for ions proved much easier; they could be confined in electromagnetic traps (Paul 1990). Moreover, long-term confinement of ions made it possible to put forward the concept of laser cooling of ions, based on irradiation in long-wave sidebands of their spectral lines and subsequent reemission of shorter-wavelength photons (Wineland and Dehmelt 1975). It was in this way that the first experiment on the laser cooling of ions down to temperatures of a few millikelvins was performed (Neuhauser *et al.* 1978; Dehmelt 1990).

The idea of using laser radiation for cooling free neutral atoms (Hansch and Schawlow 1975), which was realized in the first experiments with beams of Na atoms (Andreyev *et al.* 1981, 1982; Phillips and Metcalf 1982), led to the development of methods for the *laser control of the position (trapping) and velocity (cooling) of atoms* and the emergence of a new avenue of inquiry in atomic physics, wherein laser light first cooled atoms in laser fields of various configurations and then finally and purposefully controlled their motion. The attainment of submicrokelvin temperatures in an ensemble of atoms resulted in the manifestation of their wave properties, for their de Broglie wavelength became commensurable with the wavelength of the coherent light used for their cooling. In that case, the behavior of the atomic ensemble becomes nonclassical and is determined by quantum statistics. The Bose atoms (those with an integer spin) in the ensemble make a phase transition into a *Bose–Einstein condensate* (Anderson et al. 1995), while the Fermi atoms (those with a half-integer spin) make a transition into a Fermi "fluid" (DeMarco and Jin 1999), although they are actually dilute quantum gases.

Another line in the laser control of atoms and molecules can be called the *photoselective control* of atomic–molecular photoprocesses. I started working in this area in 1969. I recall my talk with Professor A. Schawlow during an evening stroll in May 1972 in Montreal (the regular IQEC (International Quantum Electronics Conference) was being held there at the time). I told him about the first results of our work on the stepwise photoselective ionization of atoms and the vibrationally mediated photodissociation of molecules for the purpose of isotope separation using lasers. He took great interest in this work and suggested that I should write a paper on the subject for *Science* (Letokhov 1973a). This work led, first, to a proposal for the laser separation of nuclear isomers (Letokhov 1973b) that later was successfully implemented in experiments at the ISOLDE-CERN facility, and, second, to the discovery of infrared (IR) multiphoton resonance (isotope-selective) photodissociation of polyatomic molecules (Ambartzumian et al. 1974, 1975b), which formed the basis for the construction of the first industrial plant for the laser separation of the isotopes ^{13}C and ^{14}C. Laser methods for the *noncoherent control of photoionization and photodissociation processes* have found successful application in science and technology.

One more trend in laser control is based on the use of the property of coherence of the laser light. To effect *coherent laser control*, it is necessary that not only the light, but also the atom (or molecule) should be in a coherent state during the interaction. For atoms in a beam or in a low-pressure gas, the phase relaxation time of their wave functions depends on spontaneous decay or on collisions and can be comparatively long (from 10^{-3} to 10^{-9} s). It was for precisely this reason that the main experiments on coherent interaction were conducted with atoms. These experiments led in the final analysis to the discovery of new effects, such as coherent population trapping (Arimondo 1996), electromagnetically induced transparency (Harris 1997), and the slow-light effect (Hau et al. 1999; Kash et al. 1999).

The situation with molecules is much more involved for several reasons. First, for polyatomic molecules, the intramolecular relaxation processes that occur on a subpicosecond timescale are essential. It was for exactly this reason that the first successful experiments were conducted on the *noncoherent laser control* of polyatomic molecules with intermolecular selectivity. Second, the phase relaxation time in a condensed medium is also on a subpicosecond scale because of the interaction between the quantum system and its surroundings. Therefore, it was only the creation of relatively simple and inexpensive femtosecond lasers that made it possible to set about realizing the ideas of the *coherent laser control of unimolecular processes* (Tannor and Rice 1985; Brumer and Shapiro 1986; Judson and Rabitz 1992), particularly the

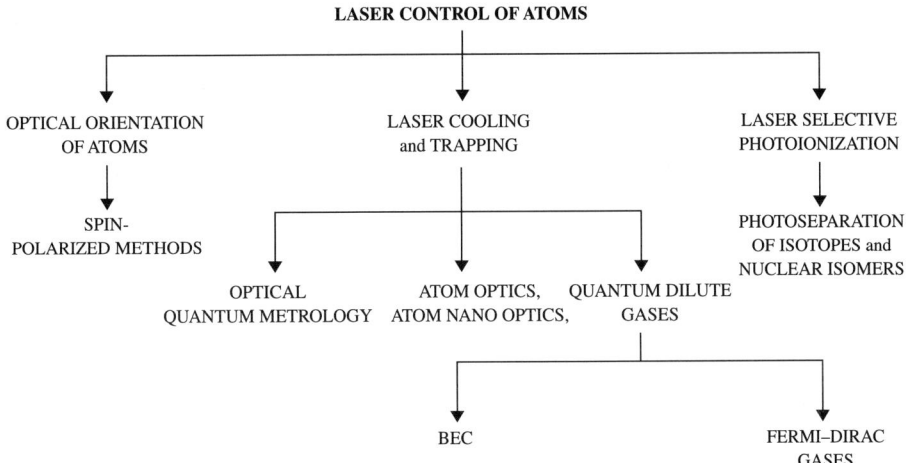

Fig. 1.4 Main trends in the laser control of atoms (Chapters 4–8).

coherent control of the photoionization and photofragmentation channels, and to perform successful experiments.

Figure 1.4 illustrates for the reader's convenience the main trends in the laser control of atoms. The first trend is associated with laser photoionization control of atoms, which has found application both in basic research (laser detection of rare atoms and laser separation of nuclear isomers) and in technology (laser isotope separation). However, the main trend is the laser cooling and trapping of atoms, which has led to (1) the discovery of the Bose–Einstein condensation of ultracold atoms, (2) methods of atom optics and possibly atom nanooptics, and (3) new precision measurement techniques, such as optical frequency standards with an accuracy much better than $\Delta\nu/\nu = 10^{-15}$ and atom interferometry methods. Deeply cooled atoms exhibit the properties of de Broglie waves, whose wavelength becomes comparable to that of the coherent light used for cooling. In that case, an ensemble of cooled atoms can become coherent, that is, form a Bose–Einstein condensate. The interaction of such coherent matter with coherent light is the next stage of the *coherent control of an atomic ensemble*.

Figure 1.5 illustrates the methods for the noncoherent and coherent control of molecules by lasers. The first stage of development was associated with the use of nanosecond and picosecond lasers for effecting noncoherent control, namely, by the vibrationally mediated photodissociation (IR multiphoton dissociation) of polyatomic molecules and the multiphoton ionization of molecules. This stage yielded important practical results: (1) laser isotope separation and (2) laser mass spectrometry of large biomolecules. The next stage (the coherent control of molecules) promises the solution of some problems which cannot be solved by laser noncoherent control, for example the separation of molecular enantiomers and the switching of molecular fragmentation channels. In essence, this will mean a transition from methods providing intermolecular selectivity to methods ensuring intramolecular (mode) selectivity, thanks to the progress of femtosecond laser technology.

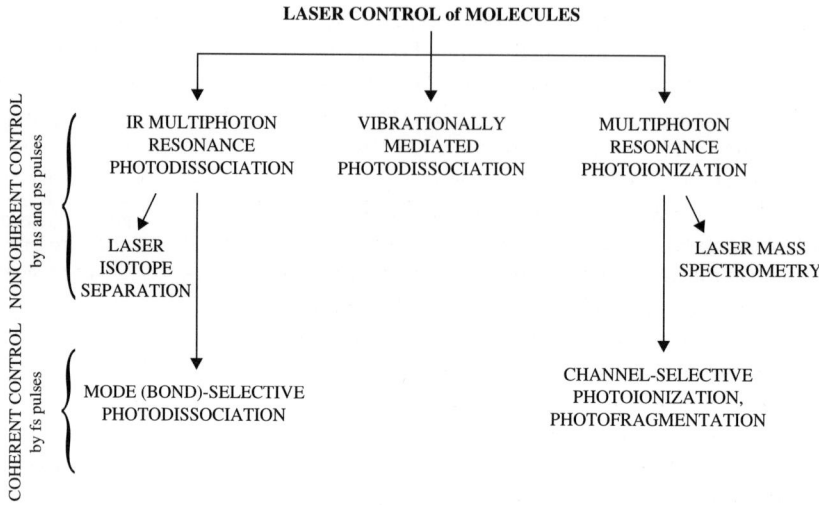

Fig. 1.5 Main methods for the laser noncoherent and coherent control of molecules by lasers (Chapters 9–12).

1.3 On the aims of this book

The goal of this book is to present in a "coherent" way the problems of the laser control of matter at the atomic–molecular level, namely, control of the velocity distribution of atoms and molecules (saturation Doppler-free spectroscopy); control of the absolute velocity of atoms (laser cooling); control of the orientation, position, and direction of motion of atoms (laser trapping of atoms, and atom optics); control of the coherent behavior of ultracold (quantum) gases; laser-induced photoassociation of cold atoms, photoselective ionization of atoms; photoselective multiphoton dissociation of simple and polyatomic molecules (vibrationally or electronically excited); multiphoton photoionization and mass spectrometry of molecules; and femtosecond coherent control of the photoionization of atoms and photodissociation of molecules.

Though the main material of the book relates to atoms and molecules, the ideas of laser control are very fruitful and are applicable to both the simplest objects (electrons) and dielectric particles. For this reason, these applications are also considered briefly in the last chapter. The possibility of the laser control (reflection, for example) of electrons was considered as far back as 1933 (the Kapitza–Dirac effect: Kapitza and Dirac 1933), and after the advent of the laser it was proposed to accelerate electrons to relativistic energies in an intense light wave. But it is only now, when great progress has been made in the development of ultrahigh-intensity femtosecond laser technology, that these ideas are coming to the forefront of investigations. The possibility of the laser control of the position of microbodies was proposed by A. Ashkin in 1970. This approach has gradually led to the development of a very efficient tool for microbiological investigations, known as optical tweezers, and it was impossible not to consider this tool in this book.

All of these methods are developing rapidly but more or less separately from one another. The material of the book brings together various trends in the development of the laser control of atoms and molecules and summarizes an enormous number of scattered publications. However, the references cited are restricted to the pioneering, key papers, major reviews, and books in the particular fields.

Naturally, the entire huge bulk of this rapidly developing domain of laser physics and technology cannot be considered in detail in a relatively small book. For this reason, my treatment of the material is in many cases inevitably very brief and will serve rather as an introduction to the subject to the reader, who can refer for a more in-depth study to the specialized reviews and monographs cited in each chapter. In essence, each chapter can be considered as a short essay, but taken together, Chapters 3–13 give the reader an idea of the amazing capabilities that laser light has to control atoms and molecules. Therefore, one should treat this book as a first introduction and only then proceed to a more detailed reading.

The book is addressed to readers familiar with the elements of physics, spectroscopy, and quantum mechanics. It can be used as a handbook for graduate students, for scientists entering the field, and in some aspects, for science historians as well. A reading of the book will allow readers to familiarize themselves with the frontier level of research and its developments.

2
Elementary radiative processes

In this introductory chapter, we consider the elementary radiative processes for both free atomic particles (atoms, ions, and molecules) and particles interacting with a monochromatic laser field. For more advanced considerations, the reader is advised to refer to the classical textbooks (Loudon 1973; Sargent *et al.* 1974; Allen and Eberly 1975; Sobelman 1979; Haken 1981; Knight and Allen 1983; Meystre and Sargent 1990; Cohen-Tannoudji *et al.* 1992; Foot 2004). Extensions of this material are treated in subsequent chapters, wherein various light field configurations, atomic particles, and the processes of interaction between atomic particles and laser fields are considered in detail.

2.1 Spontaneous emission

An excited atom (or molecule) can make a transition to a lower state and emit a photon with an energy equal to the transition energy. This radiative transition occurs spontaneously, as a quantum jump (Bohr jump), in a random direction, via the allowed decay channels (Fig. 2.1(a)). The instant of time when the spontaneous transition occurs is also random. Consequently, complete information about the process can be obtained by averaging the results of a large number of independent measurements of the energy $E = \hbar\omega$ (or the frequency ω or the wavelength $\lambda = 2\pi c/\omega$) of the emitted photon and of the delay time between the excitation of the atom and its decay from the excited state $|e\rangle$.

Let the atom, at the instant of time $t=0$, be in the upper state $|e\rangle$. All the lower states will be designated as by $|i\rangle$. The decay of the state $|e\rangle$ to the lower states $|i\rangle$ can be described by a decrease in the probability $P_e(t)$ that the atom is in the state $|e\rangle$ at the instant t, which obeys the exponential law

$$P_e(t) = \exp\left[-\left(\sum_i \gamma_{\text{spont}}^{e \to i}\right) t\right], \tag{2.1}$$

where the quantity $\gamma_{\text{spont}}^{e \to i}$ is identified with the *rate* of the respective spontaneous transition $|e\rangle \to |i\rangle$. Thus, the average *lifetime* of the atom in the state $|e\rangle$ is

$$\tau_{\text{spont}}^{(e)} = \left(\sum_i \gamma_{\text{spont}}^{e \to i}\right)^{-1}. \tag{2.2}$$

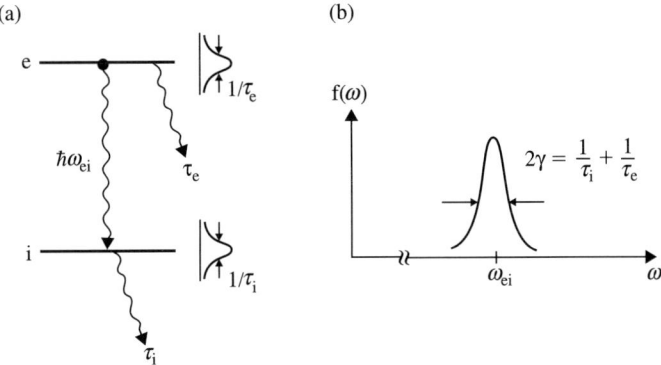

Fig. 2.1 (a) Spontaneous decay of an excited state $|e\rangle$ to a lower state $|i\rangle$; (b) the radiative (natural) width of the spectral line of the transition $|e\rangle \to |i\rangle$ is determined by the sum of the widths of the initial and final quantum states $|e\rangle$ and $|i\rangle$.

When methods of trapping and cooling single ions by means of laser radiation (see Chapters 5 and 6) made their appearance, it became possible to observe quantum jumps directly. Figure 2.2 presents the results of an experiment performed at the laboratory of H. Dehmelt (Nagourney et al. 1986) that demonstrates such jumps of a trapped Ba$^+$ ion, along with a graph of the average lifetime of the ion in its excited quantum state, obeying the exponential law (2.1).

$\gamma_{\text{spont}}^{e\to}$ is a constant of any given atomic particle that depends on the mutual properties of the states $|e\rangle$ and $|i\rangle$. The strongest transitions are those which correspond to a nonzero off-diagonal matrix element of the dipole moment operator. The dipole moment operator $\hat{\mathbf{d}} = \sum_j e_j \mathbf{r}_j$ depends on the charges e_j of the electrons and the nucleus (or nuclei) forming the given atom (or molecule) and their radius vectors $\mathbf{r}_j$. Simply stated, the off-diagonal matrix element

$$\mathbf{d}_{ei} = \langle e|\mathbf{d}|i\rangle = \sum_j e_j \int \Psi_e^*(\mathbf{r}_j) \mathbf{r}_j \Psi_i(\mathbf{r}_j)\, \mathbf{dr}_j \tag{2.3}$$

is determined by the overlap of the wave functions Ψ_e and Ψ_i of the states $|e\rangle$ and $|i\rangle$ involved in the transition. But even if the overlap of the wave functions is strong, the off-diagonal matrix element can be zero because of their mutual symmetry properties. If $\mathbf{d}_{ei} \neq 0$, the rate of the spontaneous transition $|e\rangle \to |i\rangle$ is, to a high degree of accuracy, given by

$$A_{ei} = \gamma_{\text{spont}}^{e\to i} = \frac{4\omega_{ei}^3}{3\hbar c^3}|\mathbf{d}_{ei}|^2 = \frac{32\pi^3}{3\hbar \lambda_{ei}^3}|\mathbf{d}_{ei}|^2, \tag{2.4}$$

where $\hbar$ is Planck's constant, c is the velocity of light, ω_{ei} is the transition frequency (in rad/s), and $\lambda_{ei} = 2\pi c/\omega_{ei}$ is the radiation wavelength corresponding to the transition. The above expression includes two equivalent quantities, A_{ei}, the Einstein coefficient defining the rate (or probability per unit time) of spontaneous emission into a solid angle of 4π steradians, and $\gamma_{\text{spont}}^{e\to i}$, the total width of the energy level $|e\rangle$, determined by

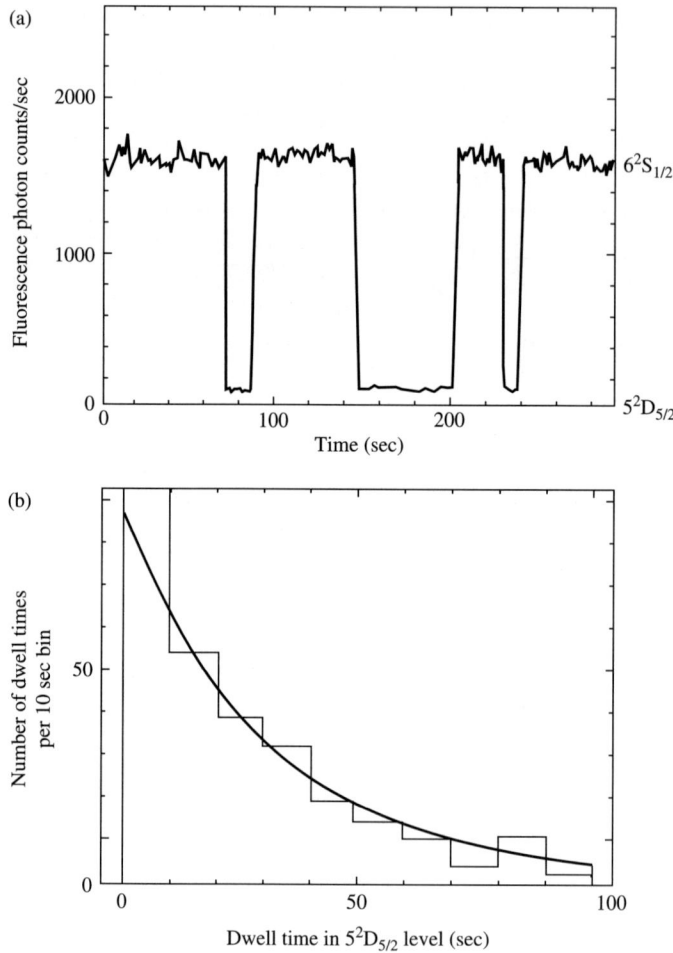

Fig. 2.2 Experimental observation of quantum jumps: (a) a typical trace of the 493 nm fluorescence from the $6^2P_{1/2}$ level of a trapped B$^+$ ion; (b) histogram of the distribution of dwell times in the shelf level $5^2 D_{5/2}$ for 203 "off" times. A fitted theoretical (exponential) distribution for a metastable lifetime of 30 s is superposed on the experimental histogram. (Reprinted from Nagourney et al. 1986, with courtesy and permission of the American Physical Society.)

the inverse of its lifetime $\tau_{\text{spont}}^{(e)}$ as regards the spontaneous transition to the lower level i under consideration. If $\mathbf{d}_{ei}=0$, one should also consider another two characteristics of the system of charges forming the given atom or molecule, namely, the magnetic dipole moment and the electric quadrupole moment. The parameter determining the smallness of the emission rate of magnetic-dipole and electric-quadrupole radiation, compared with that of electric-dipole radiation, can, as a rule, be estimated from the square of the fine-structure constant $\alpha = e^2/\hbar c \approx 1/137$ (where e is the electron charge). This estimation comes from the fact that this small parameter is defined by the square of the ratio between the mean velocity of the charges in the atom (or molecule) and the velocity of light, or the square of the ratio between the characteristic size of the atom (or molecule) and the transition wavelength. In any case, for the optical spectra of atoms and molecules, this is a small quantity, which allows one to consider the transitions corresponding to $\mathbf{d}_{ei} = 0$ as *forbidden*, and to consider excited states having no dipole-allowed downward transitions as *long-lived* or *metastable*.

Dipole moments can be measured conveniently in Debye units (1 Debye unit is equal to 10^{-18} erg$^{1/2}$ cm$^{3/2}$). This unit is of the same order of magnitude as the product of the electron charge and the characteristic size of the atom (or molecule) and can be used for qualitative estimation of the rates of allowed spontaneous electronic transitions. One can remember that $\gamma_{\text{spont}}^{e\to i} \approx 0.5 \times 10^8 \text{s}^{-1}$ for $\mathbf{d}_{ei} = 1$ Debye and $\lambda_{ei} = 2.5 \times 10^{-5}$ cm $= 2500$ Å $= 250$ nm. This estimate gives an approximate value of the rate of a spontaneous transition corresponding to single-quantum excitation in atoms (or molecules) with not very large principal quantum numbers. Vibrational transitions in molecules in the infrared region of the spectrum ($\lambda \sim 3\text{--}30\,\mu$m) are usually characterized by both a relatively large wavelength and a small dipole moment (around 0.1 Debye unit), which is due to the low vibration amplitude ($\sim 10^{-9}$ cm $= 0.1$ Å) of the nuclei in molecules in comparison with the nuclear separation ($\gtrsim 1$Å). Taken together, these two circumstances result in the fact that the rates of spontaneous vibrational transitions in molecules are usually in the range of $1\text{--}10^3$ s^{-1}.

What is the *spectrum* of spontaneous emission for an individual transition $|e\rangle \to |i\rangle$? For a finite lifetime of the excited state, this spectrum cannot be reduced to a spectrum only at the central frequency of the transition of the emitting atom (or molecule), but has a certain characteristic width that is usually defined as the full width at half maximum (FWHM). This radiative *width* $\Delta\omega_{\text{rad}}^{ei}$ is called the *natural width*. It is equal to the sum of the inverse lifetimes of the states $|e\rangle$ and $|i\rangle$. If the lower state is the ground state or a metastable state, the width of the spectrum is governed only by the lifetime of the excited state $|e\rangle$. The shape of the spectrum is described by the *Lorentzian* function (Fig. 2.1b)

$$f(\omega) = \frac{1}{2\pi} \frac{\Delta\omega_{\text{rad}}^{ei}}{(\omega - \omega_{ei})^2 + \gamma^2}; \quad \int f(\omega)\,d\omega = 1,$$

$$\Delta\omega_{\text{rad}}^{ei} = 2\gamma = \left(\tau_{\text{spont}}^{(e)}\right)^{-1} + \left(\tau_{\text{spont}}^{(i)}\right)^{-1}.$$

(2.5)

The function $f(\omega)$ in the above expression is normalized so as to make its integral over all frequencies equal to unity. Note that the natural width of the spontaneous-emission line for the transition between the levels indicated in Fig. 2.1 may be greater than $\gamma_{\text{spont}}^{e\to i}$ because of the decay of the excited state to other states via some channels other than the transition $|e\rangle \to |i\rangle$.

2.2 Stimulated absorption and emission

When considering the conditions of equilibrium between atomic particles and thermal radiation (black-body radiation), Einstein introduced another two elementary radiative processes whose rate depended on the radiation intensity. This was a phenomenological approach based on two fundamental laws: (1) Planck's distribution describing the relationship between black-body radiation and frequency, and (2) the Boltzmann distribution for the probability of population of a quantum level of energy E in an atom. Equilibrium could be attained in a natural way if one introduced, in addition to spontaneous emission of radiation with a rate of A_{ei} (s^{-1}), stimulated upward and downward transitions with rates which were proportional to coefficients B_{ie} and B_{ei}, respectively (Fig. 2.3):

$$\left(\frac{g_i}{g_e}\right) B_{ie} = B_{ei}, \tag{2.6}$$

$$\left(\frac{\hbar \omega_{ei}^3}{\pi^2 c^3}\right) B_{ei} = A_{ei}. \tag{2.7}$$

The coefficients A_{ei}, B_{ei}, and B_{ie} came to be known as the Einstein coefficients for spontaneous emission and for stimulated emission and absorption of radiation, respectively. The rates of spontaneous and stimulated emission A_{ei} and W_{ei} are connected by the expression

$$W_{ei} = A_{ei} \frac{4\pi^3 c^2}{\omega_{ei}^2} J(\omega) = A_{ei} \frac{4\pi}{c} \frac{J(\omega)}{\rho(\omega)}, \tag{2.8}$$

where $J(\omega)$ is the spectral intensity of the radiation in photons/cm^2, and $\rho(\omega)$ is the spectral density of the number of degrees of freedom (or modes) (1.1). The coefficient 4π can quite naturally be explained by the fact that spontaneous radiation in free space is emitted randomly into all modes, while the rate of stimulated emission depends only on the total radiation intensity within the limits of the spontaneous-emission line, no

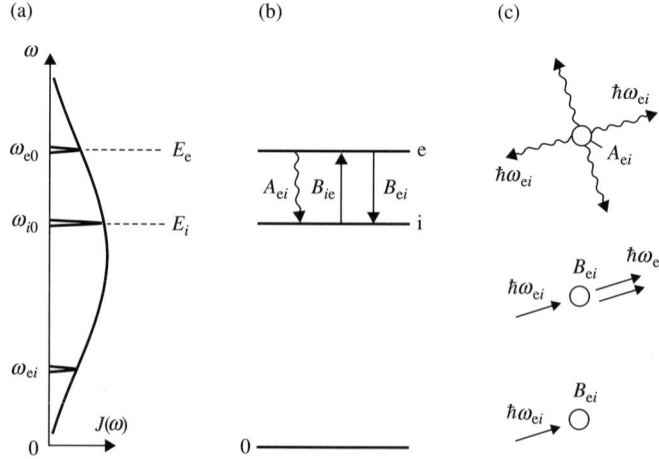

Fig. 2.3 Radiative processes of spontaneous emission and stimulated emission and absorption between quantum levels $|e\rangle$ and $|i\rangle$, which control the equilibrium (Boltzmann) population of levels due to radiative interaction with the equilibrium (Planck) distribution $J(\omega)$ of the photon energies of black-body radiation. Spontaneous emission happens in all directions, but stimulated emission occurs in the direction of the incident radiation.

matter what the radiation direction, that is, on the number of modes involved in the stimulated-emission process.

The physical meaning of the phenomenological relations in eqns (2.6)–(2.8) was established by Dirac in his quantum theory of radiation (see Dirac 1958), wherein the following relation was derived for the probability of spontaneous and stimulated emission in a given mode of free space:

$$W_{ei} \propto \langle n \rangle + 1, \qquad (2.9)$$

where $\langle n \rangle$ is the number of photons (the occupation number) in the mode. The first term in (2.9) describes stimulated emission, and its rate is proportional to the number of photons in the mode, that is, the radiation intensity in the mode. The second term is independent of the number of photons and describes spontaneous emission. A detailed quantitative consideration of eqns (2.6)–(2.9) can be found in several textbooks (Loudon 1973; Sargent et al. 1974; Allen and Eberly 1975; Sobelman 1979; Haken 1981; Knight and Allen 1983; Meystre and Sargent 1990; Cohen-Tannoudji et al. 1992; Foot 2004).

The classical relations (2.6) and (2.7) have been studied and found convenient in investigating black-body radiation, for example in the case of radiative processes for atomic particles in stellar atmospheres. In the case of lasers, where stimulated processes take place under the effect of monochromatic, directional radiation, it is more convenient to operate with the rates W_{ei} and W_{ie} of stimulated processes, expressed in s^{-1}. In that case, the rate W_{ei} is defined by the following simple expression:

$$W_{ei} = \int J(\omega) \sigma_{ei}(\omega) \, d\omega, \qquad (2.10)$$

where the spectral intensity $J(\omega)$ is expressed in photons/cm^2 s rad s^{-1} (or photon/cm^2), ω is expressed in rad/s, and the radiative-transition cross section (cm^2) is related to the spontaneous transition rate A_{ei} (s^{-1}) by a simple relation

$$\sigma_{ei}(\omega) = \frac{\lambda_{ei}^2}{2\pi} A_{ei} f(\omega), \qquad (2.11)$$

where the normalized Lorentzian shape of the spontaneous-emission line $f(\omega)$ is described by eqn (2.5). In the case of monochromatic radiation with intensity I (photons/cm^{2}s) and frequency ω with spectral width $\delta\omega \ll$ spectral line width, the rate of stimulated emission is given by $W_{ei} = I\sigma_{ei}(\omega)$. This formula is very convenient for practical estimations. Rates of stimulated emission W_{ei} and absorption W_{ie} are related by eqn (2.6).

Note that if the radiative width ω_{ei} of the spontaneous-emission line is governed only by the spontaneous radiative transition $|e\rangle \to |i\rangle$, then at the center of the spectral line ($\omega_{ei} = \omega_0$),

$$\sigma_{ei}(\omega_0) = \frac{\lambda_{ei}^2}{2\pi}. \qquad (2.12)$$

In terms of the radiative-transition cross section $\sigma(\omega)$ (cm^2), one can understand the simple physical meaning of the condition for the rate of the stimulated (induced) transition to become equal to that of its spontaneous counterpart. One can say qualitatively that the spontaneous and stimulated transition rates are quantities of the same order of magnitude when a single photon within the spectral line passes through an area λ^2 in size during the lifetime of the excited state. But this is an ideal case, where the spectral line suffers no broadening other than its own radiative broadening. If the actual spectral transition width $\Delta\omega_{ei}$ exceeds $\Delta\omega_{rad}^{ei}$, then instead of eqn (2.12), the following simple relation holds true:

$$\sigma_{ei}(\omega_0) = \frac{\lambda_{ei}^2}{2\pi} \frac{A_{ei}}{\Delta\omega_{ei}}. \qquad (2.13)$$

Here, for simplicity's sake, the degeneracy of the quantum levels of a "two-level" system in a real atom is not taken into account. Consideration of the degeneracy of levels due to the angular momentum of an atom in a given quantum state (s,p,d...) and the polarization of radiation (linear, circular, or natural) is essential (see, for example, Foot 2004). This effect is fundamental for the optical orientation of atoms and nuclei (Chapter 4) and the laser cooling of atoms by the polarization gradient method (Chapter 5).

The central frequency of the spontaneous or stimulated emission (or absorption) spectrum of an atom is shifted relative to the transition frequency because of the translational degree of freedom of the atomic motion, owing to two effects: the Doppler effect and the recoil effect.

2.3 Recoil effect and Doppler effect

The first simple, and at the same time fundamental, conclusions about momentum and energy exchange between an atom and a resonant light field can be drawn by proceeding from the conservation laws. Let us assume that an atom having a velocity $\mathbf{v}_0$ absorbs or emits a photon with a wave vector $\mathbf{k}$ and a frequency $\omega = kc$. The atom is considered to be a simple two-level quantum system, with the states $|e\rangle$ and $|i\rangle$ separated by an energy interval of $\hbar\omega_0$. When the atom moves with a nonrelativistic velocity, the momentum and energy conservation laws, applied to the case of an elementary act of photon absorption (plus sign) or emission (minus sign), give the following equations:

$$M\mathbf{v}_0 \pm \hbar\mathbf{k} = M\mathbf{v}, \qquad (2.14)$$

$$\tfrac{1}{2}M\mathbf{v}_0^2 \pm \hbar\omega = \tfrac{1}{2}M\mathbf{v}^2 \pm \hbar\omega_0, \qquad (2.15)$$

where $\mathbf{v}$ is the velocity the atom acquires after absorbing or emitting a photon, and M is the mass of the atom.

Considering eqn (2.14), one can see that in the nonrelativistic approximation the photon momentum is the same in any reference frame. Indeed, we can rewrite eqn (2.14) in the form

$$\pm\hbar\mathbf{k} = M(\mathbf{v} - \mathbf{v}_0). \qquad (2.16)$$

The fact that the right-hand side of this equation is independent of the choice of the reference frame immediately proves the invariance of the photon momentum under nonrelativistic velocity transformations.

The invariance of the photon momentum shows that, in the nonrelativistic-velocity approximation, the modulus of the recoil momentum is the same for stimulated absorption, stimulated emission, and spontaneous emission. Indeed, the moduli of the recoil momentum for all three processes coincide in the reference frame of the atom at rest:

$$\hbar|k_{\text{abs}}| = \hbar|k_{\text{em}}| = \hbar|k_{\text{sp}}| = \frac{\hbar\omega_0}{c}. \tag{2.17}$$

But in this case, as noted above, they must coincide in any other nonrelativistic reference frame as well. In connection with these remarks, one should emphasize specially the difference in transformation properties between the wave vector and the radiation frequency for the processes of interaction between a nonrelativistic atom and resonant radiation. Indeed, under a nonrelativistic velocity transformation, the wave vector retains its value. At the same time, the radiation frequency changes according to the first-order Doppler effect. If, for example, ω_0 is the frequency of radiation emitted by a source that is motionless in the laboratory reference frame, then the radiation frequency in the rest frame of the atom ω' will be

$$\omega' = \omega_0 - \mathbf{k} \cdot \mathbf{v}_0, \tag{2.18}$$

where $\mathbf{v}_0$ is the velocity of the atom in the laboratory frame.

Similarly, if the frequency of a spontaneously emitted photon is ω_0 in the rest frame of the atom, then in the laboratory frame (where the velocity of the atom is $\mathbf{v}_0$) the photon frequency will be

$$\omega_{\text{sp}} = \omega_0 + \mathbf{k}_{\text{sp}}\mathbf{v}_0, \tag{2.19}$$

whereas the wave vector $\mathbf{k}_{\text{sp}}$ has the same value in both reference frames. Consider now some specific features of the energy exchange between a nonrelativistic atom and a resonant photon. To this end, we write down from eqns (2.14) and (2.15) the change in the kinetic energy of the atom due to the photon recoil,

$$\Delta E_{\text{kin}} = \tfrac{1}{2}M(\mathbf{v}^2 - \mathbf{v}_0^2) = \pm\hbar\mathbf{k} \cdot \mathbf{v}_0 + R, \tag{2.20}$$

and the energy of the absorbed or emitted photon,

$$\hbar\omega = \hbar\omega_0 + \hbar\mathbf{k} \cdot \mathbf{v}_0 \pm R, \tag{2.21}$$

where

$$R = \frac{\hbar^2 k^2}{2M} \tag{2.22}$$

is the recoil energy. Note that the plus sign in eqns (2.20) and (2.21) refers to the absorption of a photon, and the minus sign to the emission of a photon.

According to eqn (2.20), the change in the kinetic energy of the atom is made up of the Doppler shift of the photon energy and the recoil energy. These two terms play quite different roles. This can be clearly seen by going over to the reference frame in which the atom is initially at rest. Assuming that $\mathbf{v}_0 = 0$, we have, in the atomic rest frame, the following relations instead of eqns (2.20) and (2.21):

$$\Delta E_{\text{kin}} = R, \qquad (2.23)$$

$$\hbar\omega' = \hbar\omega_0 \pm R, \qquad (2.24)$$

where $\omega' = \omega - \mathbf{k} \cdot \mathbf{v}_0$ is the photon frequency in the atomic rest frame. Thus, in the atomic rest frame, the absorption or emission of a photon with a momentum of $\hbar\mathbf{k}$ always increases the kinetic energy of the atom by an amount equal to the recoil energy R. In the reference frame where the velocity of the atom is $\mathbf{v}_0$, the change in the kinetic energy is due to both the transfer of the momentum $\hbar\mathbf{k}$ to the atom and the shift of the photon energy as a result of the Doppler effect.

The influence of the recoil effect on the photon frequency shift depends on whether the photon is absorbed or emitted. According to eqn (2.23), in the rest frame of the atom, the energy of an absorbed photon is higher than the energy of the quantum transition by an amount equal to the recoil energy that is converted into kinetic energy of the absorbing atom. When a photon is emitted, the recoil effect, on the contrary, decreases the photon energy by an amount equal to the recoil energy. The recoil energy itself has the same value in both the photon absorption and the photon emission process, owing to the invariance of the modulus of the recoil momentum.

Example. A quantitative estimate of the energy and momentum exchange during the process of absorption or emission of a photon can be obtained by considering, as an example, a sodium atom absorbing radiation with a wavelength of $\lambda = 5890$ Å in resonance with the transition from the ground state 3S to the excited state 3P (the yellow line of atomic sodium). The change in the velocity of the sodium atom upon absorption or emission of a single photon at the given wavelength is $v_{\text{rec}} = \hbar k/M = 3$ cm/s. The recoil energy, in units of frequency, is $R/h = h/2M\lambda^2 = 25$ kHz. It should be noted, for comparison, that the mean thermal velocity of sodium atoms at the ambient temperature of $T = 300$ K is $v_0 = 5 \times 10^4$ cm/s. At such a velocity, the Doppler shift of the photon frequency is $\Delta\nu = kv/2\pi = 850$ MHz. The natural linewidth of the 3S $\to$ 3P transition in atomic sodium is equal to 10 MHz.

This example clearly illustrates the smallness of the recoil effect due to the absorption or emission of an optical photon. For the usual thermal atomic velocities, the typical relative change in atomic velocity (or atomic momentum) caused by the absorption or emission of a single optical photon is $v_{\text{rec}}/v_0 \simeq 10^{-4}$.

The shift of emission and absorption spectral lines due to the recoil effect is also small, making it rather difficult to observe the recoil effect in the optical region of the spectrum. For optical atomic transitions, the ratio between the recoil energy and the atomic transition energy, $R/\hbar\omega_0$, typically ranges from 10^{-10} to 10^{-11}. The recoil effect in the optical region of the spectrum has been experimentally resolved only in ultranarrow infrared absorption lines of molecules by means of saturation laser spectroscopy (Chapter 3).

So, if an excited atom at rest emits a photon, it acquires a momentum in a direction opposite to the photon emission direction, the magnitude of this momentum being exactly equal to that of the photon momentum $\hbar\omega/c = \hbar k$. The acquired velocity of the atom after emission is

$$v_{\rm rec} = \frac{\hbar\omega}{Mc}, \qquad (2.25)$$

where M is the mass of the atom, and its kinetic energy will be equal to the recoil energy defined by eqn (2.22). Because the recoil effect is inevitably present, the frequency of the emitted photon is somewhat lower than the transition frequency. The relative difference between these frequencies is $(\omega - \omega_{ei})/\omega \approx -\hbar\omega/2Mc^2$, which, for the optical region of the spectrum, yields only a small quantity of around 10^{-12}. However, when the atom interacts repeatedly with a laser wave, this effect can accumulate and substantially change the atomic velocity (Chapters 5–8).

If the excited atom is moving and the projection $v_\parallel$ of its velocity onto the direction in which it emits a photon is such that $Mv_\parallel^2/2 \gg R$, the difference between the radiation frequency and the transition frequency is much greater. The shift of the frequency of the emitted photon with respect to the transition frequency is called the linear Doppler shift and is defined by the expression

$$\omega - \omega_{ei} = \Delta\omega_{\rm D} \approx \omega_{ei}\frac{\mathbf{v}\cdot\mathbf{n}}{c}, \qquad (2.26)$$

where $\mathbf{v}\cdot\mathbf{n}$ is the projection of the velocity of the atom onto the projection of the direction of photon emission. When we are observing radiation emitted by an ensemble of atoms distributed over a range of velocities, this shift determines the broadening of the spectral line of the transition, which is referred to as the *Doppler broadening*. For a Maxwellian distribution of atom velocities at room temperature, the value of the relative Doppler broadening $\Delta\omega_{\rm D}/\omega_{ei} \approx v/c$ is typically around 10^{-6}. If the motion directions of the atoms are equiprobable, the Doppler shift (2.26) is linear in the projection of the velocity of the atom, and causes no shift of the center of the emission line.

The recoil effect, however, is not the only cause of a shift of the line center. Using the relativistic laws of conservation of energy and momentum yields the following exact formula for the Doppler shift:

$$\omega = \omega_{ei}\frac{(1 - v^2/c^2)^{1/2}}{1 - (v_\parallel/c)}. \qquad (2.27)$$

Expanding the above expression into a series in powers of v/c, we obtain, in addition to the linear shift proportional to $v_\parallel/c$, also a quadratic Doppler shift

$$\frac{\Delta\omega_{\rm D}}{\omega_0} = \frac{\omega - \omega_{ei}}{\omega} \approx \frac{v_\parallel}{c} - \frac{v^2}{2c^2}. \qquad (2.28)$$

The line center should be slightly shifted toward lower frequencies, as is the case with the recoil effect, provided there is no specified atomic motion direction.

2.4 Resonant excitation of a two-level system free from relaxation

The brief description of the interaction between atomic particles and photons presented in Sections 2.1–2.3 was based on energy and momentum conservation laws. This was quite sufficient for the description of incoherent interaction effects due either to the nonmonochromatic character of the radiation or to the incoherent state of the atomic-particle ensemble. With the advent of lasers, the situation has changed, for laser light is highly coherent, has a high intensity, and is capable of producing coherence in matter. For this reason, the next step is, naturally, to consider the quasi-classical interaction between atomic particles and intense laser light, which requires no quantization of the light field, the only quantized system being the atomic medium.

The laser control of atoms and molecules is most effective when the particles are isolated from external influences. The necessary degree of isolation differs widely between different laser control methods. For coherent control techniques, even an accidental displacement of the phase of the wave function must be excluded. For incoherent laser control techniques, it is sufficient to exclude relaxation of the quantum-state populations of the atomic particles. One should therefore consider first the interaction of an isolated simple (two-level) quantum system with a coherent light field (Fig. 2.4(a)).

2.4.1 Schrödinger equation

The state of a quantum system with two energy levels E_1 and E_2 in an external laser-light field is fully described by the wave function Ψ, governed by the Schrödinger equation

$$i\hbar \frac{\partial}{\partial t}\Psi = \hat{H}\Psi, \tag{2.29}$$

where $\hat{H}$ is the system's Hamiltonian. This includes the unperturbed Hamiltonian of the system, H_0, and the operator of interaction between the system and the light field, $\hat{V}$:

$$\hat{H} = \hat{H}_0 + \hat{V}. \tag{2.30}$$

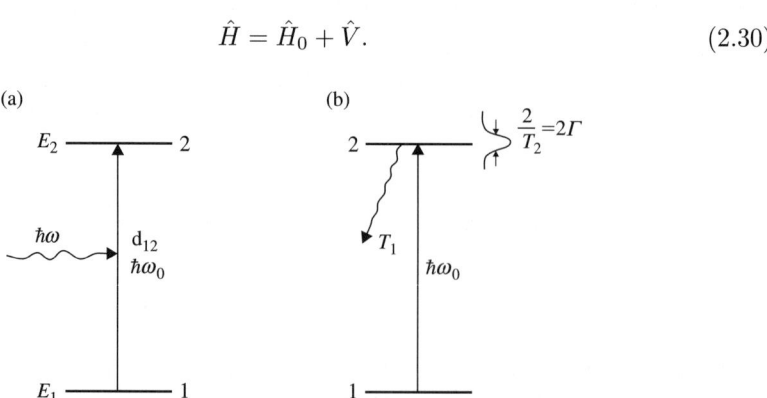

Fig. 2.4 Resonant excitation of (a) an ideal, relaxation-free two-level system and (b) a two-level system with a population relaxation time T_1 and a phase relaxation time T_2.

In the case of an electric dipole interaction between the light field and the particle, the interaction operator $\hat{V}$ has the form

$$\hat{V}(t) = -\mathbf{d} \cdot \mathbf{E}(t), \tag{2.31}$$

where $\mathbf{d}$ is the electric dipole moment operator of the particle, and $\mathbf{E}$ is the electric field strength of the light wave.

According to the superposition principle, the wave function Ψ of the two-level quantum system interacting with the field may be written in a general form as

$$\Psi = a_1(t)\Psi_1 e^{-i(E_1/\hbar)t} + a_2(t)\Psi_2 e^{-i(E_2/\hbar)t}, \tag{2.32}$$

where Ψ_1 and Ψ_2 are the wave functions of the stationary states of the system with energies E_1 and E_2, and $a_1(t)$ and $a_2(t)$ are the probability amplitudes of the states 1 and 2, respectively. The temporal evolution of the probability amplitudes obeys the set of equations obtained by substituting eqn (2.32) into the initial Schrödinger equation (2.29) (Dirac 1958; Loudon 1973; Allen and Eberly 1975):

$$i\hbar \frac{da_1}{dt} = V_{21} a_2 e^{-i\omega_0 t},$$
$$i\hbar \frac{da_2}{dt} = V_{12} a_1 e^{+i\omega_0 t}, \tag{2.33}$$

where $\omega_0 = (E_2 - E_1)/\hbar$ is the resonance transition frequency, and V_{12} and V_{21} are the matrix elements of the field–system interaction operator defined by the expression

$$V_{12}(t) = \int \Psi_1^* V \Psi_2 \, dq, \quad V_{12} = V_{21}^*. \tag{2.34}$$

Here, q stands for the coordinates of the quantum system, and the asterisk denotes the complex conjugate.

2.4.2 Evolution of the probability amplitude. Rabi frequency

For most laser spectroscopy applications involving stimulated transitions of quantum systems between energy levels, the laser-light wave may be represented in the form of a linearly polarized coherent wave with frequency ω and amplitude E,

$$\mathbf{E}(t) = \mathbf{e} E \cos \omega t = \tfrac{1}{2} \mathbf{e} E (e^{i\omega t} + e^{-i\omega t}). \tag{2.35}$$

In this case, the interaction energy $V_{12}(t)$ defined by eqn (2.34) has the following simple form:

$$V_{12} = -d_{12} E \cos \omega t, \quad V_{21}^* = V_{12}, \tag{2.36}$$

where $d_{12} = \mathbf{e} \cdot \mathbf{d}_{12}$ is the projection of the dipole matrix element on the polarization vector of the light wave. In an isotropic medium the vector $\mathbf{d}_{12}$ is randomly oriented relative to the vector $\mathbf{e}$, and accordingly can be averaged over all possible orientations:

$$d^2 = \langle (\mathbf{e} \cdot \mathbf{d}_{12})^2 \rangle = \tfrac{1}{3} |d_{12}|^2. \tag{2.37}$$

In the case of resonant interaction, the light-field frequency ω is assumed to be close to the transition frequency ω_0, so that the frequency difference $\Delta = \omega - \omega_0$ is much smaller than the frequency of the field:

$$|\Delta| = |\omega - \omega_0| \ll \omega_0, \omega. \tag{2.38}$$

Also, the energy of interaction between a light field and a two-level system is usually much lower than the quantum transition energy:

$$dE \ll \hbar\omega_0. \tag{2.39}$$

This is only natural, for otherwise the interaction of the field with the quantum system would no longer be resonant. In this case, all the time-dependent terms can be greatly simplified by treating them only by reference to fast motions of the type $e^{i\omega_0 t}$. With what is known as the *rotating-wave approximation* (Feynman et al. 1957), the probability amplitudes a_1 and a_2 oscillate only at the slow frequency Δ. Under these conditions (which generally hold true), we have, instead of eqn (2.33), the following set of simple equations for the probability amplitudes:

$$\begin{aligned} i\hbar\dot{a}_1 &= -\tfrac{1}{2}dE e^{i\Delta t} a_2, \\ i\hbar\dot{a}_2 &= -\tfrac{1}{2}dE e^{-i\Delta t} a_1. \end{aligned} \tag{2.40}$$

These are easy to solve.

Let the particle be definitely in the lower level at first, that is,

$$|a_1(t=0)|^2 = 1, \quad |a_2(t=0)|^2 = 0, \tag{2.41}$$

and let a light field with a constant amplitude,

$$E(t) = \begin{cases} E\cos\omega t, & t \geq 0, \\ 0, & t < 0, \end{cases} \tag{2.42}$$

be switched on at the moment $t=0$.[1] The solution of the set of equations in eqn (2.40) can then easily be found:

$$\begin{aligned} a_1(t) &= \left\{ \cos\left[\left(\frac{\Omega}{2}\right)t\right] - i\left(\frac{\Delta}{\Omega}\right)\sin\left[\left(\frac{\Omega}{2}\right)t\right] \right\} e^{i(\Delta/2)t}, \\ a_2(t) &= i\left(\frac{dE}{\hbar\Omega}\right)\sin\left[\left(\frac{\Omega}{2}\right)t\right] e^{-i(\Delta/2)t}, \end{aligned} \tag{2.43}$$

where Ω is the generalized Rabi frequency, defined by the expression (Rabi 1937)

$$\Omega^2 = \Delta^2 + \left(\frac{dE}{\hbar}\right)^2 = \Delta^2 + \Omega_0^2. \tag{2.44}$$

Here, $\Omega_0 = dE/\hbar$ is the *Rabi oscillation frequency* in the case of exact resonance ($\Delta = 0$).

---

[1] Let us emphasize that use is made here of a time-stepped switching of the resonant field that is quite applicable to the problem at hand. In a number of cases, the main features of the switching of the field are of importance, specifically in the case of the optimal shape of the laser pulse for the coherent control of femtosecond laser-induced chemical processes (Chapter 12).

2.4.3 Excitation probability

The squared probability amplitude $|a_2(t)|^2$, that is, the probability of finding the two-level particle in the upper state 2 at the moment t, is

$$|a_2(t)|^2 = \left(\frac{dE}{\hbar\Omega}\right)^2 \sin^2\left(\frac{\Omega}{2}t\right) = \left[1 - \left(\frac{\Delta}{\Omega}\right)^2\right] \sin^2\left(\frac{\Omega}{2}t\right). \qquad (2.45)$$

So, the probability that the particle is excited to the upper level, W_{12}, is given by

$$W_{12}(t) = \sin^2\left[\frac{1}{2}(\Delta^2 + \Omega^2)^{1/2}t\right] L\left(\frac{\Delta}{\Omega}\right), \qquad (2.46)$$

where $L(x)$ is the Lorentzian line form factor, normalized to unity at maximum,

$$L(x) = \frac{1}{(1+x^2)}. \qquad (2.47)$$

In the case of resonance ($\Delta = 0$),

$$W_{12}(t) = \sin^2\left(\frac{\Omega_0 t}{2}\right) = \frac{1}{2}(1 - \cos\Omega_0 t), \qquad (2.48)$$

while, under conditions far off resonance ($\Delta \gg \Omega$), the following approximate expression for W_{12} is valid:

$$W_{12}(t) = \left(\frac{\Omega}{\Delta}\right)^2 \sin^2\left(\frac{\Delta}{2}\right) t = \left(\frac{\Omega}{2\Delta}\right)^2 (1 - \cos\Delta t). \qquad (2.49)$$

In the case of exact resonance (see Fig. 2.5), complete excitation of the particle to the upper level can be achieved during a time $t_{\text{inv}} = \pi/\Omega_0 = \pi\hbar/dE$. However, this is

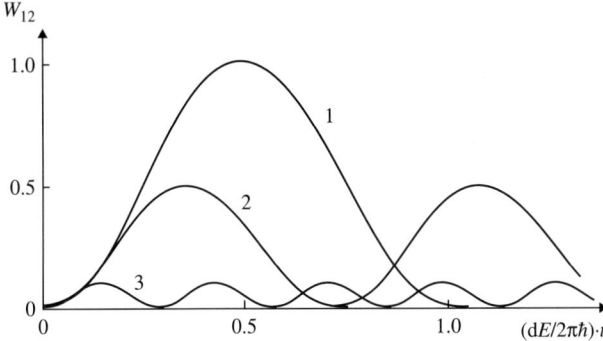

Fig. 2.5 Coherent oscillations of the excited-level population in an ideal, relaxation-free two-level system in a resonant laser-light field switched on at the moment $t = 0$, at various values of the detuning Δ of the field frequency ω from the exact resonance frequency ω_0: (1) $\Delta = \omega - \omega_0 = 0$; (2) $\Delta = dE/\hbar$; and (3) $\Delta = 3dE/\hbar$.

in principle impossible to do when the field frequency ω is detuned from the resonance. To properly consider the case of detuning from resonance ($\Delta \neq 0$), we must first of all go ouside the bounds of the highly idealized model of an isolated two-level particle interacting with an ideal coherent light field.

2.5 Resonant excitation of a two-level system with relaxations

2.5.1 Population relaxation and phase relaxation

A real two-level particle is inevitably subject to the decay of its excited state, owing to at least spontaneous emission (Fig. 2.4(b)). Giving consideration to the decay of the excited state immediately makes the absorption spectral line of the transition $1 \to 2$ have a finite width even in the absence of an external field. In actual fact, the interaction of the particle with its surroundings, which can be considered as a thermal bath, can additionally shorten the lifetime of the excited state. In the general case, the relaxation of the population of the excited level to its equilibrium state is called *longitudinal relaxation* and is characterized by a longitudinal relaxation time T_1 (Bloch 1946).

The interaction of the particle with its surroundings causes a "random shift" of the phase of the particle's wave function in each of its steady states, which is not necessarily accompanied by the decay of the particle to the lower level. The mean time of such a "phase relaxation" is denoted by T_2, and the relaxation itself is frequently referred to as *transverse relaxation* (Bloch 1946). The phase relaxation has no effect on the relaxation of the population of levels, but it broadens the spectral line of the $1 \to 2$ transition. The homogeneous half-width Γ of the Lorentzian in eqn (2.47) is related to the time T_2 by a simple relation at $T_1 \gg T_2$:

$$\Gamma = \frac{1}{T_2}. \tag{2.50}$$

2.5.2 Density matrix equations

The excitation of a quantum system under conditions of population and phase relaxation cannot be described by the probability amplitudes $a_i(t)$. Instead, the time evolution of the particle is described by the combinations $a_i^*(t)a_j(t)$ averaged over the ensemble. This is a standard procedure in quantum mechanics, according to which the quantity $\langle a_i^*(t)a_j(t)\rangle$ is an element $\rho_{ji}(t)$ of a density matrix (see, for example, Sargent et al. 1974):

$$\rho_{ji}(t) = \langle a_i^*(t)a_j(t)\rangle. \tag{2.51}$$

The population probability of state i introduced above is described by a diagonal element of the density matrix:

$$n_i(t) = \langle a_i^*(t)a_i(t)\rangle = \rho_{ii}(t). \tag{2.52}$$

The population evolution of levels 1 and 2 in an external resonant laser-light field is described by equations that follow from the Schrödinger equation after the appropriate averaging in accordance with eqn (2.51) and include terms accounting for the relaxation of the level population:

$$\frac{dn_1}{dt} = \left(\frac{i}{\hbar}\right) \mathbf{E}\left(\mathbf{d}_{12}\rho_{21} - \mathbf{d}_{21}\rho_{12}\right) - \left(\frac{1}{T_1}\right)(n_1 - n_1^0),$$

$$\frac{dn_2}{dt} = -\left(\frac{i}{\hbar}\right) \mathbf{E}\left(\mathbf{d}_{12}\rho_{21} - \mathbf{d}_{21}\rho_{12}\right) - \left(\frac{1}{T_1}\right)(n_2 - n_2^0),$$

(2.53)

where n_i^0 is the population probability of level i in the absence of the laser-light field, and the values of T_1 are equal for each level. The change of the level population depends on the off-diagonal elements $\rho_{12}(t)$ and $\rho_{21}(t)$, which describe the polarizability of the two-level particle, that is, the high-frequency dipole moment $\mathbf{P}(t)$ induced in the particle by the laser-light field,

$$\mathbf{P}(t) = \mathbf{d}_{12}\rho_{21} + \mathbf{d}_{21}\rho_{12}. \qquad (2.54)$$

The evolution of the off-diagonal matrix elements is described by equations that also follow from the Schrödinger equation after the appropriate averaging and introduction of the phase relaxation time T_2, that is, the time of the decay of the particle polarization:

$$\frac{d\rho_{12}}{dt} = i\omega_0\rho_{12} + \left(\frac{i}{\hbar}\right) \mathbf{E} \cdot \mathbf{d}_{12}(n_2 - n_1) - \left(\frac{1}{T_2}\right)\rho_{12},$$

$$\frac{d\rho_{21}}{dt} = -i\omega_0\rho_{21} - \left(\frac{i}{\hbar}\right) \mathbf{E} \cdot \mathbf{d}_{21}(n_2 - n_1) - \left(\frac{1}{T_2}\right)\rho_{21}.$$

(2.55)

2.5.3 Polarization and population equations

The set consisting of eqns (2.53) and (2.55), which looks rather cumbersome, can be reduced to a simpler form by using the rotating-wave approximation and introducing, instead of the four elements $\rho_{ij}(t)$, the physically observable quantities. We introduce the difference in population probability between levels 2 and 1 in the form

$$N(t) = n_2(t) - n_1(t), \qquad (2.56)$$

which, by virtue of the condition $n_1 + n_2 = 1$, is related to the level populations by

$$n_1 = \tfrac{1}{2}(1 - N), \quad n_2 = \tfrac{1}{2}(1 + N). \qquad (2.57)$$

Let us next represent the polarization oscillating with the frequency Ω of the light field in eqn (2.35) in the form

$$\mathbf{P}(t) = \mathbf{e}P(t)\cos\left[\omega t + \varphi(t)\right]. \qquad (2.58)$$

Then, instead of the set of equations (2.53) and (2.55), the following simple set of equations can be obtained with the rotating-wave approximation:

$$\frac{d}{dt}N + \left(\frac{1}{T_1}\right)(N - N_0) = -\left(\frac{1}{\hbar}\right)PE\sin\varphi,$$

$$\frac{d}{dt}P + \left(\frac{1}{T_2}\right)P = \left(\frac{d^2}{\hbar}\right)NE\sin\varphi,$$

$$(\omega - \omega_0)P = \left(\frac{d^2}{\hbar}\right)NE\cos\varphi,$$

(2.59)

where $N_0 = N(t=0)$. We consider now the response of the two-level system to a switching-on of the light field of the form of eqn (2.42) in the simple case of exact resonance:

$$\omega = \omega_0. \tag{2.60}$$

In this case, the change of the level population difference N is described by a simple equation with constant coefficients:

$$\frac{d^2}{dt^2}N(t) + \left(\frac{1}{T_2} + \frac{1}{T_1}\right)\frac{d}{dt}N(t) + \left[\Omega_0^2 + \frac{1}{(T_1T_2)}\right]N(t) = N_0\left[\frac{1}{T_1T_2}\right]. \tag{2.61}$$

The behavior of the two-level system depends materially on the relation between two parameters: the Rabi frequency $\Omega_0 = dE/\hbar$ and the parameter $(1/T_2 + 1/T_1)$. In most cases, phase relaxation takes its course much faster than level population relaxation ($T_2 \ll T_1$), and therefore, to simplify expressions, the quantity $(1/T_2 + 1/T_1)$ will hereinafter be considered equal to $1/T_2$.

2.5.4 Coherent and incoherent interactions

In the case of a sufficiently strong light field, where

$$2\Omega_0 = 2\left(\frac{dE}{\hbar}\right) \gg \frac{1}{T_2} = \Gamma, \tag{2.62}$$

the change of the level population is described by the approximate expression

$$N = N_0 \exp\left(\frac{-t}{2T_2}\right) \cos\Omega_0 t. \tag{2.63}$$

In this case, the populations of states 1 and 2 oscillate at the Rabi frequency, as shown in Fig. 2.6. The only difference is that the oscillations decay exponentially during the phase relaxation time T_2. If condition (2.62) is satisfied, the interaction of the two-level system with the laser-light field is said to be *coherent*.

Let the light-field amplitude now fails to satisfy condition (2.62):

$$2\Omega_0 \ll \frac{1}{T_2}. \tag{2.64}$$

In this case, the evolution of the population difference N is described by the expression

$$\frac{N}{N_0} \simeq \frac{1}{1+G} + \left[\frac{G}{1+G}\right]\exp\left[-\left(\frac{t}{T_1}\right)(1+G)\right], \tag{2.65}$$

where G is the dimensionless saturation parameter, defined as

$$G = \Omega_0^2 T_1 T_2 = \left(\frac{dE}{\hbar}\right)^2\left[\frac{1}{T_1T_2}\right] = \sigma(\omega_0)IT_1, \tag{2.66}$$

with $I = (c/8\pi)(E^2/\hbar\omega)$ being the radiation intensity (photons/cm² s), $d^2 = |\mathbf{d}_{21}|^2/3$, and $\sigma(\omega)$ is the cross section for the stimulated transition between levels 1 and 2 at a

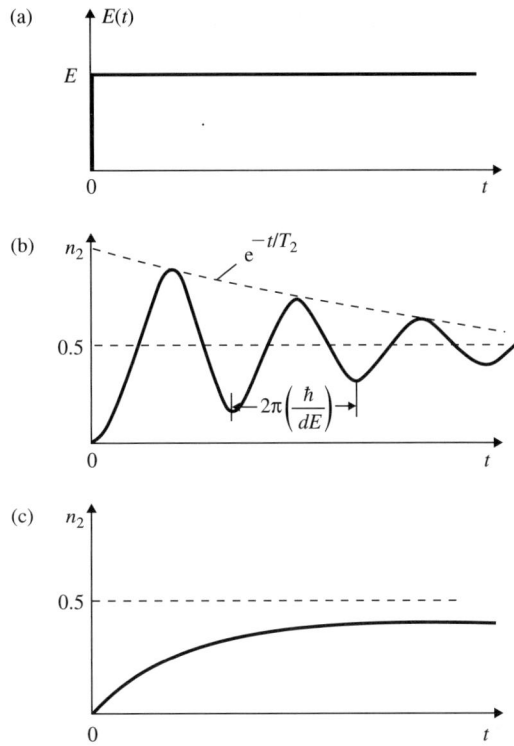

Fig. 2.6 Response of a two-level system to (a) a pulsed resonant laser-light field $E(t)$, (b) coherent interaction at $2(dE/\hbar) \gg 1/T_2$, and (c) incoherent interaction at $2(dE/\hbar) \ll 1/T_2$ (n_2 is the relative population of the excited level, and E is the light-field amplitude).

light-field frequency ω equal to ω_0:

$$\sigma(\omega) = \left(4\pi\omega_0 d^2/c\hbar\Gamma\right) L\left(\frac{\omega-\omega_0}{\Gamma}\right) = \frac{\lambda^2}{2\pi}\frac{A_{21}}{\Gamma} L\left(\frac{\omega-\omega_0}{\Gamma}\right), \qquad (2.67)$$

where $L(x) = (1+x^2)^{-1}$ is the Lorentzian line shape (eqn 2.47). This expression is another form of eqn (2.11) for the more general case of a homogeneous spectral width 2Γ. The populations of the two levels here tend to become equal in an aperiodic fashion. If condition (2.64) is satisfied, the interaction of the two-level quantum system with the laser-light field is said to be *incoherent*. The qualitative distinction between the evolutions of the population of, for example, the upper level, n_2, in the case of coherent and incoherent interaction is illustrated in Fig. 2.6.

2.5.5 Rate equation approximation. Saturation power and energy

With the incoherent interaction approximation, the set of equations (2.59) is greatly simplified. In this case, the polarization P of the two-level system depends in a

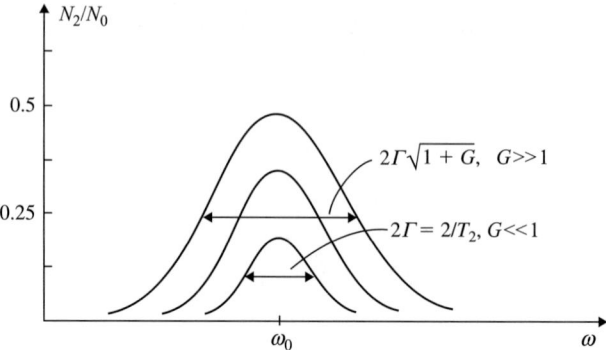

Fig. 2.7 Excitation resonance broadening of a two-level system with increasing laser radiation intensity (ω is the field frequency, ω_0 is the central transition frequency, $\Gamma = 1/T_2$ is the homogeneous spectral-line half-width, and G is the saturation parameter).

steady-state manner on the light-field amplitude E:

$$P = \left(\frac{d^2}{\hbar \Gamma}\right) NE \sin \varphi. \tag{2.68}$$

Therefore, the interaction of the system with the light field can be described by a simple *rate* equation following from eqns (2.59) and (2.68):

$$\frac{dN}{dt} + \left(\frac{1}{T_1}\right)(N - N_0) = -2\sigma(\omega) IN, \tag{2.69}$$

where $\sigma(\omega)$ is the radiative transition cross section at the light-field frequency ω, and I is the radiation intensity (in photons/cm² s). The relative populations of each level, n_1 and n_2, can be found by solving eqn (2.69) with the relation $n_1 + n_2 = 1$.

For instance, under steady-state conditions, when the laser-pulse duration τ_p is much longer than the population relaxation time T_1, the difference in the level populations is defined by the very simple relation

$$N(t) = \frac{N_0}{1+G} = N_0 \left(1 + \frac{I(t)}{I_{\text{sat}}}\right)^{-1}, \tag{2.70}$$

where I_{sat} is the absorption *saturation intensity*, given by

$$I_{\text{sat}} = \frac{1}{2\sigma(\omega) T_1}, \quad \tau_p \gg T_1. \tag{2.71}$$

When the saturation is strong enough ($I(t) \gg I_{\text{sat}}$), there evidently occurs equalization of the level populations. This is only natural, for there then occurs the establishment of equilibrium between the two-level system and a light field having a very high effective radiation temperature ($kT_{\text{rad}} \gg \hbar\omega_0$).

In the opposite case of excitation by a short light pulse ($\tau_p \ll T_1$), we can omit the longitudinal relaxation term in eqn (2.69) and obtain the following simple relation for the change in the population difference:

$$N(t) = N_0 e^{-\Phi/\Phi_{\text{sat}}}, \tag{2.72}$$

where Φ is the laser pulse fluence, defined as

$$\Phi(t) = \int_{-\infty}^{t} I(t') \, dt', \tag{2.73}$$

and Φ_{sat} is the absorption saturation fluence, given by

$$\Phi_{\text{sat}} = \frac{1}{2\sigma(\omega)}, \quad \tau_p \ll T_1. \tag{2.74}$$

Again, when the saturation is sufficiently strong ($\Phi \gg \Phi_{\text{sat}}$), there occurs equalization of the populations of the two levels.

2.5.6 Power broadening

Of great importance in spectroscopy is the width of the absorption line under conditions of saturation, which cannot be avoided if the probability of exciting an atom is to be maximized. Solving the set of equations (2.59) in the steady-state case, we can obtain the dependence of the difference in the level populations on the frequency detuning $\Delta = \omega - \omega_0$:

$$\frac{N(\omega)}{N_0} = 1 - \frac{\Gamma^2 G}{(\omega - \omega_0)^2 + \Gamma^2(1+G)}. \tag{2.75}$$

According to eqn (2.52), the probability of excitation $n_2(\omega)$ or $N_2(\omega)$ to the upper level is

$$n_2(\omega) = \frac{1}{2} \left\{ \frac{\Gamma^2 G}{(\omega - \omega_0)^2 + \Gamma^2(1+G)} \right\}, \tag{2.76}$$

where, for the sake of definiteness, the particle is assumed to be initially in the lower level. Figure 2.7 shows the frequency dependence of the upper-level excitation probability at various degrees of saturation. As can be seen, the *power broadening effect* of the excitation spectrum is manifest in the graph. As follows from eqn (2.76), the effective half-width of the excitation spectrum is given by (Karplus and Schwinger 1948)

$$\Delta\omega_{\text{eff}} = \Gamma\sqrt{1+G} = \left(\frac{1}{T_2}\right)\left[1 + \left(\frac{d^2 T_1 T_2}{\hbar^2}\right)E^2\right]^{1/2}. \tag{2.77}$$

The broadening of spectral lines due to excitation by a strong light field should be taken into account when one is interpreting the results yielded by the various spectroscopic techniques based on the use of stimulated quantum transitions.

2.5.7 Field splitting of levels

The broadening of the excitation spectrum in a strong light field is amenable to a simple physical explanation, especially if we consider the case of a strong field where a particle oscillates between two levels. For the sake of simplicity, we consider the case of exact resonance ($\Delta = 0$). Then, according to eqn (2.43), the probability amplitudes $a_k(t)$ oscillate as

$$a_1(t) = \cos\left(\frac{\Omega_0}{2}\right)t, \quad a_2(t) = \sin\left(\frac{\Omega_0}{2}\right)t, \qquad (2.78)$$

while the squared probability amplitudes $|a_k(t)|^2$ describing the level populations oscillate of the frequency Ω_0.

In accordance with eqn (2.32), the wave function of the two-level system may be represented, considering eqn (2.79), in the form

$$\Psi = \frac{1}{2}\sum_{k=1,2}\exp\left[-i\left(\frac{E_k}{\hbar} + \frac{\Omega_0}{2}\right)t\right] + \exp\left[-i\left(\frac{E_k}{\hbar} - \frac{\Omega_0}{2}\right)t\right]. \qquad (2.79)$$

It is at once evident from the above expression that the particle's oscillation at a frequency Ω_0 between the two energy levels E_k results in a *field splitting* of each level into two sublevels $E_k \pm (\Omega_0/2)\hbar$, spaced a distance $\hbar\Omega_0$ apart. Such split energy levels, referred to as quasi-energy levels, have a common origin associated with the properties of the periodic wave function of a two-level system in an external periodic field of constant amplitude (Zel'dovich 1973).

Field splitting of energy levels (see Fig. 2.8) can also be observed in transitions from level 1 or 2 to some third level. As a result, of this splitting, the spectral line of the $2 \rightarrow 3$ transition is split into two lines (Fig. 2.8(b)). This phenomenon was first observed in microwave spectroscopy (the Autler–Townes effect (Autler and Townes 1955)). In the optical region of the spectrum and in the presence of phase relaxation, such a splitting can apparently be observed only if condition (2.62) is satisfied, that is, when the response of the two-level system to the laser-light field is truly oscillatory.

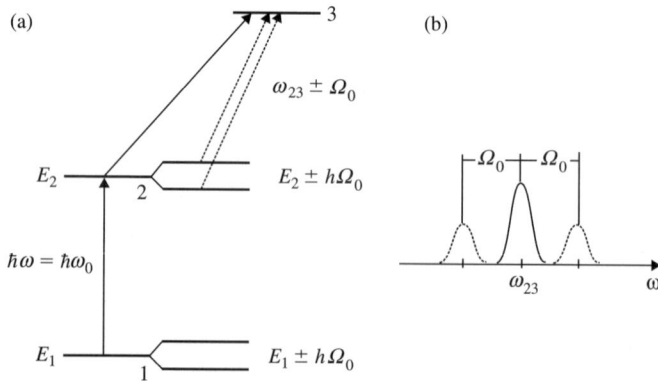

Fig. 2.8 Level splitting of a two-level system in a strong resonant light field, which can be observed in a transition to some third level.

The quantum levels of a real atom or molecule are usually degenerate. In that case, the description of the interaction between a field and degenerate two-level system becomes more complicated. In particular, there is no longer any simple, graphic picture of the particle's Rabi-frequency oscillations between the two levels. There are, instead, the particle's oscillations between individual sublevels with frequencies of their own, which combine to smooth out the oscillations of the net level populations.

However, in the case of incoherent interaction, the picture looks simpler, for there are no level population oscillations at all. In that case, to describe the evolution of level populations, use can be made as before of rate equations, such as eqn (2.61), but the quantities n_1 and n_2 should now be understood to be the total populations of each level. The difference in level populations given by eqn (2.48) should then be modified for naturally polarized light as follows:

$$N(t) = n_2(t) - \left(\frac{g_2}{g_1}\right) n_1(t). \tag{2.80}$$

In the case of strong saturation ($G \gg 1$), there occurs equalization of not the total level populations n_k but of the sublevel populations, that is,

$$\frac{n_1}{g_1} \simeq \frac{n_2}{g_2}. \tag{2.81}$$

Under the condition $g_2 > g_1$, the number of excited atoms can be much higher than in the initial state. For instance, for an allowed S $\to$ P dipole transition ($J_1 = 0$, $J_2 = 1$), $g_1 = 2J_1 + 1 = 1$ and $g_2 = 2J_2 + 1 = 3$, and so 75% of the particles can be excited, provided that the saturation is strong enough.

2.6 Radiation-scattering processes

The laser control of atoms and molecules usually uses the processes of resonant absorption and emission of photons. But the processes of scattering of photons by atoms, molecules, macroparticles, and free electrons (Chapter 13) also prove very useful in a number of cases. These processes are fairly diverse, and concrete information about them can be found elsewhere in the text. Here we shall restrict ourselves to a general description only.

All scattering processes can be subdivided into elastic and inelastic processes (Fig. 2.9(a)). In the case of elastic scattering, the atomic particle and the photon being scattered exchange only their momenta, without a change of the internal energy of the atomic particle. In such scattering, the scattered photon changes the direction of its momentum, and so does the scattering particle. When this process is repeated many times, the motion direction and the velocity of the scattering particle can be changed (Chapter 5). In the case of inelastic scattering, the energy of the photon being scattered and the quantum state of the scattering atomic particle are changed (Figs. 2.9(b) and (c)). This process is referred to as Raman scattering. If an atomic particle is raised to an excited state upon inelastic scattering of a photon, the process is called Stokes scattering, and conversely, if an excited particle drops to a lower energy level upon inelastic scattering of a photon, the process is called anti-Stokes scattering.

A scattering processes possess two general properties. First, as the frequency of the scattered photon approaches a resonance frequency, the scattering cross section

34 Laser control of atoms and molecules

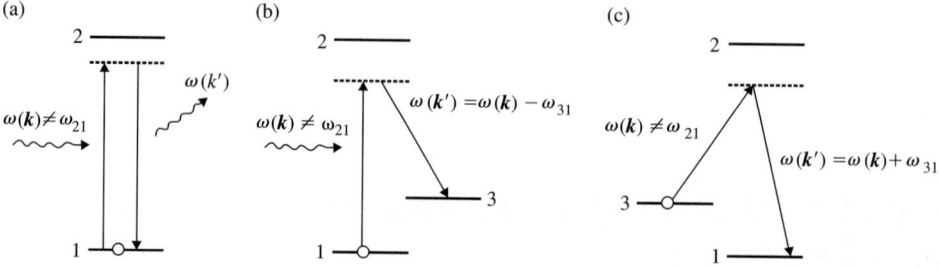

Fig. 2.9 Elementary processes of elastic scattering (a) and of nonelastic (Raman) scattering: (b) Stokes and (c) anti-Stokes Raman scattering.

increases, the frequency dependence of the cross section having a Lorentzian shape. For example, when the frequency of the scattered photon exactly coincides with the resonance frequency, the scattering cross section reaches its maximum, defined by the following expression (Loudon 1973):

$$\sigma_{\text{scatt}} = \frac{\lambda^2}{2\pi}. \qquad (2.82)$$

That is, it coincides with the maximum resonant absorption cross section given by eqn (2.12). In that case, the processes of resonant absorption and subsequent spontaneous decay are inseparable from the resonant scattering process. Here we introduce the notions of *coherent* and *incoherent* scattering. In coherent scattering, the phases of the photon in the initial and the final state (after scattering) are correlated, and, conversely, in incoherent scattering, these phases are absolutely uncorrelated. Thus, in the case of exact resonance, where the photon energy is imparted to the atomic particle for the lifetime of the excited state and then is reemitted in the form of spontaneous radiation, the resonant scattering process is incoherent. However, when the frequency of the photon being scattered differs from the resonance frequency by an amount greater than the natural width of the excited level of the scattering atomic particle, the photon cannot excite the latter because of the difference in energy between them and the impossibility of compensating for this difference in the isolated system "photon + atomic particle." For this reason, such a quasi-resonant scattering process is coherent.

Secondly, each type of spontaneous scattering process has a stimulated counterpart of its own. In other words, if there is a strong external field that has the same direction and frequency as the scattered photon, with a mode occupation number $|n_{\text{scatt}}\rangle \gg 1$, the scattering probability grows higher in accordance with Dirac's law (eqn 2.9). The process we have now is a two-quantum transition of the atom from one state into another in a two-frequency laser field. In this process, both energy and momentum, are imparted to the atom, which is also of interest from the standpoint of laser control.

3
Laser velocity-selective excitation

3.1 Doppler broadening of optical spectral lines

A moving particle (an atom or molecule) emits or absorbs radiation that is not exactly at the quantum transition frequency $\omega_0 = \omega_{21}$ between two energy levels E_1 and E_2, which is determined by the Bohr quantization condition

$$\hbar\omega_0 = E_2 - E_1, \qquad (3.1)$$

where $\hbar$ is Planck's constant, but at a frequency shifted because of the Doppler effect (Fig. 3.1(a)). The spectral line of a single particle, $S(\omega)$, is shifted by an amount that depends on the projection of the particle velocity $\mathbf{v}$ on the direction of observation $\mathbf{n}$:

$$S_{\mathbf{v}}(\omega) = S\left(\omega - \mathbf{n}\frac{\mathbf{k}\cdot\mathbf{v}}{c}\right). \qquad (3.2)$$

In a gas, particles move in all possible directions. For this reason, the Doppler shift differs between individual particles. At thermal equilibrium, all directions are equiprobable, that is, the velocity distribution of the particles is isotropic. Therefore, the projection of the particle velocity on any direction ($v = \mathbf{n}\cdot\mathbf{v}$) is given by the Maxwell distribution

$$W(v) = \frac{1}{\sqrt{\pi}u}\exp\left[-\left(\frac{v}{u}\right)^2\right], \quad u = \left(\frac{2kT}{M}\right)^{1/2}, \qquad (3.3)$$

which has the symmetrical form of a Gaussian curve (Fig. 3.1(b)). As a result, the spectral line of an ensemble of particles has a symmetrical profile with its center at the transition frequency ω_0 (Fig. 3.1(c)).

In the simple case where the broadening due to the Doppler effect and that caused by particle collisions are statistically independent, the shape of the spectral line $S(\omega)$ of the whole ensemble of particles is defined by the convolution of the line shape of an individual particle and the Doppler-shift distribution, that is, by the distribution of the projections of the atomic velocities onto the observation direction. The full width at half maximum (FWHM) of a spectral line due to the Doppler effect, $\Delta\omega_D$, is given by

$$\Delta\omega_D = \frac{2\omega_0}{c}\left(2\ln 2\frac{kT}{M}\right)^{1/2} = 7.163\times 10^{-7}\left(\frac{T}{A}\right)^{1/2}\omega_0, \qquad (3.4)$$

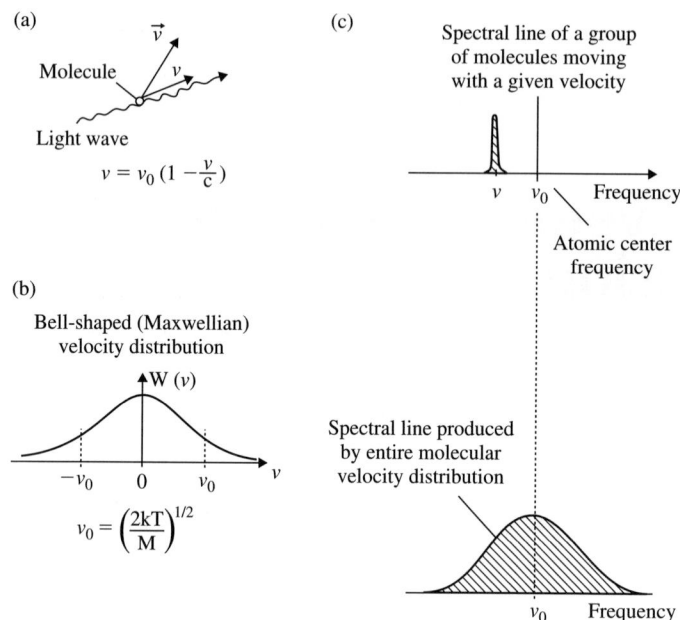

Fig. 3.1 Influence of the Doppler effect on the shape of a spectral line. (a) Doppler shift due to a molecule moving with velocity **v**. The Doppler effect shifts the emission frequency from ν_0 to ν. (b) Thermal distribution of particle velocities. (c) Corresponding spectral lines. The upper curve is the response of particles moving with velocity **v** as in (a). The lower curve is the response due to atoms over the entire thermal distribution.

where M and A denote the mass and the atomic weight of the particle, k is the Boltzmann constant, and T is the absolute temperature (in K). For atoms and molecules with an atomic weight of $A = 100$ at room temperature, the Doppler spectral linewidth $\Delta\omega_D$ is approximately $10^{-6}\omega_0$. A Doppler-broadened spectral line is essentially a set of a large number of much narrower absorption or emission spectral lines of particles with different velocities. Therefore, Doppler broadening is often referred to as *inhomogeneous* broadening. *Homogeneous* broadening means that the width of a spectral line does not depend on the specific velocity of every particle.

Consider the interaction between molecules and a monochromatic light wave with a frequency $\omega_0 = 2\pi\nu_0$ close to the frequency of a quantum transition between two molecular levels. Such a wave can interact only with molecules moving almost transverse to the light wave, since such molecules have a very small Doppler frequency shift (Fig. 3.2). Other molecules cannot interact with the light wave, because of the Doppler shift. If the wave frequency is not close to the central frequency, the molecules that move transverse to the beam do not interact with the field. Molecules whose velocity projection onto the direction of the light wave is about $v_{res} = (\nu - \nu_0)c/\nu_0$ are in resonance with the wave (Fig. 3.2(b)). Such a velocity is needed to compensate the detuning of the light wave frequency relative to the quantum-transition frequency by the Doppler shift. When $\nu > \nu_0$, the light wave interacts with molecules

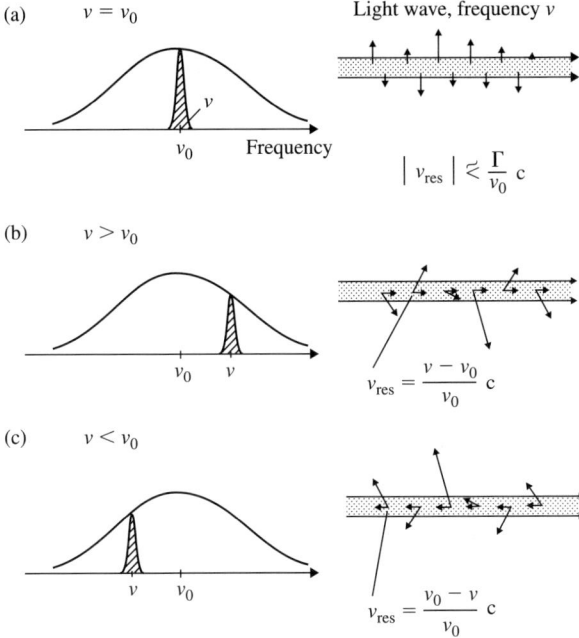

Fig. 3.2 Velocity groups of atoms or molecules resonating with an applied monochromatic field. (a) The monochromatic field is tuned to the atomic center frequency ($\nu = \nu_0$). The resonant atoms are those that have no velocity component in the direction of the light wave. (b) In this case, the light frequency is tuned above the atomic center frequency ($\nu > \nu_0$), so that the atoms that are Doppler shifted into resonance move in the same direction as the light wave. (c) In this case, the light frequency is tuned below the atomic center frequency ($\nu < \nu_0$), so that the atoms that are Doppler shifted into resonance move in the direction opposite to the propagation direction of the light wave.

that have positive velocity projections onto the wave propagation direction and, conversely, when $\nu < \nu_0$, it interacts with those which have negative velocity projections (Fig. 3.2(c)). Therefore, the wave can interact only with molecules that have the appropriate Doppler shift. Since the spectral linewidth of each of these molecules is the homogeneous width, a group of such molecules occupies a narrow spectral range within the Doppler profile; its center is at the field frequency ν and its width is equal to the homogeneous width.

So, a monochromatic traveling laser wave of frequency ω interacts only with a group of particles that have a certain velocity projection $\mathbf{v}$ on the direction of the wave vector $\mathbf{k}$ within the limits of the homogeneous width:

$$|\omega - \omega_0 + \mathbf{k} \cdot \mathbf{v}| \lesssim \Gamma. \tag{3.5}$$

It is this group of particles that can be selectively excited to the upper state. This possibility forms the basis for the laser control of the velocity distribution of particles at desired quantum levels.

3.2 Homogeneous broadening mechanisms

The width of a velocity-selective excitation resonance is determined by the homogeneous width 2Γ. Table 3.1 provides a quantitative understanding of the various mechanisms contributing to the homogeneous width.

The first and the most fundamental effect is the *radiative* broadening due to the spontaneous decay of an excited state (the half-width of the spectral line is $\Gamma = \gamma_{\text{rad}} = \gamma$). A spectral line due to spontaneous decay has a Lorentzian shape (eqn 2.5). The lifetime of an excited atomic or molecular state depends on the oscillator strength of the transition and the radiation wavelength. For electronic transitions of atoms and molecules in the visible region of the spectrum, $1/\gamma_{\text{rad}} \simeq 10^{-8}$ s. For metastable atomic and vibrational molecular levels, the radiative lifetime may be much longer, namely $1/\gamma_{\text{rad}} \simeq 10^{-1}$–$10^{-5}$ s.

Particle collisions contribute much to the homogeneous broadening. Each collision shifts the phase of the periodic motion of an electron in an atom, or that of the vibrations of the nuclei in a molecule. Such random shifts give rise to a quasi-periodic process that replaces the periodic process that defines the state of the atom or molecule. This quasi-periodic process occurs as a sequence of coherent trains with an average duration of τ_{coll}, which is the average time between successive collisions of one particle with the others. Besides a phase change, collisions can also cause a given atomic or molecular state to decay. The collision-induced line shape in the simple case where either a phase shift or a level decay takes place in the course of a collision is

Table 3.1 Homogeneous broadening mechanisms of spectral lines

Type	Origin	Line width, $\Delta\nu$	Range of values
Natural broadening	Spontaneous decay of an excited state	$\frac{1}{2\pi\tau}$ $\tau = $ natural lifetime	Atoms: 10^5–10^7 Hz Molecules: 10–10^3 Hz
Lorentz (collision) broadening	Interparticle collisions	$(\pi\tau_{\text{coll}})^{-1}$ $\tau_{\text{coll}} = $ mean time between collisions	3×10^3–3×10^4 Hz (at 1 mTorr pressure)
Wall-collision broadening	Particle collisions with the walls of the sample cell	$v_0/2\pi L$ $v_0 = $ mean velocity $L = $ cell diameter	10^3–10^4 Hz
Transit-time broadening	Transit of particles through light beam	$v_0/2\pi a$ $v_0 = $ mean velocity $a = $ beam diameter	10^3–10^4 Hz
Power broadening	High intensity of laser beam induces high rate of transition	$d_{12}E/2\pi\hbar$ $d_{12} = $ transition dipole moment $E = $ strength of laser field $\hbar = $ Planck's constant	10^4–10^5 Hz (for 1 mW/cm^2 intensity)

also Lorentzian. The collisional half-width at half maximum $\Delta\omega_{\text{coll}}$ is defined by the frequency of the particle collisions, that is,

$$\Delta\omega_{\text{coll}} = \frac{2}{\tau_{\text{coll}}}, \tag{3.6}$$

where the average time between collisions is determined by the concentration of the particles (cm^{-3}), the velocity-averaged collision cross section σ_{coll} (cm^2), and the average particle velocity v:

$$\tau_{\text{coll}} = \frac{1}{n\langle\sigma_{\text{coll}} \cdot v\rangle_v}. \tag{3.7}$$

Even a relatively weak interaction of particles, where their velocity and direction of motion vary only slightly, is sufficient to cause phase shifts. Therefore, the cross section for collisions that cause line broadening is usually much larger than the gas-kinetic cross section that determines strong particle-to-particle interaction. The collisional broadening $\Delta\nu_{\text{coll}} = \Delta\omega_{\text{coll}}/2\pi$ typical of molecules ranges between 3 and 30 MHz at a gas pressure of 1 Torr. This gives some 10^{-7} to 10^{-8} s for the average time between collisions causing broadening. At such a pressure, the collisional broadening for, say, vibrational transitions in molecules is thousands of times the natural width. Since the collisional line broadening is proportional to the gas pressure, pressures below 10^{-3} Torr should be used to reduce the collisional broadening to the natural linewidth. But at low gas pressures one more mechanism comes into play, whose contribution to the line broadening is significant.

At low gas pressures, the mean free path length of the molecules, Λ, which is related to their mean velocity v_0 and the mean free time τ_{coll} by the relation $\Lambda = v_0 \tau_{\text{coll}}$, increases and can, in principle, become comparable to or even greater than the diameter of the gas cell. In that case, the time between collisions τ_{coll} is determined not by particle–particle collisions but by collisions between particles and the walls of the gas cell. When the transverse dimension of the cell is a few centimeters, the particle–wall collisions cause a broadening of the order of 10^4 Hz. This value is rather small, but it exceeds the natural width of molecular transitions in the infrared region of the spectrum.

When moving particles interact with a beam of limited diameter, the resonance is broadened by one more effect, apart from those already mentioned. A molecule that has a velocity v_0 crosses a light beam with diameter a in a time of $\tau_{\text{tr}} = a/v_0$. The light beam may be regarded as a measuring tool with which the molecules interact in a finite time interval $\Delta t = \tau_{\text{tr}}$. The energy of a transition between levels cannot be evaluated to better than $\Delta E = \hbar/\Delta t$. This corresponds to an uncertainty in the transition frequency $\Delta E/\hbar = 1/\tau_{\text{tr}}$, that is, to a spectral line broadening due to the finite time of flight through the light beam (transit-time broadening),

$$\Delta\omega_{\text{tr}} = \frac{1}{\tau_{\text{tr}}}. \tag{3.8}$$

At relatively high gas pressures, when the free-path length Λ is much shorter than the light-beam diameter, this effect is not significant in comparison with collisional broadening, but at low pressures it may become the main mechanism broadening the long-lived transitions.

3.3 Doppler-free saturation spectroscopy
3.3.1 Saturation of a Doppler-broadened spectral line

Thus, a monochromatic light wave with a small divergence, that is, a light field of high spatial and temporal coherence, can interact with a small proportion of the atoms or molecules contributing to a Doppler-broadened transition. The field can accordingly change the state of this small proportion of the particles and *discriminate* them distinctly from the rest of the particles, whose velocities are far from the resonance condition (eqn 3.5). Let the intensity of the light field be sufficient to excite a considerable proportion of the particles to the upper state. In the simple case of a steady-state excitation under incoherent interaction conditions (eqn 2.64), the degree of excitation is governed by what is known as the degree of saturation (eqn 2.66). The total probability W_{12} of a stimulated transition of a particle with a velocity $\mathbf{v}$ to the excited level under the effect of a traveling wave $E\cos(\omega t - \mathbf{k}\cdot\mathbf{r})$ is

$$W_{12}(\mathbf{v}) = \frac{G}{2}\frac{\Gamma^2}{(\Delta - \mathbf{k}\cdot\mathbf{v})^2 + \Gamma^2(1+G)}, \qquad (3.9)$$

where $\Delta = \omega - \omega_0$ is the detuning of the wave frequency ω from the frequency of the particle transition ω_0. The probability of a transition of the particle to the upper level is determined by the saturation parameter G and the deviation of the particle's velocity from the resonance velocity.

The predominant excitation of particles moving with a particular velocity alters the equilibrium particle velocity distribution at each of the levels of the transition (Fig. 3.3). The velocity distribution of particles at the lower level, $n_1(\mathbf{v})$, develops a shortage of particles moving with velocities close to the resonance condition (eqn 3.5):

$$\mathbf{n}_1(\mathbf{v}) = n_1^0(\mathbf{v}) - W_{12}(\mathbf{v})\left[n_1^0(\mathbf{v}) - n_2^0(\mathbf{v})\right], \qquad (3.10)$$

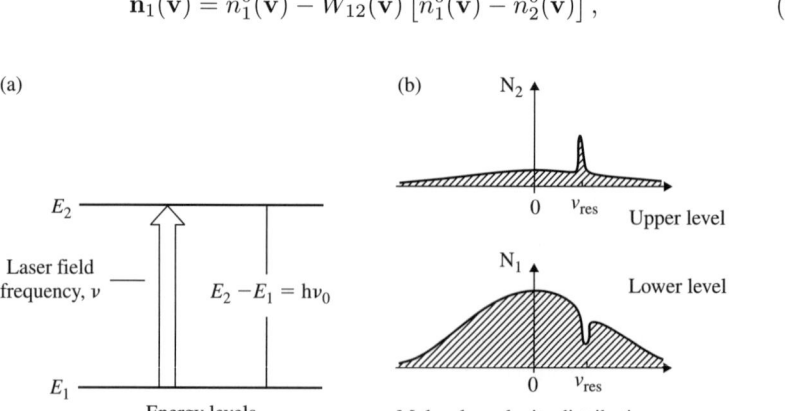

Fig. 3.3 Changes in the particle velocity distribution for two levels of a transition under the action of a traveling laser wave of frequency ν: (a) level diagram; (b) distribution of the Z component of the velocity of the particles at the lower and upper levels of the transition ($v_{\text{res}} = (\nu - \nu_0)c/\nu_0$ = resonant projected velocity).

where $n_1^0(\mathbf{v})$ and $n_2^0(\mathbf{v})$ denote the initial equilibrium velocity distributions of particles at the lower and the upper level, respectively. For the upper level, the velocity distribution, on the other hand, has an excess of particles with resonant velocities:

$$n_2(\mathbf{v}) = n_2^0(\mathbf{v}) + W_{12}(\mathbf{v})\left[n_1^0(\mathbf{v}) - n_2^0(\mathbf{v})\right]. \tag{3.11}$$

Figure 3.3 shows the distribution of the projections of the particle velocities on the light-beam direction for the lower and upper levels. When there is no light wave, these distributions are symmetrical, whereas in the presence of a strong light wave a "hole" develops in the velocity distribution for the lower level and a "peak" in that for the upper level. This hole (or peak) develops at a certain particle velocity depending on the light field frequency:

$$v = v_{\text{res}} = \frac{\omega - \omega_0}{\omega} c. \tag{3.12}$$

The depth of the hole (or the height of the peak) is determined by the saturation parameter G. The width $\Delta\omega$ of the hole (or peak) is equal to the homogeneous width 2Γ, allowing for power broadening, which, according to (eqn 2.76) or (3.9), also depends on G.

Thus, the light wave changes the velocity distribution of the particles at the various levels, that is, it makes the distribution essentially anisotropic. Indeed, it causes the Doppler-broadened line to be distorted. A "hole" appears in the Doppler profile because of the particles that are excited to the upper state, and the width of the hole directly determines the homogeneous transition width, which may be many orders of magnitude narrower than the Doppler width. To obtain such a narrow structure within the Doppler profile, the light wave should satisfy three conditions:

(1) monochromaticity, or high temporal coherence;
(2) directionality, or high spatial coherence;
(3) an intensity sufficient to saturate the transition.

These conditions can only be met by laser radiation. Therefore, it is quite clear why it was inevitable that the development of quantum electronics would give birth to saturation laser spectroscopy free from Doppler broadening.

3.3.2 The Lamb dip and inverted Lamb dip

The first gas laser (Javan et al. 1961) operated at a wavelength of $\lambda = 1.15\,\mu\text{m}$ with a neon–helium mixture at a pressure of about 1 Torr. In that case, the homogeneous linewidth was much narrower than the Doppler width (Table 3.1) and therefore the line broadening had an inhomogeneous nature. In such a case, the light wave interacts only with particles that are in resonance with it. Therefore, a strong light wave that causes amplification saturation "burns" a "hole" (Bennett hole) (Bennett 1962) at the wave frequency in the Doppler profile of the amplification line. In the laser cavity, there is a standing light wave that may be represented as a superposition of two counterrunning waves of the same frequency. In that case, each wave burns its own "hole." Because these two waves run in opposite directions, there arise two holes symmetrical about the center of the Doppler profile (Fig. 3.4). In essence, the laser field absorbs the energy from two groups of amplifying particles that have opposite velocities. As the

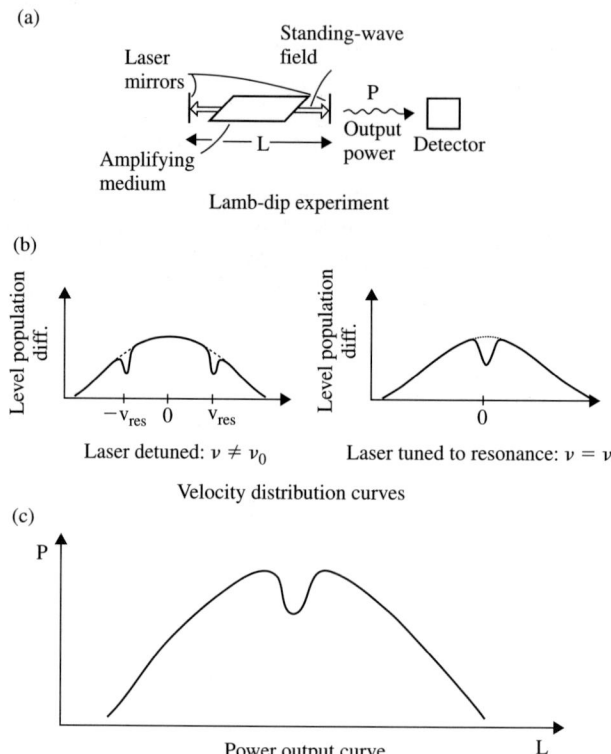

Fig. 3.4 Lamb dip experiment: (a) experimental arrangement; (b) velocity distribution curves. Note that the saturated velocity groups can overlap only when the laser frequency is tuned to the center of the Doppler profile. (c) Power output curve. The laser intensity is plotted as a function of the laser frequency or the fine tuning of the separation between the laser mirrors. The narrow dip in the center is the Lamb dip.

laser frequency is tuned in toward the center of the Doppler profile, the two holes coincide and the standing light wave interacts with only one group of particles. This results in a resonant decrease of power at the center of the Doppler amplification line. This effect, now called the "Lamb dip," was first considered by Lamb in his theory of the gas laser (Lamb 1964). Experimental observations of the effect were first reported in 1963 (McFarlane et al. 1963; Szoke and Javan 1963).

The width of the dip at the center of the Doppler amplification line is equal to the homogeneous linewidth 2Γ which may be considerably smaller than $\Delta\omega_D$. This fact has opened up strong possibilities for spectroscopy within the Doppler profile. Such a method has been used in experiments on measuring collisional broadening and isotope shifts and on stabilizing the laser frequency at the center of the amplification line. However, the spectral resolution in these experiments was limited by the homogeneous width $2\Gamma_{\text{hom}}$ of the amplified spectral transition between excited states, which was relatively large.

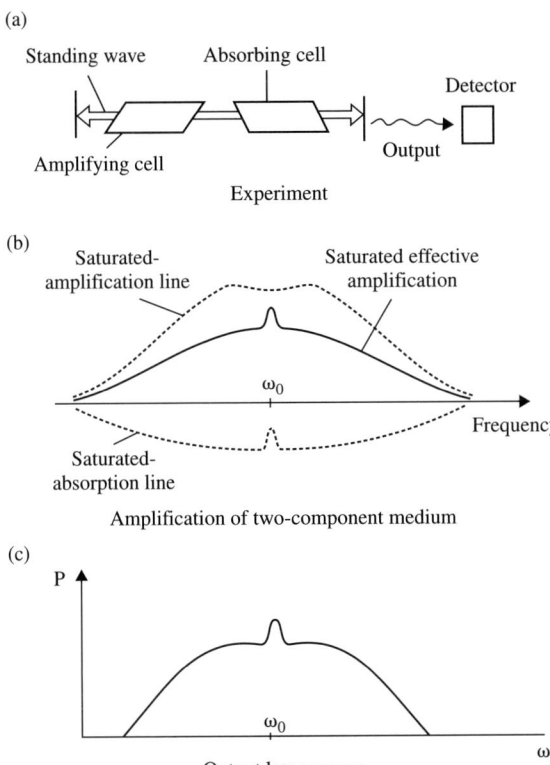

Fig. 3.5 Production of inverted Lamb dip in a gas laser with an intracavity low-pressure absorption cell: (a) experimental arrangement; (b) frequency dependence of saturated total amplification of two-component medium; (c) power output as a function of laser-field frequency.

The situation has changed greatly since attention was turned to observation of the Lamb dip in resonant-absorption media. The first suggestions of using nonlinear absorption resonances were made independently by three research groups (Lee and Skolnick 1967; Letokhov 1967; Lisitsyn and Chebotayev 1968), who also performed experiments of this type. These investigators proposed to insert a resonant-absorption cell containing gas at low pressure into the laser cavity (Fig. 3.5(a)). Saturated absorption in a standing light wave gives rise to a narrow Lamb dip at the center of the Doppler-broadened absorption line. As a result, the efficient saturated amplification by the two-component medium in the laser acquires a narrow peak at the center of the absorption line (Fig. 3.5(b)), and the laser output power exhibits a narrow peak (Fig. 3.5(c)), often called the 'inverted Lamb dip." The virtues of this method are that the absorbing gas, at a low pressure and with the appropriately selected particle species and transition, may have a very narrow homogeneous width, of the order of 10^4 to 10^6 Hz (Table 3.1). Particular emphasis was placed upon this essential feature of infrared molecular transitions by Letokhov (1967) and Lisitsyn and Chebotayev (1968). There are two circumstances of importance for the application of this method in spectroscopy and in laser frequency stabilization. First, the dip in the absorption line can be narrower by a factor of 10^{-2} or 10^{-3} than the dip in the amplification line. Actually, absorption, in contrast to amplification, can take place by means of transitions from the ground state or a state close to the ground state, to an excited long-lived state. As a result, the radiative width can be negligible. Because the ground-state population in the absence of excitation is rather large, absorption can be observed in a gas at a very low pressure, when the collisional width is also small. Second, because of the low pressure and the absence of excitation in the gas, the center of the spectral absorption line can be rather stable. Specifically, such an experiment with CH_4 molecules was proposed where the rotational–vibrational transition $P(7)$ in the ν_3 band of CH_4 was coincident with the emission line of the He–Ne laser at $\lambda = 3.39\,\mu\mathrm{m}$. This experiment was performed by Hall (Barger and Hall 1969), and the resonance obtained was only $0.3\,\mathrm{MHz}$ in width, that is, 10^3 times as narrow as the Doppler absorption linewidth of CH_4. Figure 3.6 presents the results of Hall's experiment at the National Bureau of Standards with a He–Ne laser and an absorption cell filled with methane at low pressure. Experiments of this type played an important role in developing He–Ne/CH_4 lasers with ultrahigh frequency stability.

To obtain a narrow saturation resonance at the center of an absorption line requires not a standing light wave, but only a strong running wave and a weak counterrunning wave (Fig. 3.7). This strong running wave excites molecules whose velocity projections onto the wave propagation direction are $v = (v - v_0)c/v_0$. Because the counterrunning wave has the same frequency but is opposite in direction, it interacts with molecules whose projected velocities are equal but opposite to those of the molecules interacting with the strong wave. If the frequency of the waves does not coincide with that of the center of the Doppler line, ν_0, the weak probe wave is not responsive to the strong wave.

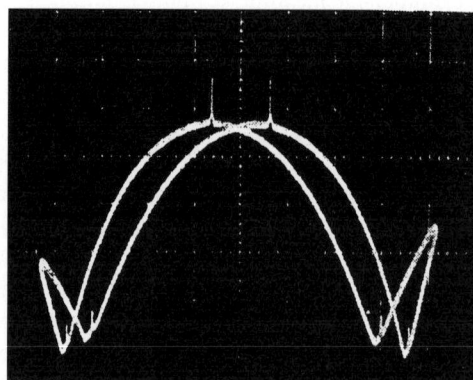

Fig. 3.6 Observation of an inverted Lamb dip in a He–Ne laser at $3.39\,\mu\mathrm{m}$ with an intracavity CH_4 absorbing cell. Output power versus laser-cavity tuning, with double scan. The saturated absorption peak has an amplitude of about 2% with a $400\,\mathrm{kHz}$ width, and is nicely centered on the ^{20}Ne Doppler gain curve. (Reprinted from Barger and Hall 1969, with courtesy and permission of the American Physical Society.)

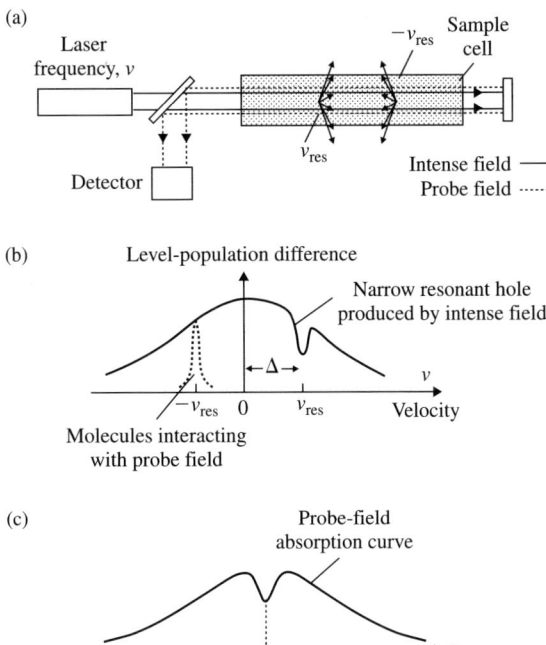

Fig. 3.7 Observation of a narrow saturation resonance by use of a single strong coherent traveling wave and a countertraveling weak probe wave. (a) Experimental arrangement. A small part of the intense wave is reflected back through the cell. The attenuation of this weak wave is studied as a function of the laser-field frequency. (b) Molecular velocity distribution, showing velocity groups that interact resonantly with the strong wave and the probe wave. (c) Absorption of probe wave as a function of frequency.

However, when the frequency is coincident with the center frequency of the Doppler profile, the weak probe wave interacts with molecules whose absorption has already been reduced by the strong counterrunning wave. Consequently, the absorption of the probe wave has a resonant minimum equal in width to the homogeneous width and centered exactly on the Doppler-broadened absorption line. This method has been demonstrated in experiments using a CO_2 laser operating at 10 μm and SF_6 molecules (Basov et al. 1969), and now it is universally accepted in laser saturation spectroscopy.

3.3.3 Saturation spectroscopy on coupled transitions

Real atoms and molecules have multilevel structures. The simplest and most interesting are three-level configurations (Fig. 3.8). The narrow "hole" in the velocity distribution of a set of multilevel particles that results when some of them are excited by a coherent light wave occurs for coupled transitions as well. For example, because of the peak in the velocity distribution at the upper level (Fig. 3.3), the absorption line corresponding to the transition from this level to a higher energy level also has a resonance peak with a width much smaller than the Doppler width (Fig. 3.9).

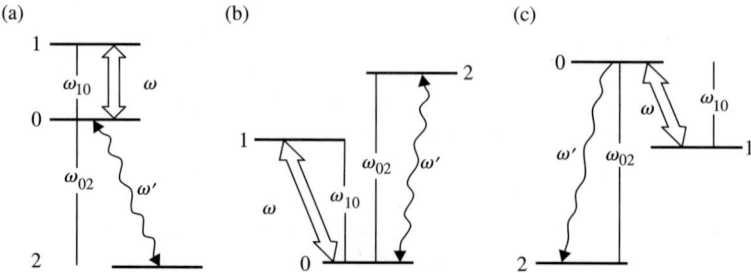

Fig. 3.8 Energy-level configurations for two coupled transitions: (a) cascade configuration; (b, c) bent V and Λ configurations.

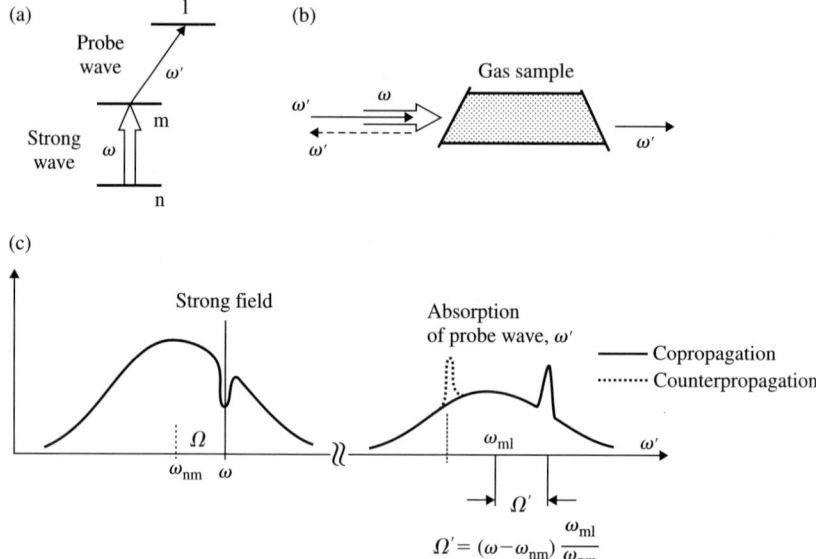

Fig. 3.9 Observation of narrow saturation resonances associated with coupled transitions: (a) energy-level diagram; (b) detection scheme; (c) absorption of probe wave for coupled transition $m-1$ as a function of its frequency ω'.

The line shape for the resulting transition can be conveniently observed by using an additional probe wave traveling in the same direction as the strong wave. This method of Doppler-free spectroscopy for coupled transitions was proposed and experimentally accomplished by Schlossberg and Javan (1966). These authors demonstrated that the shape of the absorption line for the coupled transition is determined not only by effects due to the change in the level populations; quantum coherence also gives rise to much finer effects compared with the simple change of the velocity distribution at the third level.

For coupled atomic transitions whose homogeneous linewidth depends mainly on radiative decay, saturation spectroscopy with a two-frequency field makes it possible to observe narrow resonances with a linewidth smaller than the homogeneous width. An atom absorbs photons $\hbar\omega$ and $\hbar\omega'$ from two unidirectional traveling waves of frequencies ω and ω' and makes a two-quantum transition from a level n to a level l (see Fig. 3.9) (Notkin et al. 1967). In this case, the resonance width of, say, the transition $m \to l$ is determined by the levels n and l, rather than m and l. The contribution of the level m is small if the difference between the wave vectors of the two waves is small. When $\mathbf{k}_\omega = \mathbf{k}_{\omega'}$, the resonance width is equal to the radiative width of the "forbidden" transition $n \to l$. Popova et al. (1969) introduced a classification of the effects causing changes in the absorption or emission spectrum of a gas placed in an external monochromatic field resonating with a coupled transition. The first effect is the formation of a nonequilibrium velocity distribution of the atoms; the second is a splitting of the atomic levels; and the third is a nonlinear quantum interference effect due to coherence, which is caused by the action of the strong field on the atoms. A great number of studies have been devoted to the three-level saturation spectroscopy method. Many of the results obtained have been summarized in a monograph (Letokhov and Chebotayev 1977).

The three-level saturation spectroscopy of Λ- and V-level configurations flourished in the subsequent years and led to the discovery of numerous new effects, such as "coherent population trapping" (Arimondo 1996), electromagnetically induced transparency (Harris 1997), and "lasing without population inversion" (Kocharovskaya 1992). These effects are beyond the scope of the present book. The control of quantum coherence and interference in laser-driven three-level systems has been treated in detail in the excellent reviews cited above.

3.3.4 Dispersion saturation spectroscopy

A change in the absorption of a medium in a resonant light field must be accompanied by a change in the refractive index of the medium, for the absorption variation $\Delta\chi$ (cm^{-1}) is related to the refractive-index dispersion Δn by the Kramers–Kronig relation. In the particular case of a Lorentzian profile

$$L\left(\frac{\omega - \omega_0}{\Gamma}\right) = \left[1 + \left(\frac{\omega - \omega_0}{\Gamma}\right)^2\right]^{-1} \tag{3.13}$$

with an absorption linewidth of 2Γ, this relation has the following simple form (Brillouin 1960):

$$\Delta n = -\frac{\Delta\chi}{2k}\frac{(\omega - \omega_0)}{\Gamma}L\left(\frac{\omega - \omega_0}{\Gamma}\right), \tag{3.14}$$

where $k = \omega/c$, and ω_0 is the absorption line center. Therefore, when the laser field induces an absorption saturation resonance, a corresponding resonance must develop in the *dispersion* of the medium being saturated. In accordance with eqn (3.14), the shape of the saturated dispersion resonance (in the case of weak saturation) must look like that of the first-order derivative of the absorption saturation resonance.

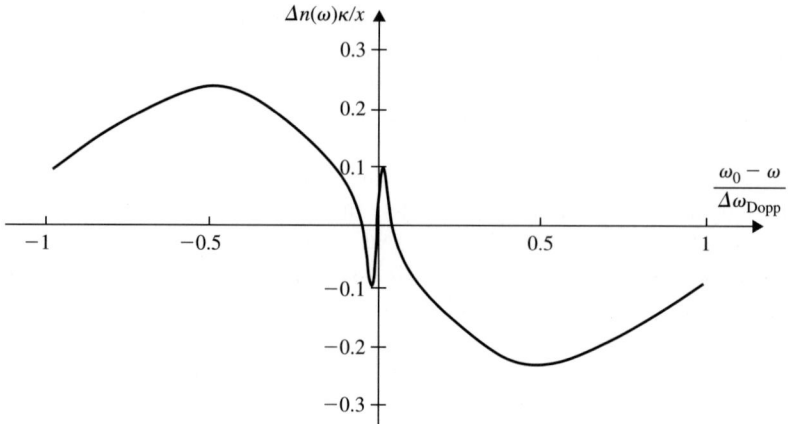

Fig. 3.10 Frequency dependence of the refractive-index change $\Delta n(\omega)$ in a standing laser wave for a moderate degree of saturation of a Doppler-broadened absorption line.

Figure 3.10 presents a typical function $\Delta n(\omega)$. It can be seen that the dispersion curve is a superposition of two curves. The relatively wide component curve describes the anomalous refractive-index dispersion within the limits of the Doppler profile. The maximum refractive-index variation due to linear absorption is reached when the field frequency is tuned off resonance by an amount approximately equal to the Doppler half-width:

$$\Delta n_{\max} = (2\pi e)^{-1/2}\frac{\chi_0}{k} \quad \text{at } |\omega - \omega_0| = \frac{\Delta\omega_D}{2} \qquad (3.15)$$

The narrow curve of reversed sign is due to the refractive-index dispersion in the saturated absorption region. The maximum dispersion value here is reached at a frequency detuning equal to the half-width of the narrow saturated absorption resonance:

$$\Delta n_{\max}^{\text{sat}} = G(4\pi)^{-1/2}\frac{\chi_0}{k} \quad \text{at } |\omega - \omega_0| = \Gamma. \qquad (3.16)$$

The relative dispersion $\Delta n_{\max}^{\text{sat}}/\Delta n_{\max} = G\sqrt{e/2}$ in the narrow resonance region grows linearly with increasing saturation parameter (at a low degree of absorption saturation).

The refractive-index variation $\Delta n_{\max}^{\text{sat}}$ in the narrow resonance region is of the same order of magnitude as $\Delta n_{\max}$. However, such absorption saturation occurs in a very narrow frequency region ($2\Gamma \ll \Delta\omega_D$) near the center of the Doppler profile. This effect substantially influences the group velocity of light in a medium. The group velocity of light in a dispersive medium is defined by the equation (Brillouin 1960)

$$v_{\text{gr}} = \frac{c}{n(\omega) + \omega\partial n/\partial\omega}, \qquad (3.17)$$

where c is the velocity of light in a vacuum. By using eqn (3.17), one can represent the group velocity of light in the narrow resonance region ($|\omega - \omega_0| \, \Gamma$) as

$$\frac{v_{\rm gr}}{c} = \left[n(\omega_0) + G \frac{c\chi}{\pi^{1/2} \Gamma} \right]^{-1}. \tag{3.18}$$

Because the magnitude of the parameter $\partial n / \partial \omega$ in the narrow resonance region is high, the group velocity of light here may be lower by many orders of magnitude than the velocity c of light in a vacuum ("slow light"), as demonstrated by Kash et al. (1999).

The refractive-index dispersion can be used to observe absorption lines. This approach is also applicable to the observation of narrow dispersion saturation resonances within the Doppler profile. The first experiments along these lines were performed by Borde et al. (1973) by means of a ring interferometer containing an iodine-vapor-filled nonlinear-absorption cell.

The concept of *polarization saturation spectroscopy*, suggested by Wieman and Hansch (1976), is close to that of interferometric saturation spectroscopy. Both methods succeed in attaining the same goal—to strongly suppress the background noise level and accordingly improve the signal-to-noise ratio. The essence of polarization spectroscopy is as follows. A strong circularly polarized light field causes absorption saturation, and a corresponding optical orientation of particles and an anisotropy in the absorption cell. A linearly polarized probe light beam passes through the absorption cell and through crossed polarizers. Under normal conditions, the transmission of the probe beam is zero. However, in the vicinity of the center of the Doppler profile, the strong polarized wave causes an optical anisotropy, so that the plane of polarization of the probe beam is rotated and the crossed polarizers let the beam pass through the cell and produce a transmission signal.

Polarization spectroscopy can be considered as a version of the interferometric observation of saturation resonances. The linearly polarized probe wave can be treated as a superposition of two circularly polarized waves, with left- and right-hand rotation of the polarization vector. If the propagation conditions of these two left- and right-hand polarized probe beams in the absorption cell are identical, they combine at the exit of the cell to form a linearly polarized beam with the same direction of polarization vector. Such a beam cannot pass through the crossed polarizers. The strong circularly polarized laser beam gives rise to two effects. First, there develops an optical dichroism, that is, the absorption values for the left- and right-hand polarized beams, χ^+ and χ^-, in the cell become different. Second, there occurs an optical birefringence, that is, the refractive indices n^+ and n^- for these beams also become different. All this, of course, takes place near the center of the Doppler profile, where the probe beam and the strong laser beam interact with the same particles. The polarization method has proved very useful in "labeling" the transition that is being saturated in a molecular spectrum, which has materially simplified the identification of spectral lines in complex spectra (Teets et al. 1976).

3.4 Ultrahigh spectral resolution

Saturation spectroscopy makes it possible to attain very high spectral resolution by means of ultranarrow spectral saturation resonances whose width is five to six orders of magnitude smaller than the Doppler width. Tunable lasers capable of an emission linewidth much smaller than the width of the narrow saturation resonances of interest have allowed numerous fundamental experiments with atoms and molecules to be performed. Here I must launch into a short digression on the problem of ultranarrow spectral resonances and their application.

3.4.1 Ultranarrow spectral resonances and ultrastable-frequency lasers

Concurrent with the progress of laser saturation spectroscopy, there have also been developed other Doppler-free spectroscopic methods, such as (a) a two-photon spectroscopy technique using counterpropagating laser waves (Vasilenko et al. 1970) and (b) a separated-light-field technique (Baklanov et al. 1976a,b)—a modification of the Ramsey spatially-separated-microwave-field method (Ramsey 1950, 1987) to cover the optical region. These methods do not use the saturation velocity selection that forms the basis of saturation spectroscopy, but are also capable of ultrahigh spectral resolution (see the excellent review in Chebotayev 1985). The two-photon Doppler-free spectroscopy technique has proved especially valuable for fundamental experiments with hydrogen atoms (ultrahigh-precision measurements of the Rydberg constant, the Lamb shift, etc. (Hansch 1989)). The next breakthrough occurred with the advent of the methods for cooling and trapping atoms and ions by means of electromagnetic fields that are considered in Chapters 5 and 6.

The ultranarrow saturation resonances and ultranarrow absorption lines of cooled atoms and cold trapped ions have been successfully used in developing ultrastable-frequency lasers, which caused a true revolution in the optical metrology of lengths, frequencies, and times (see the reviews by Hall et al. 2001 and Hollberg et al. 2001). These questions, however, are beyond the scope of this book. The advent of high-precision atomic clocks opened up the possibility of test experiments in fundamental physics. Specifically, ultranarrow saturation resonances using saturation velocity selection of slow molecules and rotational–vibrational transitions offer unique possibilities of performing fundamental experiments on parity violation.

3.4.2 Ultrahigh-resolution saturation spectroscopy based on selection of slow molecules

Rotational–vibrational transitions in molecules have a natural linewidth of the order of a few hertz, which means that they are potentially capable of extremely narrow saturation resonances. Naturally, to attain this goal, it is necessary to eliminate the other causes of spectral broadening listed in Table 3.1. First, the absorbing gas should be used at a very low pressure, for the collisional broadening is over 1 Hz at 10^{-7} Torr for most molecules, and the absorption cell should be a few tens of meters long. The next limiting factor is the broadening due to the finite time the molecules interact with the light beam, that is, the finiteness of the light-beam diameter a, which is very difficult to make greater than 30 cm. For example, a 3.39 μm standing He–Ne laser wave 30 cm in diameter interacting with CH_4 molecules moving with an average thermal velocity of v_0 at 300 K produces a saturation resonance with a width of $\Delta\nu_{tr} = v_0/a \simeq 300$ Hz. Therefore, to reduce the transit-time broadening, it is advisable to use slow molecules. Cooling the gas, for example, to 77 K will reduce v_0 and hence $\Delta\nu_{tr}$ by merely one-half. The laser-cooling techniques that operate very efficiently with atoms are inapplicable to molecules. A natural possibility is to use the original method for narrowing saturation resonances by way of predominant saturation of slow molecules in the Maxwell distribution of the projection of the molecular velocity onto a specified direction, $W(u) = (u/v_0^2)\exp(-u^2/v_0^2)$ (Rautian and Shalagin 1970). To this end, it is necessary to reduce the saturation parameter for the molecules moving with the mean thermal velocity v_0 so as to make the laser field saturate only the molecules moving with a slow thermal velocity v_{sl}, their time of interaction with the field in the course of their flight across the light beam (with a diameter of a) being longer. To attain this goal, it is necessary that the resonance saturation time (i.e. the duration of the $\pi/2$ pulse in Fig. 2.5) for the slow molecules should approximately coincide with the time it takes for them to cross the beam:

$$\tau_{\pi/2} = \pi \frac{\hbar}{d_{12}E} \simeq \frac{a}{v_{sl}} \gg \frac{a}{v_0}. \tag{3.19}$$

One should then operate with low light intensities that cause no saturation of the absorption by molecules with the mean thermal velocity v_0. Of course, use should be made of high-sensitivity methods for detecting ultranarrow resonance signals. The intensity of such signals is very low because of the need to use low-pressure gases (at some 10^{-6}–10^{-7} Torr) to reduce collisional broadening, and the small proportion of slow molecules in the thermal molecular-velocity distribution.

The method of saturation selection of slow molecules was successfully implemented in experiments with a 3.39 μm He–Ne laser and CH_4 molecules (Bagayev et al. 1991) and with a 10.6 μm CO_2 laser and OsO_4 molecules (Chardonnet et al. 1994). The best results for the spectral resolution of saturation spectroscopy so far were obtained in those experiments. Specifically, in the case of a He–Ne laser with a beam diameter of 30 cm and an 8 m long absorption cell filled with CH_4 molecules at a pressure of 2×10^{-7} Torr, ultranarrow saturation resonances with a half-width of 60 Hz (!) were obtained. This corresponds to the saturation selection of slow molecules with an effective temperature of $T_{\text{eff}} = 8$ mK. In this case, the ultranarrow saturation resonances of the slow molecules are also free from the shift due to the second-order Doppler effect, $\Delta/\omega_0 = -(1/2)(v_{sl}/c)^2$. That is, this shift in the above experiment (Bagayev et al. 1991) was negligible: $\Delta = -6$ mHz. In the experiments with OsO_4 molecules at 10^{-6} torr filling an 18 m long absorption cell irradiated by a light beam 3.5 cm in diameter (Chardonnet et al. 1994), the results obtained were as follows: resonance half-width 230 Hz, $T_{\text{eff}} = 0.6$ K, and $\Delta = -6$ mHz.

Saturation spectroscopy with selection of ultraslow molecules is a unique method for studying rotational–vibrational transitions in polyatomic molecules, with a spectral resolution of $\Delta\nu/\nu_0 = 10^{-15}$–$10^{-16}$. A spectral resolution as high as this is necessary in investigations into the fundamental effect of parity violation in chiral molecules.

3.4.3 Parity violation tests on chiral molecules

Immediately following the discovery of weak neutral currents in high-energy physics, Rein (1974) and Letokhov (1975a) independently predicted the effect of splitting of electronic energy levels in the two mirror-image enantiomers of chiral molecules. This phenomenon was termed the parity-violation energy difference (PVED) effect. For chiral molecules, one can consider two states—the right and left image states, with wave functions $|\Psi^R\rangle$ and $|\Psi^L\rangle$, respectively. If tunneling barrier of the potential-energy

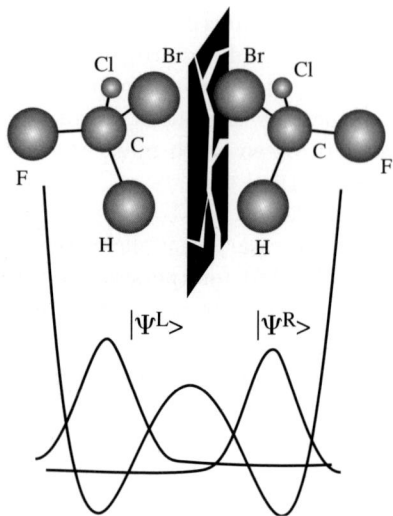

Fig. 3.11 Because of parity violation in electroweak interactions, the right and left enantiomers of CHFClBr are not mirror images.

curve is very high, the enantiomorphic left- and right-handed molecules are stable. If the parity is violated, the enantiomers can no longer be mirror images of each other (Fig. 3.11). According to the simple estimates made by Letokhov (1975a), the PVED effect lies in the region $\Delta\nu/\nu_0 \simeq 10^{-16}$–$10^{-17}$, and because the associated energy is potential energy, the relative changes in the electronic, vibrational, and rotational frequencies in the right- and left-handed molecules should be comparable:

$$\frac{\Delta E_e^{PV}}{E_{el}} \simeq \frac{\Delta P_{vib}^{PV}}{E_{vib}} \simeq \frac{\Delta P_{rot}^{PV}}{E_{rot}}. \tag{3.20}$$

This relation forms the basis for the search for the PVED effect in exactly those transitions for which ultranarrow saturation resonances were obtained.

Early experiments on the search for the PVED effect by means of laser saturation spectroscopy were conducted by Kompanets et al. (1976) at a spectral resolution of 10^{-8}. The left- and right-handed enantiomers of CHFClBr molecules, contained in separate gas cells, were the subject of the investigation. Of course, the above value of spectral selectivity was too low for the effect to be detected, but the experiment posed a challenging problem to laser spectroscopy. It was only 20 years later that experiments were performed by Daussy et al. (1999) using the method of saturation spectroscopy, with selection of slow molecules at a spectral resolution of the order of 10^{-11}. But the accuracy of measurement of the level splitting was much higher, at the level of 10^{-13}, for the centers of resonances in the separate gas cells could be determined much more accurately than the widths of the resonances. The saturation resonance frequencies of the separate enantiomers were compared and found to be identical within 13 Hz ($\Delta\nu/\nu < 4 \times 10^{-13}$). Theoretical estimates predict much lower PVED values for CHFClBr, of the order of 10 mHz, which so far is beyond the capabilities of saturation spectroscopy.

However, there is a way to significantly decrease the requirements on the ultimate spectral resolution of saturation spectroscopy, based on the use of molecules with heavy atoms, because parity violation effects increase with nuclear charge approximately as Z^5 (Zel'dovich et al. 1977). Therefore, heavy atoms attached to a carbon center should enhance the PVED effect in enantiomers. A search for such molecules has yielded encouraging results. Relativistic four-component electronic-structure calculations, including parity-odd electroweak interactions, give a large energy difference of about 0.2 Hz for the C–F stretching mode in the enantiomers of $PH_3AuCHFCl$ and $ClHgCHFCl$ (Bast and Schwerdtfeger 2003). These organometallic compounds are therefore ideal candidates for future high-resolution interdisciplinary experiments combining the capabilities of many sciences, such as chemical synthesis, enantiomer separation, and saturation laser spectroscopy, for the study of the fundamental laws of nature.

4
Optical orientation of atoms and nuclei

A century ago Einstein introduced the notion of the linear momentum of a photon, $h\mathbf{k}$, and analyzed its transfer to an atom (and back) in the establishment of thermodynamic equilibrium between black-body radiation and matter. In addition to a linear momentum, the photon also possesses an angular momentum, $\hbar$, directed parallel (or antiparallel) to its propagation direction (the OZ axis in Fig. 4.1), depending on its polarization direction (left- or right-handed). This angular momentum is due to the spin of the photon, equal to 1.[1] Half a century later, Kastler (1950) considered the use of the transfer of the angular momentum of a polarized photon to the angular momentum of an atom as a method of optical orientation of atoms. In essence, this was the first method to control atomic motion by means of low-intensity light, developed even before the advent of the laser; it which found widespread application in investigating and then using spin polarization effects in atomic physics. But even after lasers had made their appearance, the method of optical orientation of atoms (or pumping of certain magnetic sublevels in atoms) proved very important, specifically in the laser cooling of atoms (Chapter 5).

4.1 Optical orientation of atoms

Consider for the sake of simplicity the resonant interaction between a circularly polarized photon and an atom that has a single valence electron beyond a completed electron shell possessing zero orbital angular momentum (i.e. an alkali metal atom). The only optical electron in the ground state has an angular momentum of $\hbar/2$ due to its spin equal, to 1/2, which can be directed along a spatial quantization axis (OZ axis in Fig. 4.1) and can have two projections on it, namely, $-1/2$ or $+1/2$. These are the magnetic (Zeeman) sublevels m, which are degenerate, that is, they have the same energy, in the absence of a magnetic field. To simplify the explanation of the idea, we consider first the case of zero nuclear spin.

4.1.1 Zero nuclear spin

In the first excited state, the orbital angular momentum is $l = h$ and can be oriented either antiparallel to the spin angular momentum or parallel to it, forming two states,

---

[1]Strictly speaking, the total angular momentum of a photon is the sum of the spin angular momentum $\vec{s}$ and the orbital angular momentum l (Allen et al. 1999), so that the total angular momentum of a photon may take on large values in the range from $-n\hbar$ to $+n\hbar$. However, under ordinary conditions, where light is only circularly polarized, the orbital angular momentum is given by $l = 0$.

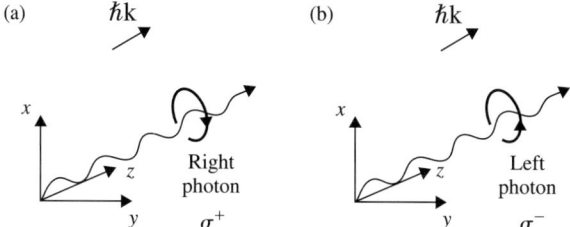

Fig. 4.1 Orientation of angular (spin) momentum of photon with momentum $\hbar k$, with right (a) and left (b) polarization.

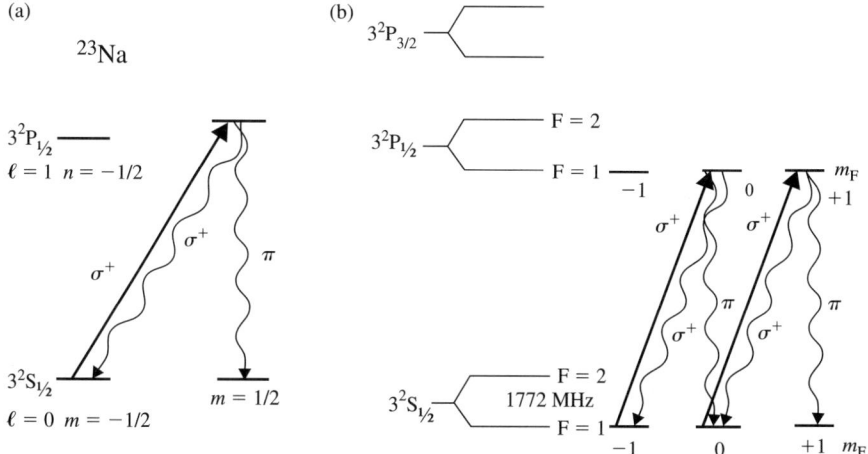

Fig. 4.2 Optical orientation of Na by circularly polarized σ^+ photons using the quantum transition $3^2S_{1/2} - 3^2P_{1/2}$ (D$_1$ line): (a) without nuclear-spin effect; (b) with the nuclear spin $I = 3/2$ of ^{23}Na.

with a total angular momentum of $j = 1/2$ (in units of $\hbar$) (antiparallel orientation) and $j = 3/2$ (parallel orientation). Because of the interaction between the spin and orbital angular momenta of the electron, these two quantum states have different energies (the $3^2P_{1/2}$ and $3^2P_{3/2}$ fine-structure sublevels for the Na atom, Fig. 4.2(b)). The radiative transitions between the ground state $3^2S_{1/2}$ and the first excited states correspond to the D$_1$ and D$_2$ resonance lines at 5896 and 5890 Å.

Now consider the transfer of the angular momentum of a photon to an atom, using as an example resonant transitions between the ground state $3^2S_{1/2}$ and the excited state $3^2P_{1/2}$. The selection rules for the allowed dipole transitions require that the energy and angular momentum of the system "atom + photon" be conserved (as well as the linear momentum and the parity). Since the angular momentum of the photon equals 1, the absorption of a σ^+ photon by the atom must cause the projection of its angular momentum onto the quantization (observation) axis OZ to change by an amount $\Delta m = +1$, whereas the absorption of a σ^- photon must cause it to change by $\Delta m = -1$ (Fig. 4.2(a)). The atom in the excited state $3^2P_{1/2}$ spontaneously emits a photon in a random direction and returns to its ground state, but not necessarily to the initial magnetic sublevel $m = -1/2$. When the spontaneously emitted photon is a σ^+ photon, the atoms returns to its initial magnetic sublevel with a probability of 2/3, and when it is a "linearly polarized π-photon," the atom returns to the magnetic sublevel $m = +1/2$ with a probability of 1/3.

Going over from classical to quantum mechanical terminology for describing polarized light, one should bear in mind two different situations in the case above. In the case of a single atom, one can effect *multiply repeated excitation* with σ^+ photons and observe that it returns to the $m = -1/2$ sublevel in 2/3 of the cases and to the sublevel $m = +1/2$ in 1/3 of the cases. In the case of excitation of *many atoms* with σ^+ photons, one would observe that 2/3 of the atoms had returned to the magnetic sublevel $m = -1/2$, and 1/3 of them to the sublevel $m = +1/2$. Thus, if the atoms were initially distributed equally between the two magnetic sublevels of the state $3^2S_{1/2}$, that is, they had equiprobable angular momentum orientations, then after the "optical pumping" there would be twice as many atoms in the state $m = -1/2$ as in the state $m = +1/2$. Herein lies the essence of the optical orientation of atoms with circularly polarized light. The simplest case of optical orientation of Na atoms using the D_1 line can be extended similarly to the other fine-structure transition $3^2S_{1/2} \leftrightarrow 3^2P_{3/2}$ in Na and also to other suitable atoms (K, Rb, and others).

Experiments on the optical orientation of atoms in vapors were performed in the "prelaser" epoch with the aid of the resonance spontaneous-emission lines of the same atoms as those subject to optical pumping. The weak intensity of the spontaneous radiation notwithstanding, a high degree of optical orientation was attained. The success of these experiments was due to two factors at least. First, the orientation of atoms required no alteration of the atomic energy. Secondly, the disorientation of the atoms, that is, their collisional relaxation, between the magnetic sublevels of the ground (usually an S) state turned out to be fairly low on a per-collision basis (Dehmelt 1957). The process of accumulation of atoms in a magnetic sublevel with the maximum $|m|$ value takes only a few optical-pumping cycles to reach a steady state. Therefore, optical pumping is an effective method for producing and sustaining a nonequilibrium distribution of a population N_m among the magnetic sublevels of the ground state of a set of atoms. Since the different m-states correspond to different orientations of the spin of the optical electron, this phenomenon is frequently called *spin polarization*.

The degree of polarization can be defined as the extent of the accumulation of atoms, with density N_m, in the states with the maximum m value (Cohen-Tannoudji and Kastler 1966):

$$P = \frac{1}{J} \frac{\sum_m m N_m}{\sum_m N_m}, \qquad (4.1)$$

where J is the total angular momentum of the atom ($J = 1/2$ in the case of the Na atom in the ground state shown in Fig. 4.2) and N_m is the population of sublevel m. For example, the degree of polarization attained as a result of a single cycle of optical pumping of Na atoms using the transition $3^2S_{1/2} \to 3^2P_{1/2}$ is $P = 1/3$. When the optical pumping cycle is repeated a sufficiently great number of times, one attains complete polarization, either $P = +1$ (accumulation of atoms in the sublevel with the maximum value of $+m$) or $P = -1$ (accumulation of atoms in the sublevel with the maximum value of $-m$).

Experimentally, the optical orientation of atoms in the ground state is usually observed by means of the change of the polarization of their spontaneous radiation consequent upon optical excitation.

4.1.2 Nonzero nuclear spin. Optical orientation of nuclei

For atoms with a nonzero nuclear spin, optical pumping with a circularly polarized light makes it possible to orient the nucleus also. There are two cases of orientation of atoms in the ground state: (1) the angular (spin) momentum of the electron and the nuclear spin are both other than zero (^{23}Na, ^{35}K, ^{85,87}Rb, and ^{133}Cs), so that the hyperfine coupling of the angular momenta of the electron and the nucleus allows the concurrent orientation of the atom and the nucleus; and (2) the angular momentum of the electron is zero (two coupled optical electrons), but the spin angular momentum of the nucleus is nonzero (^{199,201}Hg), In that case, the angular momentum of the atom is due only to that of the nucleus. Nevertheless, optical pumping with polarized radiation orients both the atom and the nucleus. This illustrates the efficiency of optical radiation in orienting atoms. For example, when a magnetic field is used for this purpose, because of the difference in magnetic moment between an electron and a nucleus, the field necessary to orient nuclei is thousands of times as strong as that required to orient atoms. At the same time, however, for a circularly polarized photon, these two cases are absolutely identical. It should be added here that the spin angular momentum of a photon is independent of frequency. Optical and microwave fields carry the same angular momentum on a per-photon basis (!). Consider the above two cases using as examples the ^{23}Na and ^{199}Hg atoms.

^{23}*Na atom with a nuclear spin of* $I = 3/2$. The hyperfine splitting of the ground-state levels is shown in Fig. 4.2(b) ($\Delta\nu_{hfs} = 1772$ MHz). This means that the orientation of the optical electron, interacting with the nuclear spin with an energy of $\Delta E = h\Delta\nu_{hfs}$ orients the total angular momentum of the atom, $\mathbf{F} = \mathbf{J} + \mathbf{I}$. In other words, one can optically pump certain hyperfine-structure sublevels by using identical projections of the total angular momentum $\mathbf{F}$ with $F = 1, 2$. Because the coupling between the nuclear and electronic spins is weak, optical pumping should be carried out by means of low-intensity radiation (where the Rabi frequency is much less than the hyperfine splitting) so as not to decouple their hyperfine interaction, that is, the orientation of the atom should follow the orientation of the optical electron.

^{199}*Hg atom with a nuclear spin of* $I = 1/2$. Figure 4.3(a) presents the energy levels and transitions of the ^{199}Hg atom that are used for the optical orientation of this atom and nucleus. The case of ^{199}Hg is convenient, for it can be raised to the longer-lived triplet state 6^3P_1 by means of the intense UV line at 2537 Å. For this reason, optical orientation of Hg atoms can be observed not only in the ground state, but also in an excited electronic state (Brossel and Bitter 1952). In the case of Zeeman splitting of the magnetic sublevels of the $6S_0$ ground state, one can observe transitions between these magnetic sublevels in a radio-frequency field. This phenomenon is termed nuclear magnetic resonance (NMR), in contrast to electron spin resonance or electron paramagnetic resonance (EPR), involving transitions between sublevels differing in the spin angular momentum of the electron. Observations of nuclear magnetic and electron paramagnetic resonances have revealed remarkable effects in the behavior of both electronic and nuclear spins. Before discussing these effects, we note that a wide range

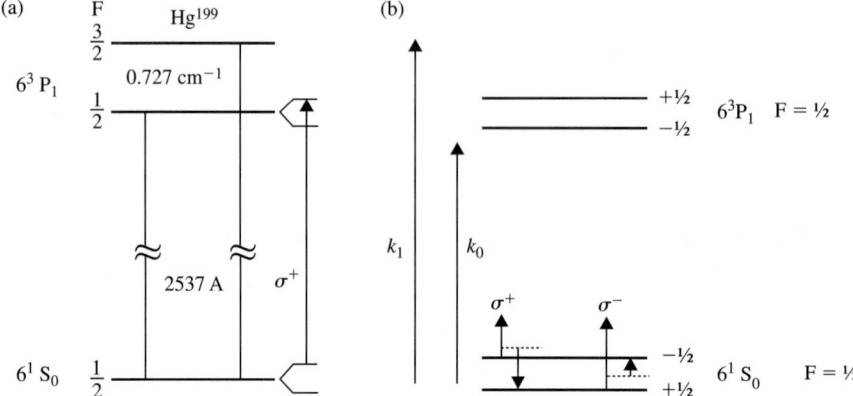

Fig. 4.3 (a) Energy scheme of the hyperfine structure of the Hg isotope ^{199}Hg ($I = 1/2$), and (b) scheme showing the shift of ground-state Zeeman levels produced by virtual transitions of frequency $k_1 > k_0$. (Adapted and modified from Cohen-Tannoudji 1961.)

of efficient nonoptical methods have been developed for the orientation of nuclei in solids at low temperatures in a static magnetic field, where the Zeeman sublevels differ noticeably even without the use of optical pumping (Abragam 1961).

On the basis of the use of the orientation of atomic and nuclear spins, Kastler formulated a generalization of the Franck–Condon principle: "In a rapid process involving the electronic configuration (a spectral transition, disorienting collision, or exchange collision), the position and *orientation* of the atomic nuclei remain unchanged" (Kastler 1966).

4.2 Radio-frequency spectroscopy of optically oriented atoms

Optical pumping with circularly polarized light results in a nonequilibrium population of the magnetic sublevels. In the absence of a magnetic field, the energies of these sublevels can differ only because of their interaction with the nuclear spin, that is, by an amount equal to the hyperfine interaction energy. This fact has opened up the possibility of observing stimulated quantum transitions between hyperfine-structure sublevels ($\Delta F = 1$) under the effect of microwave radiation, and thus measuring hyperfine-interaction energies with high precision. Experiments of this type are called "double-resonance" experiments, owing to the resonant interaction of the atom with both an optical and a radio-frequency field. If the excited electronic level is a metastable one, that is, it has a long lifetime (e.g. a few milliseconds), one then can effect optical pumping of the atom in the excited state and measure the hyperfine structure of the excited level. Such a possibility occurs, for example, when Hg atoms are excited into triplet states. In the presence of a magnetic field, the degeneracy of the magnetic sublevels vanishes and there arises a Zeeman splitting of energy levels, which can be observed by means of stimulated radio-frequency transitions between the Zeeman sublevels. Depending on the polarization direction of the photon (parallel or antiparallel to the OZ-axis), one can stimulate transitions from the lower sublevel to the upper one (stimulated absorption) or, vice versa, from the upper to the lower sublevel (stimulated emission). Stimulated spin-flip transitions also return the magnetic-sublevel populations to their equilibrium condition.

The radio-frequency transitions observed provide valuable information about the interaction between the spin-oriented atoms and their surroundings, that is, about collisions with one another, with the buffer-gas atoms, and with the container walls.

4.2.1 Slow spin relaxation of optically polarized atoms

While studying the decay with time of the amplitude of a radio-frequency resonance signal that was proportional to the degree of polarization of the atomic spins, Dehmelt (1957) discovered the effect of the slow (a fraction of a second long) relaxation of optically polarized atoms. This effect was observed with Na atoms in the presence of a buffer gas (Ar) when the diffusion of the atoms to the container walls was slowed down, which prevented them from undergoing disorientation upon collision with the walls. At the same time, the probability of relaxation of the spin of the Na atoms upon collision with Ar atoms was very low. In terms of the relaxation times T_1 and T_2 introduced in Chapter 2, this means that in the presence of a buffer gas, the longitudinal relaxation time T_1 becomes longer (the buffer-gas effect). The width of the radio-frequency resonances provides information on the transverse relaxation time T_2. Experiments conducted by Cagnac and Cohen-Tannoudji at the Ecole Normale Superiere in the early 1960s demonstrated that the width of the nuclear magnetic resonance curve lay in the region of a few hertz. Figure 4.4(a) shows the shape of some such resonances for the ^{199}Hg atom in the ground state 6^1S_0, with a nuclear spin of $I = 1/2$, for the $m_F = -1/2 \rightarrow m_F = +1/2$ transitions in the presence of a weak magnetic field that eliminated the degeneracy of those magnetic sublevels. As the intensity of the radio-frequency field is increased, broadening and splitting effects of the resonance become clearly manifest in the figure (the power-broadening and Autler–Townes splitting effects; see Chapter 2).

Investigation of the parameters of very narrow radio-frequency resonances of oriented atoms has opened up the possibility of precision measurement of the

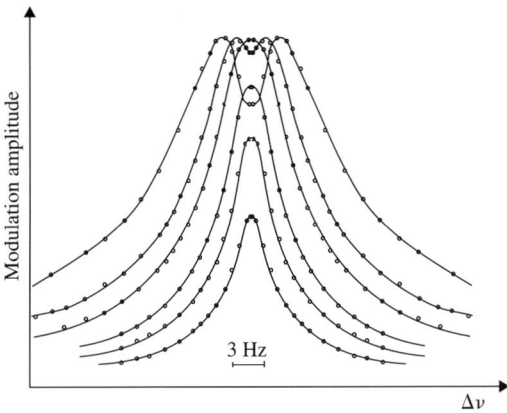

Fig. 4.4 Nuclear magnetic resonance curve of optically oriented ^{199}Hg in the ground state. (Reprinted with courtesy and permission from Cohen-Tannoudji 1962.)

4.2.2 Light shifts of atomic levels

The exceptionally narrow width of radio-frequency resonances of oriented atoms enables one to observe very small shifts of atomic levels in the light-wave field of rather weak (nonlaser) spontaneous-emission lines, that is, the Stark effect in a high-frequency light field. It is amazing that this effect was discovered without using intense laser light; however, it plays an important part in the cooling of atoms by resonant laser light (Chapter 5). Figure 4.3 explains how this effect was observed with ^{199}Hg atoms (Cohen-Tannoudji 1962).

By using polarized resonant radiation, the ^{199}Hg atoms are first oriented so as to make it possible to observe, in the $m_F = -1/2 \to m_F = +1/2$ transition, a narrow radio-frequency absorption line similar to the narrow resonance shown in Fig. 4.4. A second light beam is then directed opposite to the orienting beam, its frequency being off resonance with the optical transitions of the oriented atoms, that is, the wave vector k of the second beam obeys $k > k_0 = \omega_0/c$, where ω_0 is the optical transition frequency. This second light beam causes virtual transitions in the atoms that cause their energy levels to shift, but populate no actual quantum states. This shift is so negligible (a fraction of a hertz) that it cannot be observed in the optical region, but it stands out distinctly in the narrow magnetic resonances when the circular-polarization direction of the second beam is reversed: a σ^+ beam shifts only the $m = 1/2$ sublevel, whereas a σ^- beam shifts only the $m = -1/2$ sublevel (Fig. 4.3(b)). The results of such an experiment are presented in Fig. 4.5 (Cohen-Tannoudji 1962). A theoretical calculation of the Stark effect in a light field gives (Cohen-Tannoudji 1961)

$$\Delta\omega = \frac{1}{\Delta}\left(\frac{d_{12}E}{2\hbar}\right)^2, \tag{4.2}$$

where $\Delta = \omega - \omega_0$, E is the electric field strength of the off-resonance light beam, and d_{12} is the transition dipole moment. When the detuning Δ of the off-resonance

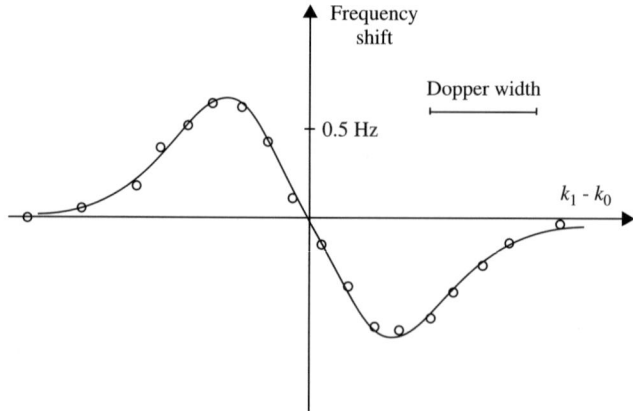

Fig. 4.5 Frequency shift produced by virtual transitions as a function of $k_1 - k_0$. (Reprinted with courtesy and permission from Cohen-Tannoudji 1962.)

light field is negative, the frequency shift is positive, that is, the energy levels of the atom are pushed apart, and the opposite occurs when the detuning is positive. Such a behavior of the energy levels of an atom corresponds to a change in its potential energy and plays an important role in the trapping and cooling of atoms by laser radiation (Chapters 5 and 6).

4.3 Spin-exchange optical pumping

Circularly polarized resonant light absorbed in an alkali-metal atom vapor in a glass cell makes the atoms highly spin polarized. Under optimum conditions, it proves possible to transfer about one-half of the total spin angular momentum of the absorbed photons to the atoms in the cell. The partial pressure of the atomic vapor absorbing light in a resonant fashion is usually very low (10^{-3} Torr) in comparison with that of the buffer gas, which can be as high as a few atmospheres. When spin-polarized atoms A collide with other particles B, they can effectively exchange their spins. A spin-exchange collision between two $^2S_{1/2}$ atoms A and B may be represented by the equation

$$A(\uparrow) + B(\downarrow) \to A(\downarrow) + B(\uparrow), \tag{4.3}$$

where the arrows indicate the direction of the electron spin. Such a spin-exchange process can be illustrated by the simple example of two colliding hydrogen atoms, which, depending on the mutual orientation of their spins, can either form an H_2 molecule in the stable singlet ground state, or not, if their spin orientations are the same (the triplet repulsive state), as shown in Fig. 4.6. Two hydrogen atoms passing within a distance of less than 4×10^{-3} cm of each other can exchange their spins with a probability of $1/2$. But one should bear in mind, as always, the nuclear spins (the proton has a spin $i = 1/2$) of the two atoms; these can be oriented both in a parallel and in an antiparallel manner, forming two modifications of the hydrogen molecule, namely, ortho-hydrogen and parahydrogen, the latter having a zero total nuclear spin. Since the proton (like the electron) is a fermion, that is, it obeys Fermi–Dirac statistics, a permutation of two protons changes the sign of the total wave function. This is manifest in the distribution

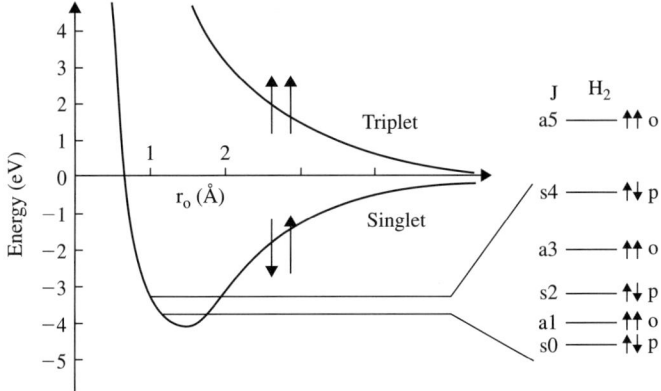

Fig. 4.6 Interaction energy of two hydrogen atoms as a function of the internuclear separation r_0, and symmetries and nuclear-spin orientations of the low rotational levels.

of rotational energy levels in orthohydrogen and parahydrogen. In parahydrogen, only the rotational levels of even angular momentum, $J = 0, 2 \ldots$, are populated, whereas in orthohydrogen, only those of odd angular momentum, $J = 1, 3 \ldots$, are populated. When molecular hydrogen is cooled, orthohydrogen gets stuck in the rotational level $J = 1$ because the ortho $\rightarrow$ para conversion is a very slow process. The radiative transition is absolutely forbidden, there being no dipole moment in the H_2 molecule, and so it is only the interaction between the magnetic moments of the protons and those of other particles that can change the mutual orientation of the protons in the hydrogen molecule. This is a vivid example of the "protection" of the nucleus provided by the electron shell against external effects during collisions.[2] This spin-exchange process has been observed in the course of interaction between optically oriented atoms and a wide variety of particles, for example, between optically polarized Na atoms and free electrons (Dehmelt, 1958) and between oriented Rb atoms and ^{3}He atoms (Bouchiat et al. 1960). The high efficiency of the spin-exchange process is due to large spin-exchange cross sections. To illustrate, the spin-exchange cross section σ for two alkali metal atoms is $\approx 2 \times 10^{-14}$ cm^2. These cross sections are an order of magnitude greater than the gas-kinetic collisional cross sections.

Most important from the standpoint of practical applications is spin-exchange optical pumping using collisional polarization transfer between optically oriented alkali metal atoms and noble gas atoms. The binary spin-exchange collisions that were observed by Bouchiat et al. (1960) are dominant at high buffer gas pressures. Practically, the polarization of noble gases is important because polarized noble gas atoms can be transported and used for purposes of nuclear magnetic resonance imaging (Albert et al. 1994).

4.4 Coherent effects and optically oriented atoms

There are several aspects of the manifestation of coherent effects, both in the optical orientation of atoms and in the use of optically oriented atoms for studies into new coherent effects in laser–atom interactions.

4.4.1 Coherent orientation of atoms

The orientation of atoms by way of their optical pumping is a *noncoherent* process involving spontaneous emission. However, it is possible to transfer angular momentum from light to atoms in a *coherent* fashion. This possibility is based on off-resonant interaction between circularly polarized light and atoms, giving rise to light-induced

---

[2] Another remarkable example is provided by rapidly rotating excited nuclei with a large nuclear spin. If the low-lying nuclear levels have a small spin, it is only high-order multipole radiative transitions with $\Delta J \gg 1$ that are allowed. The probability of such γ-transitions is extremely low. For example, an excited Hf$^{178\,m2}$ nucleus, with an energy of 2.5 MeV, has a half-life of 31 years. A kilogram of this isomer has a great store of nuclear energy (900 GJ, or about a quarter of a kiloton) that is difficult to release even with high-power laser radiation, attempts at doing so notwithstanding. Such isomeric nuclear substances are so amazing that I have even considered using negative muons for their discharge (Ivanov and Letokhov 1976). Since a meson is 200 times as heavy as an electron, it "settles" itself down on the lowermost orbit in the immediate vicinity of the nucleus and can thus exert an effect on it. In principle, having discharged one nucleus, the muon remains intact and can "settle" itself down on another nucleus, thus inducing a reaction of liberation of nuclear energy from excited isomeric nuclei similar to muon-induced cold nuclear fusion reaction. But this belongs to the field of muonic control of nuclei and not to that of laser control of atoms and molecules, which the present book is devoted to.

level shifts that depend on the magnetic quantum number (Cohen-Tannoudji and Dupont-Roc 1972). Using ^{199}Hg, ^{201}Hg, and ^{87}Rb atoms in the ground state as an example, those authors demonstrated experimentally the elimination of the degeneracy of the magnetic sublevels in these atoms, equivalent to the action of a static magnetic field of the order of 20 μG. Of course, by using laser light of a higher intensity, one can achieve much greater level shifts and much stronger "fictitious" magnetic fields, which can be used as an effective tool for spin manipulation. The most essential feature is that laser light can be focused, so that such 'fictitious' magnetic fields can be produced in local regions, which is beyond the capabilities of an ordinary magnetic field. This is of interest from the standpoint of the development of NMR methods with laser-controlled 3D spatial resolution.

4.4.2 Coherent effects with optically pumped atoms

Optical pumping produces a nonequilibrium distribution of atoms among the magnetic (Zeeman) or hyperfine-structure (HFS) sublevels. The relationships between the atomic sublevels are defined by the diagonal elements of the density matrix of the state (see Section 2.5.2). However, the phases of the wave functions of the sublevels can be interrelated, so that the off-diagonal elements of the density matrix are not equal to zero, which makes the Zeeman levels coherent. This gives rise to new effects that cannot be described within the framework of populated levels.

A still more interesting effect, referred to as coherent population trapping, arises in three-level systems, when an atom is prepared in a state that is a superposition of two states, 1 and 2 (Fig. 4.7(a)):

$$\Psi = \tfrac{1}{\sqrt{2}} \left(a_1 \Psi_1 \pm a_2 \Psi_2 \right). \tag{4.4}$$

Coherent effects in three-level systems of various configurations represent a very productive trend in coherent and atom optics (Kocharovskaya 1992), whose treatment is

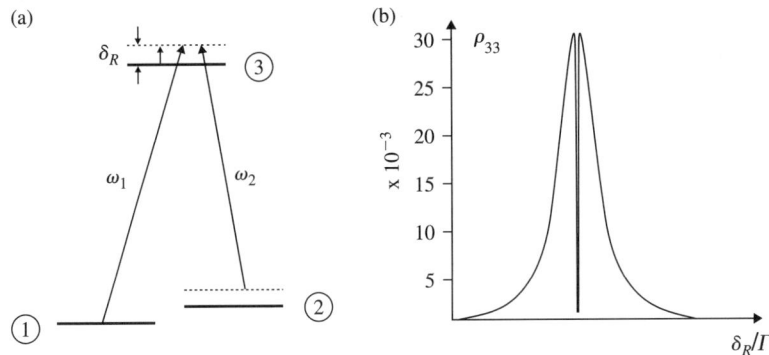

Fig. 4.7 Coherent population-trapping effect with optically oriented atoms: (a) coherent population trapping between two levels (1 and 2) and an excited state 3 in a laser field with two frequencies (ω_1 and ω_2); laser fields; (b) steady-state excited-state population ρ_{33} for a Λ-system as a function of the Raman detuning δ_R, with the typical central dip associated with the coherent population-trapping phenomenon. (Adapted from Arimondo 1996.)

beyond the scope of this book. However, the coherent population-trapping effect was discovered with optically oriented Na atoms (Alzetta et al. 1976).

An atomic system can be prepared in a superposition state by using the Raman two-photon resonance phenomenon in a two-frequency light field (Fig. 4.7(a)), with its frequencies ω_1 and ω_2 detuned relative to the exact resonance frequencies, to provide for the Raman detuning δ_R:

$$\delta_R = (\omega_1 - \omega_2) - \omega_{12}. \tag{4.5}$$

As a result of interaction with the two-frequency field, the superposition state (4.4) with the minus sign is produced between the states 1 and 2:

$$\Psi \sim \Omega_{R1}\Psi_2 - \Omega_{R2}\exp(-i\omega_{12}t)\Psi_1, \tag{4.6}$$

where Ω_{R1} and Ω_{R2} are the Rabi frequencies for the respective transitions. In this state, the atom absorbs no radiation at the frequency ω_1, nor at the frequency ω_2. This can be interpreted as a result of the destructive quantum interference of two excitation channels of the atom. Figure 4.7(b) presents the stationary excited-state population ρ_{33}^{st} of a three-level Λ-system as a function of the Raman detuning δ_R (Arimondo 1996). One can clearly see the narrow resonance minimum (dark resonance), corresponding to a zero excitation probability of the atom due to coherent population-trapping effect ($\delta_R = 0$). In that case, the laser wave with the frequency ω_2 is in exact resonance with the transition frequency, while the wave with the frequency ω_1 is subject to detuning from the resonance. It was exactly this effect that was discovered by Alzetta and coworkers (1976) in experiments with optically pumped Na atoms, where the superposition state was produced between the hyperfine-structure sublevels $3^2S_{1/2}$ ($F=1$ and $F=2$) by means of σ^+- and π-polarized radiation.

The coherent population-trapping effect occurs only in atoms that have a zero projection of their velocity onto the propagation direction, that is, under exact resonance conditions. Therefore, the atomic velocity selectivity of this effect can be used to select atoms with ultralow velocities (Aspect et al. 1988) for use with counterpropagating laser waves.

4.5 Applications of optically pumped atoms

The production of a nonequilibrium distribution of a population of magnetic or HFS sublevels in the ground state has found widespread application in high-precision atomic clocks and ultrasensitive magnetometers, and lately in noble-gas-based magnetic imaging. Let us very briefly consider these applications.

4.5.1 Atomic clocks

Atomic clocks represent one of the basic applications of atom physics, for the precise measurement of time is one of the most important needs of present-day civilization. The most familiar are atomic clocks using a microwave transition in ^{133}Cs. In 1967, an international standard was introduced for the second: 1 second = 9 192 631 770 cycles of the standard ^{133}Cs transition. Cesium atomic clocks use the magnetic sorting of sublevels in Cs and the method of spatially separated fields (Ramsey 1987) to obtain highly stable, narrow microwave resonances. Cesium clocks have demonstrated a stability of up to 1 part in 10^{13}, or one second in 300 000 years.

Laser control of Cs atoms is used to improve the microwave clock in two ways. First, the sorting of Cs atoms by their magnetic sublevels can be conveniently effected by way of optical (laser) orientation. Compact laser diodes have proved especially convenient for this purpose. Secondly, use can be made of the laser cooling of Cs atoms, which can potentially reduce both the width of the microwave resonance, because of the finite atom–field interaction time (see Section 3.2 and Table 3.1), and the frequency shift due to the quadratic Doppler effect (Letokhov and Minogin 1981b). The marriage of the laser-cooling technique and the Ramsey spaced-fields technique was successfully realized in the method of the Cs fountain using laser-cooled Cs atoms, which made it possible to reduce the width of the microwave resonance to about 1 Hz and attain a record-high frequency stability of 10^{-15} (Clairon et al. 1995). The progress in this field is so impetuous that atomic clocks built around ultracold Cs and Rb atoms will soon be used in space stations, particularly to search for variations of fundamental constants.

The rubidium atomic clock employs a transition between the ground-state HFS sublevels in the ^{87}Rb isotope at a frequency of $\nu_0 = 6384$ MHz. Optical excitation is used to pump the ^{87}Rb atoms to one of the ground-state sublevels and to detect changes in the level-population difference upon interaction of the atoms with microwave radiation in a microwave cavity. The light emitted by an ^{87}Rb lamp is passed through a filter cell containing a vapor of ^{85}Rb atoms to excite ^{87}Rb atoms in an absorption cell filled with a buffer gas (a mixture of light noble gases). The buffer gas lengthens the time of interaction between the ^{87}Rb atoms and the microwave radiation by reducing the rate of their collisions with the walls of the absorption cell. The microwave resonance width here is typically about 500 Hz.

Rubidium atomic clocks feature high frequency stability, but because of magnetic-field perturbations in the absorption cell, collisions between the ^{87}Rb atoms and buffer gas particles, and other disturbances, the scatter in their resonance frequency is of the order of 10^{-10}. For this reason, Rb clocks are being widely used as portable atomic clocks requiring calibration from atomic clocks of much higher precision (i.e. Cs atomic clocks). Highly sophisticated microwave-resonance Cs and Rb atomic clocks have now been developed using laser orientation of the atoms, both in atomic beams and in gas cells. It should be added that apart from the radical narrowing of the width of the microwave resonance by way of laser cooling of atoms, another method can also be used to obtain ultranarrow resonances, based on the coherent population-trapping effect (see Section 4.4) (Knappe et al. 2001).

The next stage in the development of atomic clocks that utilize the control of atoms will be characterized by the use of optical transitions in atoms, specifically laser-cooled trapped atoms or ions, and the laser synthesis of optical and microwave radiation (Chapter 6).

4.5.2 Ultrasensitive optical magnetometry

The measurement of ultraweak magnetic fields is one more area of effective application of optically oriented atoms. Its source lies in an effect studied by Hanle as far back as 1925 in experiments on the fluorescence of Hg vapor in the presence of a magnetic field. When Hg vapor is irradiated with linearly polarized light in resonance with the $^1S_0 \to {^3P_1}$ transition in Hg (2537 Å), the depolarization of the resonant fluorescence in the perpendicular direction depends on the strength of a weak magnetic field directed normal to the plane of the optical axis of the beam. This effect can be explained by the interference of degenerate magnetic sublevels, which depends on the magnitude of their splitting by the magnetic field or, to state it in classical terms, is described by the Larmor precession of the decaying dipole in the magnetic field. When the atoms are optically oriented, the Hanle effect can be observed in the electronic ground state as well (Lehmann and Cohen-Tannoudji 1964). It was later suggested that use should be made not of the nuclear paramagnetism in the ground state, that is, the interference of magnetic sublevels, but of the electronic paramagnetism in alkali metal atoms (Alexandrov et al. 1967). The first successful experiment on the use of this method was conducted with ^{87}Rb atoms in the ground state (Dupont-Roc et al. 1969). The sensitivity attained in this experiment in measuring magnetic-field strengths was at the level of approximately $3 \times 10^{-11} \text{G}/\sqrt{\text{Hz}} = 3 \text{ fT}/\sqrt{\text{Hz}}$ (where fT means a femtotesla). Optical magnetometers have proved very useful in measuring ultraweak magnetic field, both in practical applications and in fundamental investigations. Specifically, they have allowed the magnetic fields produced by live organs to be measured.

In parallel, magnetometers based on interference phenomena in superconducting materials (SQUIDs), which reached sensitivities as high as $\sim 1 \text{fT}/\sqrt{\text{Hz}}$ (Weinstock 1996), were also being developed. These are capable of highly localized measurements, and they made it possible, for example, to map the magnetic fields produced by the human brain and to localize the underlying electrical activity (magnetoencephalography). At the same time, optical magnetometers are continuing to be developed successfully, thanks to the use of nonlinear optical phenomena and multiple-channel detection methods, and have already reached subfemtotesla sensitivity levels, with the spatial localization of measurements being only a few millimeters (Budker et al. 2000; Komins et al. 2003).

4.5.3 Medical imaging with spin-polarized noble gases

The NMR imaging that is so important in biomedical diagnostics *in vivo* is based on the polarization of protons in water. In magnetic fields of the highest strength, such a "brute force" method makes it possible to attain a degree of polarization P as high as $\approx 10^{-5}$ under normal conditions. Therefore, it is applicable only to the high-speed high-resolution NMR imaging of water. As to other molecules, the proton concentration is too low to allow the NMR imaging of biotissues, to say nothing of atmospheric gases. The optical polarization of nuclear spins is a "soft" method excellently applicable to alkali metal, Hg, and other atoms that have a nonzero nuclear spin. Its capabilities can be extended by using spin-exchange in noble gases, which, in contrast to alkali metal, Hg, Cd, and other atoms, can exist, for example, in air that is breathed. At a pressure of 1 atm, the molar concentration of xenon gas, for instance, in a biotissue is a mere 0.04% of that of H_2O. At the same time, however, ^{129}Xe, as well as other spin-1/2 noble gases, can be polarized very efficiently through spin-exchange with an optically pumped alkali metal vapor (Walker and Happer 1997).

Biological magnetic-resonance imaging using laser-polarized ^{129}Xe was demonstrated (Albert et al. 1994) in experiments that used a small (1 ppm) admixture of Rb vapor in the xenon gas. To effectively polarize the Rb atoms in a large enough volume, use was made of laser radiation at 795 nm (the D_1 line of Rb) with a power of the order of 1–2 W, that is, thousands of times higher than the intensity of ordinary sources of resonant radiation. Gas-phase collisions between the ^{129}Xe and polarized Rb atoms resulted in the transfer of angular momentum from the Rb valence electron to the ^{129}Xe nucleus (Fig. 3.8(b)). The ^{129}Xe gas was laser-polarized to at least 25% within 5–20 min., enhancing its NMR signal to 10^5 times the thermal-equilibrium value. The gas was then delivered for imaging. The large, long-lived ^{129}Xe polarization achieved in the lungs can provide a basis for imaging beyond the lungs themselves, for ^{129}Xe can be transported from the lungs to various tissues (Albert *et al.* 1994). Besides ^{129}Xe, ^{3}He gas is also very good for imaging, for it has a larger magnetic moment and a longer relaxation time.

It should be emphasized, in conclusion, that during the course of the past several years the method of magnetic-resonance imaging with laser-polarized ^{3}He and ^{129}Xe atoms has opened up entirely new possibilities for obtaining NMR information from gas in the lung and from xenon dissolved in biotissues, for example those of the lungs, heart, and brain. Optical (laser) polarization of atoms has proved a very effective method to study and map the functioning of the brain, to measure physiological parameters, and to diagnose diseases of the lungs, heart, and brain. Static and dynamic imaging of the air spaces of the lungs have already provided a much higher spatial resolution than the standard NMR techniques have. Imaging by means of ^{129}Xe dissolved in the biotissue of vital organs has a promising future in practical medicine, as well as in basic research in physiology and neuroscience (Chupp and Swanson 2000).

5
Laser cooling of atoms

One of the most remarkable achievements of laser control is the manipulation of atomic particles, their velocities, positions, etc., which is referred to as *laser cooling* and *laser trapping*. These two effects are interrelated and usually prove effective when used jointly. That this is so one can see by considering as an example the laser cooling of trapped ions and the laser trapping of laser-cooled neutral atoms (Fig. 5.1). The sequences of processes in these two situations are directly opposite. Ions can easily be trapped in an electromagnetic trap of large depth (Paul 1990) and then be cooled for a long time by laser radiation (sideband cooling) (Wineland and Dehmelt 1975; Dehmelt 1990). Neutral atoms are first slowed down and cooled by laser radiation down to millikelvin temperatures, and only then confined in an optical trap. Trapped atoms can further be cooled to microkelvin temperatures. The present chapter and Chapter 6 consider the cooling and trapping of neutral atoms, the present chapter concentrating on their cooling, and Chapter 6 on trapping. But it would be reasonable first to familiarize the reader with the history of the early ideas in this field that eventually led to the advent of a new domain of physics—the physics of ultracold atoms and molecules (Chapters 7 and 8).

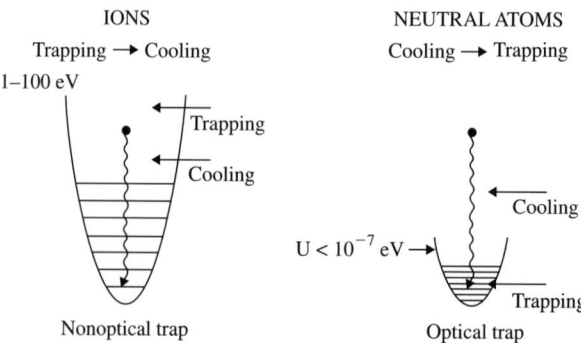

Fig. 5.1 Sequence of methods for electromagnetic cooling and trapping of ions and neutral atoms.

5.1 Introduction. History of ideas

The roots of the ideas of laser control of atomic motion lie in the continued search for effective Doppler-free laser spectroscopic techniques for single atoms. It was only after those ideas had been successfully implemented that the fundamental importance of ultracold atoms became clear. Recalling the history of those ideas, one may say that they have roots in the experiments by Ramsey and coworkers (Goldenberg *et al.* 1960). In these experiments, hydrogen atoms were trapped in a closed vessel whose internal surface was coated with a special paraffin layer. When they collided with this coating, the atoms remained with a high probability in their initial hyperfine-structure state. The vessel was placed inside a microwave cavity. The size of the vessel, a, and of the cavity was chosen to be close to the wavelength $\lambda = 21$ cm of the microwave transition between the hyperfine-structure levels of the hydrogen atom. Thanks to the fact that the free-flight length L of the atoms satisfied the condition

$$L \leq \lambda, \tag{5.1}$$

the motion of the atoms was localized within a small volume $V \leq \lambda^3$. As a result of the localization of atoms, there took place an elimination of the Doppler broadening of spectral lines in what is known as the Lamb–Dicke limit (Dicke 1953).

It seemed very tempting to try to find a way to localize atoms in a micron-size region of space and extend thus the approach to the optical spectral region. Since it was practically impossible to make such small cavities, the natural idea was conceived of localizing atoms in the nodes or antinodes of a standing laser wave, that is, in regions of the size of the optical wavelength (Letokhov 1968). To localize atoms in the inhomogeneities of a standing laser wave, use could be made of the dipole gradient force. Of course, the kinetic energy of a thermal atom far exceeds the height of the potential barrier produced by the gradient force. For this reason, it was only the trapping of thermal atoms moving almost parallel to the wavefront of the standing laser wave, that is, the 1D trapping of atoms, that was discussed in the first proposal (see Fig. 5.2).

In the same period, it was understood that the trapping of atoms by laser light might give birth to what is now called particle-trapping spectroscopy (Letokhov 1975*b*). This would be an important supplement to the Doppler-free laser spectroscopy techniques developed earlier, namely standing-wave absorption saturation spectroscopy

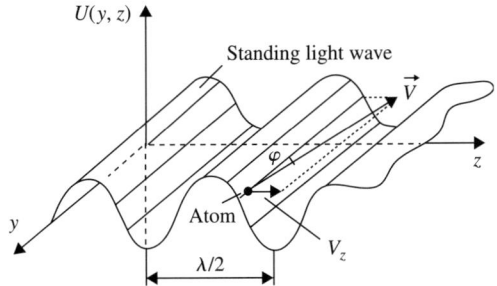

Fig. 5.2 Scheme of the 1D trapping and channeling of atoms in the periodic potential $U(y, z)$ of a standing laser wave. (From Letokhov 1968.)

and standing-wave two-photon spectroscopy (see Chapter 4). In contrast to these nonlinear spectroscopy techniques, particle-trapping spectroscopy is completely free from the transit-broadening effect resulting from the finite particle–field interaction time (Fig. 5.3).

Despite the promising applications that trapped atoms could have in spectroscopy, the trapping of atoms by an off-resonance laser field was not immediately developed experimentally, because the methods for obtaining sufficiently cold atoms were lacking at that time. The potential wells produced by the dipole interaction of an atom with an off-resonance standing light wave, $E = 2E_0 \cos kz \cos \omega t$, have a shallow depth $U_{\text{dip}} = \alpha E_0^2$ because of the low off-resonance atomic polarizability α. Accordingly, off-resonance optical trapping can be implemented only for sufficiently cold atoms, whose temperature is limited by the condition

$$T < \alpha E_0^2 / k_{\text{B}}. \tag{5.2}$$

For example, at an intensity I of the counterpropagating traveling laser waves producing the standing laser wave of the order of $(c/8\pi)E_0^2 \cong 1\,\text{kW}\,\text{cm}^{-2}$, and for a typical atomic off-resonance polarizability $\alpha \approx 3 \times 10^{-23}\,\text{cm}^3$, the condition (5.2) is satisfied only for atoms with quite a low temperature $T < 1\,\mu\text{K}$. In the case of localization of atoms in a three-dimensional standing light wave (an optical lattice) (Letokhov 1973c), the proportion of trapped atoms at normal temperature is very small. So, the

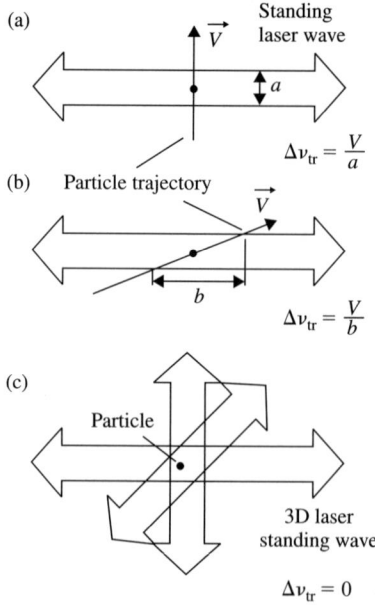

Fig. 5.3 Three methods of Doppler-free optical spectroscopy, which differ in the contribution of the transit-time broadening $\Delta\nu_{\text{tr}}$: (a) saturation spectroscopy in a standing wave; (b) two-photon spectroscopy in a standing wave; (c) particle trapping in a 3D standing wave. (From Letokhov 1975b.)

possibility of low-velocity selection of atoms from a thermal atomic beam by the photodeflection and subsequent trapping of atoms in an optical lattice was proposed.

In the mid 1970s, a major change occurred in our view of the problem of trapping atoms in electromagnetic fields. The first suggestion was put forward at that time about the possibility of deep cooling of atoms by use of resonance optical radiation red-detuned with respect to the atomic transition (Hänsch and Schawlow 1975). From a quantum mechanical point of view, the idea of optical cooling of moving atoms consists in the reduction of atomic velocities by use of photon recoil associated with the absorption of counterpropagating laser photons by the moving atoms. Recall that, owing to the Doppler effect, when the laser field is red-detuned with respect to an atomic transition, an atom predominantly absorbs counterpropagating photons. From a semiclassical point of view, the mechanism of the optical cooling of atoms consists in the retardation of atoms by the radiation pressure force, which, for red-detuned laser light, is directed opposite to the atomic velocity.

Theoretical analysis of the simplest model of the interaction of a two-level atom (Fig. 2.4, for the case $T_2 = 1/\gamma$) with counterpropagating laser beams has shown that laser cooling makes it possible to reach extremely low temperatures, five to six orders of magnitude lower than room temperature. It has been shown that in a two-level atom model, the cooling mechanism is based on single-photon absorption (or emission) processes. The minimum temperature of the atoms is reached at a red detuning equal to the natural half-width of the atomic transition line, that is, $\Delta = -\gamma$, and is determined by the natural half-width γ of the atomic transition (Letokhov et al. 1976, 1977):

$$T_\mathrm{D} = \frac{\hbar \gamma}{k_\mathrm{B}}. \tag{5.3}$$

The value of the temperature in eqn (5.3) is nowadays referred to as the *Doppler temperature* or *Doppler cooling limit*. At a typical value of the natural linewidth of an allowed transition $2\gamma = 2\pi \times 10\,\mathrm{MHz}$, the temperature T_D is of the order of $100\,\mu\mathrm{K}$. Because of the great promise that laser cooling and subsequent laser trapping of atoms held for laser spectroscopy, researchers at the Institute of Spectroscopy in Troitsk, Russia, launched experiments in this field. By the time the first successful experiment was conducted (Andreyev et al. 1981, 1982), the first theoretical work, summarized in a review of the manipulation of atoms by the light pressure force of a resonant laser (Letokhov and Minogin 1981a), had already been completed.

Concluding this brief introduction to the history of the laser cooling and trapping of atoms, we illustrate the sequence of key ideas and experiments that opened the way to the physics of ultracold atoms on the scale of the atomic temperatures and de Broglie wavelengths in Fig. 5.4. The first step is always the Doppler cooling using allowed atomic transitions, that makes it possible to reach subkelvin temperatures, starting with a normal atomic-beam temperature. The subsequent cooling can proceed by two different pathways. For atoms with a nonzero nuclear spin $(I \neq 0)$ (typical alkali metal atoms), use is made of the interaction between the atoms and circularly polarized light, with the spin angular momentum of the photon being utilized to effect polarization gradient cooling to sub-Doppler temperatures for allowed transitions. For atoms of zero nuclear spin $(I = 0)$ that have no magnetic sublevels and are insensitive to

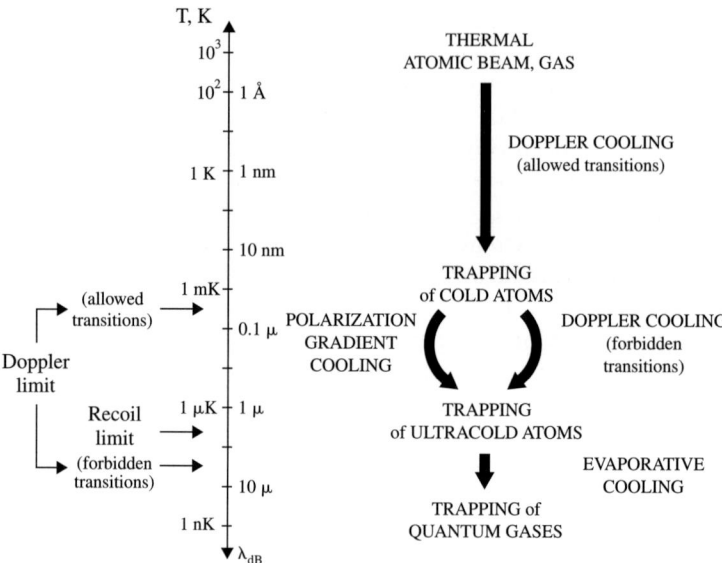

Fig. 5.4 The roads to ultracold-atom physics on the scales of temperature (10^3–10^{-9} K) and de Broglie wavelength λ_{dB}.

magnetic field, use is made of the same Doppler-cooling concept but now for forbidden radiative transitions, for which the Doppler limit can reach subrecoil temperatures. This second road toward ultracold temperatures has materially extended the variety of atoms available to ultracold-matter physics.

5.2 Laser radiation force on a two-level atom

The motion of a two-level atom in a spatially inhomogeneous laser field is generally governed by the dipole gradient force, the radiation pressure force, and the diffusion of momentum. A detailed and consistent analysis of the motion of two-level atoms in light fields can be found in Minogin and Letokhov (1987) and Kazantsev et al. (1990), and here I shall restrict myself to a brief survey of the basic formulas.

The dynamics of the center of mass of an atom in a laser field are determined by the electric dipole interaction (see Section 2.4). As a result of the dipole interaction with the electric field $\mathbf{E} = \mathbf{E}(\mathbf{r}, t)$ described by the dipole interaction operator

$$V = -\mathbf{d} \cdot \mathbf{E}, \tag{5.4}$$

the atom acquires an induced dipole moment $\mathbf{d}$. The value of the induced atomic dipole moment is defined as usual by the quantum-mechanical average

$$\langle \mathbf{d} \rangle = \mathrm{Tr}(\rho \mathbf{d}), \tag{5.5}$$

where ρ is the atomic density matrix. The interaction of the induced atomic dipole moment $\langle \mathbf{d} \rangle$ with the spatially varying laser field $\mathbf{E} = \mathbf{E}(r, t)$ causes, finally, a dipole radiation force to act on the atom, which includes both the average value and quantum fluctuations.

The energy of the dipole interaction between the atom and the laser field is

$$U = \langle V \rangle = -\langle \mathbf{d} \rangle \cdot \mathbf{E}. \tag{5.6}$$

Equation (5.6) formally coincides with the classical expression for the interaction energy of a permanent dipole with the electric field $\mathbf{E}$. Accordingly, eqn (5.6) can be used directly to calculate the force $\mathbf{F}$ acting on an atom in a laser field $\mathbf{E}$:

$$\mathbf{F} = \nabla U = \nabla(\langle \mathbf{d} \rangle \cdot \mathbf{E}) = \sum_i \langle d_i \rangle \nabla E_i, \tag{5.7}$$

where the subscript $i = x, y, z$ determines the rectangular coordinates of the vectors.

Equation (5.7) gives a general expression for the radiation force on an atom moving in a laser field. From a quantum mechanical point of view, the radiation force (5.7) arises as a result of the quantum mechanical momentum exchange between the atom and the laser field in the presence of spontaneous relaxation. The change in the atomic momentum comes from the elementary processes of photon absorption and emission: stimulated absorption, stimulated emission, and spontaneous emission. The radiation force (5.7) is a function of the coordinates and velocity of the center of mass of the atom.

The basic types of radiation force can be understood using simple models of the quasi-resonant interaction of a two-level atom with the monochromatic field of a laser beam or a standing laser wave, and using simple models describing the interaction of a multilevel atom with a laser field.

5.2.1 Traveling wave. The radiation pressure and gradient forces

In the case of the dipole interaction of a two-level atom with a spatially inhomogeneous field $\mathbf{E}$ of a monochromatic laser beam defined by a unit polarization vector $\mathbf{e}$, an amplitude $E_0(\mathbf{r})$, a wave vector $\mathbf{k}$, and an angular frequency $\omega = kc$,

$$\mathbf{E} = \mathbf{e} E_0(\mathbf{r}) \cos(\mathbf{k} \cdot \mathbf{r} - \omega t), \tag{5.8}$$

the radiation force is the sum of two forces: the radiation pressure force $\mathbf{F}_{rp}$ and the dipole gradient force $\mathbf{F}_{gr}$ (Fig. 5.5):

$$\mathbf{F} = \mathbf{F}_{rp} + \mathbf{F}_{gr}. \tag{5.9}$$

The expressions for the two parts of the radiation force on a two-level atom are (Gordon and Ashkin, 1980):

$$\mathbf{F}_{rp} = \hbar \mathbf{k} \gamma \frac{G(\mathbf{r})}{1 + G(\mathbf{r}) + (\Delta - kv)^2/\gamma^2}, \tag{5.10}$$

$$\mathbf{F}_{gr} = -\frac{1}{2} \hbar (\Delta - \mathbf{k} \cdot \mathbf{v}) \frac{\nabla G(\mathbf{r})}{1 + G(\mathbf{r}) + (\Delta - kv)^2/\gamma^2}, \tag{5.11}$$

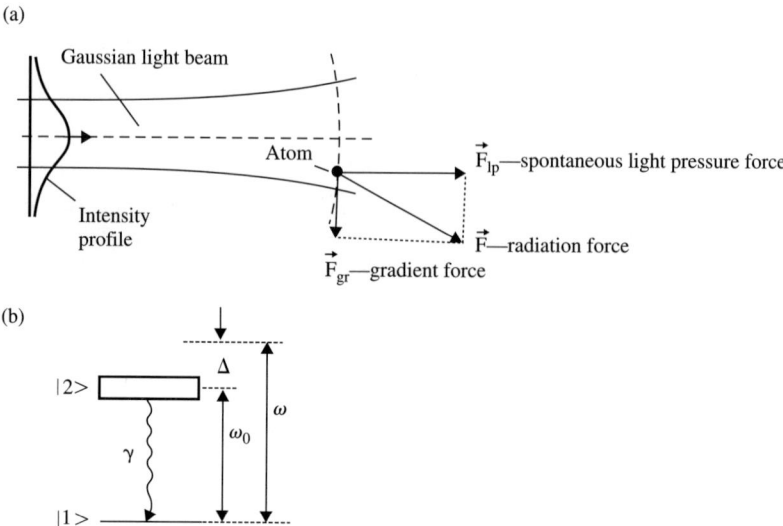

Fig. 5.5 (a) Radiation force of a Gaussian beam and (b) resonant interaction of laser light with a two-level atom.

where $G(\mathbf{r})$ is the dimensionless saturation parameter (see Section 2.5.3):

$$G(\mathbf{r}) = \frac{2\Omega^2(\mathbf{r})}{\gamma^2} = \frac{1}{2}\left(\frac{dE_0(\mathbf{r})}{\hbar\gamma}\right)^2 = \frac{I(\mathbf{r})}{I_S}. \tag{5.12}$$

In the above, $I(\mathbf{r}) = (c/8\pi)E_0^2(\mathbf{r})$ is the intensity of the laser beam at the point $\mathbf{r}$; $I_S = (c/4\pi)(\hbar\gamma/d)^2$ is the saturation intensity; $d = \mathbf{d} \cdot \mathbf{e}$ is the projection of the dipole moment matrix element of the polarization vector $\mathbf{e}$ of the laser beam; Δ is the detuning of the laser field frequency ω with respect to the atomic transition frequency ω_0, that is, $\Delta = \omega - \omega_0$; and the quantity 2γ defines the rate of spontaneous decay of the atom from the upper level $|e\rangle$ to the lower level $|g\rangle$, that is, the Einstein coefficient A. Figure 5.6 shows the dependence of the radiation pressure force and the gradient force on the projection $v_z = v$ of the atomic velocity on the propagation direction of a Gaussian laser beam for the case of strong saturation of the D-line of Na.

The radiation pressure force (5.10) results from the transfer to the atom of the photon momentum in the course of the stimulated absorption and subsequent spontaneous emission of the photon. The force is thus related to *dissipative optical processes*. The field of a *spatially inhomogeneous* laser beam can be treated as a superposition of many plane waves propagating within the divergence angle of the beam. In a field composed of many plane light waves, the momentum of the atom can also be changed by another elementary process, namely, stimulated absorption by the atom of a photon from one plane wave and subsequent stimulated emission into another plane wave. The two photons participating in this process have the same energy and differ only by their propagation direction. This process results in the gradient force, which is accordingly directed along the intensity gradient of the laser beam, as defined by eqn (5.11). The gradient force is thus related to *conservative optical processes*.

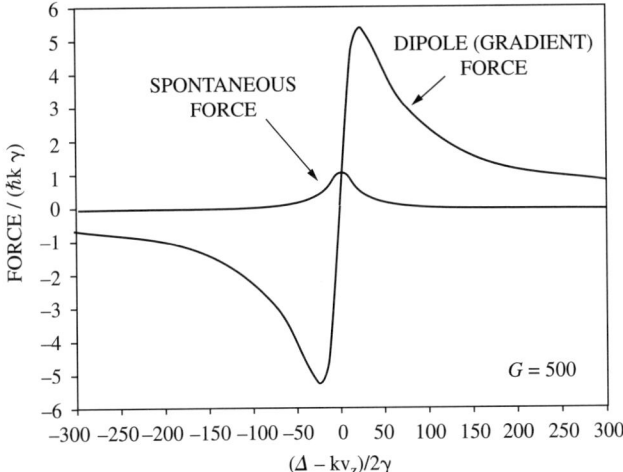

Fig. 5.6 Velocity dependence of the radiation pressure force and the gradient force for a saturation parameter $G = 500$ and the D-line of Na.

The effects of the radiation pressure force and the gradient force on an atom are essentially different. The radiation pressure force (5.10) always accelerates the atom in the direction of the wave vector $\mathbf{k}$. The gradient force (5.11) pulls the atom into the laser beam or pushes it out of the beam, depending on the sign of the Doppler shift detuning $\Delta - \mathbf{k} \cdot \mathbf{v}$. Both the radiation pressure force and the gradient force have a resonance at a velocity such that $\mathbf{k} \cdot \mathbf{v} = \Delta$, when the detuning Δ is compensated by the Doppler shift $\mathbf{k} \cdot \mathbf{v}$.

5.2.2 Standing wave. Friction force

In a single standing light wave composed of two traveling light waves, two Doppler resonances occur under weak-saturation conditions, one for each travelling wave: $\pm kv = \Delta = \omega - \omega_0$. The counterpropagating wave decelerates the atom, whereas its copropagating counterpart accelerates the atom. The combined effect of the two traveling waves acting independently is described by a dispersive curve (Fig. 5.7). At exact resonance ($kv = 0$), the spontaneous force is zero, and the slope of the curve at $v = 0$ gives the friction coefficient β for the atom, which governs its cooling rate. Summing the two partial forces in eqn (5.9), one obtains, to a first approximation in the atomic velocity, the following expression for the total radiation pressure force:

$$F_{\rm rp} = -\beta v, \qquad (5.13)$$

where β is the dynamic friction (damping) coefficient, given at weak saturation ($G \ll 1$) by

$$\beta = (4\hbar k^2) \left[\frac{G\Delta/\gamma}{1 + (\Delta/\gamma)^2} \right]. \qquad (5.14)$$

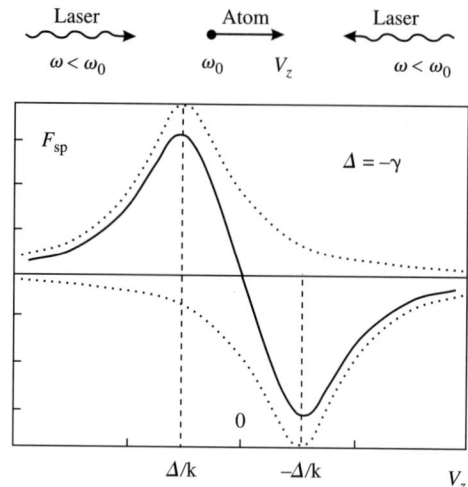

Fig. 5.7 Doppler cooling in a standing light wave, where the atom is acted upon by two radiation forces, one from each traveling light wave. When the saturation of the atomic transition is weak, the atomic-velocity dependences of the forces have a Lorentzian form (the curves marked + and −). The average total force is described by a curve of dispersive character, whose slope at $v_z = 0$ determines the friction coefficient.

The friction force in eqn (5.13) is proportional to the velocity (with a negative sign). For this reason, it causes viscous damping at a red frequency detuning of $\Delta < 0$. It is precisely this force that is responsible for "optical molasses" (see Section 5.4.3 below).

5.3 Quantum fluctuation effects. Temperature limits of laser cooling

The exchange of the momentum $\hbar k$ between an atom and a light field is of discrete character. Such an exchange takes place owing to three elementary processes (stimulated absorption, stimulated emission, and spontaneous emission) that change the atomic momentum. Stimulated absorption increases the atomic momentum, and stimulated emission decreases it, in the direction of the radiation vector k. Its direction being stochastic, spontaneous emission on average has no effect on the momentum of the atom, but ensures its relaxation to the ground state. When the atom interacts with radiation for a long period of time, a change in the atomic momentum is caused by the combined action of recoils due to stimulated and spontaneous transitions. Each time the atom is stimulated to absorb (or emit) a photon, it acquires a recoil momentum of $\hbar k = \hbar \omega_0 / c$ along the radiation wave vector ($\hbar k$ in absorption and $-\hbar k$ in emission). Because of the statistical nature of the spontaneous relaxation of the atom to the ground state, the sequence of stimulated transitions is random. In the course of spontaneous emission, whose direction fluctuates, the atom acquires a recoil momentum $\hbar \omega_0 / c$, which has a fixed magnitude but a variable direction. For this reason, the combined action of the recoils due to the stimulated and spontaneous transitions always makes the atomic momentum vary in a stochastic manner.

The quantum fluctuations in the atomic momentum (or velocity) correspond to those of the radiation pressure force of the light given in eqn (5.10). This gives rise to fluctuation heating of the atoms. This heating prevents the atoms from reaching zero temperature, which might be expected from eqn (5.13) for an atom in the field of two counterpropagating light waves. The minimum temperature of a laser-cooled two-level atom is governed by the competition between the atomic cooling and heating mechanisms (Letokhov et al. 1976, 1977; Wineland and Itano 1979):

$$T_\mathrm{D} = \frac{\hbar\gamma}{2k_\mathrm{B}}\left(\frac{|\Delta|}{\gamma} + \frac{\gamma}{|\Delta|}\right). \tag{5.15}$$

When the red frequency detuning of the laser radiation is optimal for cooling purposes, that is, when $\Delta = -\gamma$, the temperature T_D reaches the minimum value given by eqn (5.3), which usually lies in the millikelvin range. As stated earlier, the temperature T_D is usually referred to as the Doppler cooling limit.

Another ultimate temperature, determined by the recoil energy R (eqn 2.22) is defined by

$$T_\mathrm{rec} = \frac{\hbar^2 k^2}{2k_\mathrm{B} M}, \tag{5.16}$$

that is, it corresponds to the minimal recoil velocity $v_\mathrm{rec} = \hbar\omega/Mc$ (see Section 2.3). The temperature T_rec is customarily called the *recoil cooling limit*, and it lies in the submicrokelvin range.

The fluctuations of the velocity and coordinates of an atom cause it to move in a diffusive fashion. This will take place if the length of the path the atom travels during a time $\tau \simeq 1/\gamma$ is much smaller than the size of its region of interaction with the field. Such a diffusive motion of the atom is well described by the Fokker–Planck equation (Minogin 1980). Diffusive redistribution of atomic velocities has been observed in an atomic beam propagating in a counterrunning light wave (Balykin et al. 1981). In the three-dimensional case, the diffusive motion of the atom takes place in the space of both velocities and coordinates. It is similar to the motion of a particle in a viscous medium and has therefore been termed "optical molasses" (Chu et al. 1985).

In accordance with the above temperature-scale estimates (Fig. 5.4), it is convenient to consider consecutively the cooling of atoms first to the Doppler limit T_D, then sub-Doppler cooling to T_rec, and finally subrecoil cooling. Various mechanisms for cooling atoms in light fields of various configurations are considered below in the same sequence. The reader can find more detailed analysis in the book by Metcalf and van der Straten (1999).

5.4 Doppler cooling

The first successful experiments on the cooling of atoms were conducted with atomic beams irradiated by a counterrunning light wave. This is known as slowing, or one-dimensional cooling. Thereafter, experiments were performed on the transverse cooling of an atomic beam by means of standing light waves directed at right angles to the beam (2D cooling). The next natural step was the 3D cooling and trapping of atoms by way of all-round irradiation of a region containing preliminarily slowed

5.4.1 Slowing and longitudinal cooling of an atomic beam

Figure 5.8 shows the basic idea of the deceleration and longitudinal cooling of a thermal atomic beam by a counterpropagating laser beam. In the scheme of Fig. 5.8(a), a red-detuned laser beam produces the radiation pressure force (eqn 5.10), which most effectively decelerates the atoms with longitudinal velocities v_z close to the resonance velocity $v_{\rm res} = |\Delta|/k$. The deceleration of atoms whose velocities are far from the resonance velocity is less effective, for the radiation pressure force $F_{\rm rp}$ has a Lorentzian velocity dependence (Fig. 5.8(b)). As a result, the radiation pressure force both decelerates the atoms and narrows the atomic velocity distribution, that is, it produces a cooling of the atomic beam (Fig. 5.8(c)). Figure 5.9 shows the experimental velocity distribution profile of a beam of sodium atoms slowed and cooled by radiation from a dye laser in the first experiment on the laser cooling of atoms (Andreyev et al. 1981, 1982).

The above, simplest method of longitudinal laser cooling at a fixed detuning is most effective for cooling atoms moving at the resonance velocity. The efficiency of slowing and cooling naturally drops when the atoms go off resonance with the laser light owing to the velocity decrease. To maintain a high deceleration and cooling rate, some experimental techniques make use of chirping of the laser frequency (Balykin et al. 1979) or Zeeman tuning of the atomic transition frequency by an inhomogeneous magnetic field whose strength varies along the propagation direction of the atomic beam (Prodan et al. 1982). The use of these techniques makes it possible to decelerate

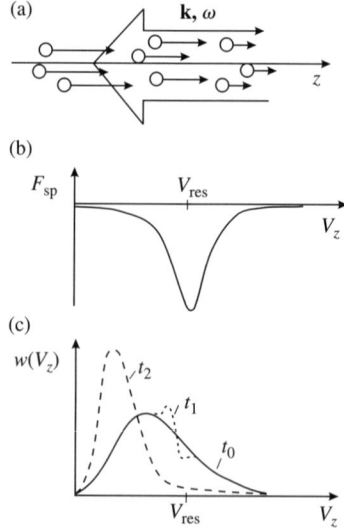

Fig. 5.8 (a) Scheme of the longitudinal slowing and cooling of a thermal atomic beam by a counterpropagating laser beam. (b) Radiation pressure force as a function of the longitudinal atom velocity. (c) Evolution of the atomic velocity distribution at times $t_2 > t_1 > t_0$.

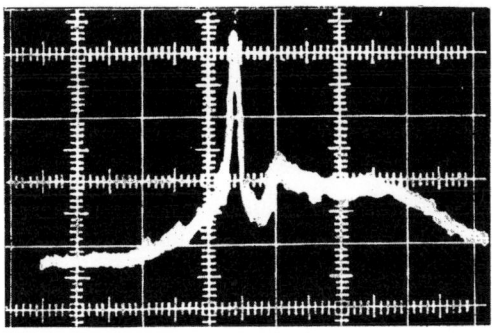

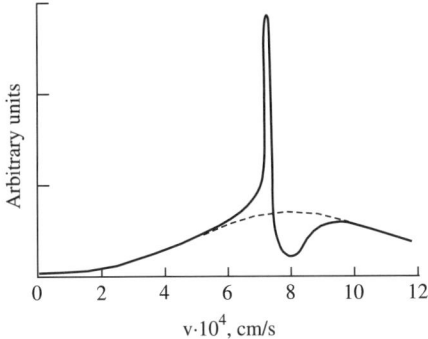

Fig. 5.9 Experimental velocity distribution of a beam of sodium atoms decelerated by a dye laser in the first experiment on the laser cooling of atoms, and a theoretical prediction of the parameters relevant to the experiment. (From Andreyev *et al.* 1981, 1982.)

an atomic beam that has an initial thermal velocity v_0 to a zero average velocity, the deceleration length being

$$l = \frac{v_0^2}{2v_{\text{rec}}\gamma} \frac{1+G}{G}, \qquad (5.17)$$

where v_{rec} is the recoil velocity and G is the saturation parameter. At a thermal velocity $v_0 \approx 10^5 \text{ cm s}^{-1}$ and a moderate saturation $G \approx 1$, the deceleration length l ranges between 10 and 100 cm. The longitudinal temperature of an atomic beam under optimal conditions can be reduced to the millikelvin range.

The Zeeman technique for compensating for the Doppler shift is to use the Zeeman shift of the atomic transition frequency while keeping constant the frequency of the cooling radiation. The Zeeman technique is applicable to cooling by means of σ^+- or σ^--polarized radiation. In a magnetic field directed along the atomic beam axis z, the laser light is in resonance with the atomic transition under the condition

$$kv(z) = -\Delta + \frac{\mu_B B(z)}{\hbar}, \qquad (5.18)$$

where $\Delta = \omega - \omega_0$ and μ_B is the Bohr magneton. To retard atoms that have an initial velocity of v_0 with an acceleration of a, the magnetic-field strength along the beam axis must have the form (Phillips and Metcalf 1982)

$$B(z) \underset{\sim}{\simeq} B_0 \left(\frac{1-z}{z_0}\right)^{1/2}, \qquad (5.19)$$

where $B_0 = kv_0\hbar/\mu_B$ is the field that causes a Zeeman shift compensating for the initial Doppler shift kv_0 of the resonance transition frequency, and z_0 is the length of the magnet. The Zeeman laser-cooling technique for an atomic beam has gained wide recognition, thanks to its capability of producing continuous flows of cold atoms.

5.4.2 2D cooling: collimation of an atomic beam

When an atomic beam is being irradiated crosswise, the dissipative force (eqn 5.10) may reduce the transverse atomic velocities, that is, collimate the beam (Balykin et al. 1984). The atomic beam (see Fig. 5.10) is in this case irradiated on all sides with an axisymmetric light field whose frequency ω is shifted toward the red relative to the atomic transition frequency ω_0; the axisymmetric field is formed by laser radiation reflected from the inner surface of a conical reflecting axicon. In the plane of the figure, this field consists of two counterpropagating light waves, the intensities of which are equal at any point of space in the axicon.

An atom moving with a transverse velocity of v_ρ in the axisymmetric light field is acted upon by a radiation pressure force, which for $\omega < \omega_0$ is directed against the radial velocity and for $\omega > \omega_0$ is directed with the velocity. These directions of the radiation pressure force are due to the fact that the atom, because of the Doppler shift of the atomic transition frequency, absorbs more effectively those photons which come from the light wave propagating counter to the radial velocity. This means that for $\omega < \omega_0$ the total force caused by the two light waves is directed against the radial atomic-velocity vector. Owing to the action of this force, for $\omega < \omega_0$ there occurs in the inner region of the axicon a rapid narrowing of the transverse velocity distribution of the atomic beam, which reduces the angular divergence of the beam and increases its atomic density, that is, it improves the beam collimation. The joint effect of radiative

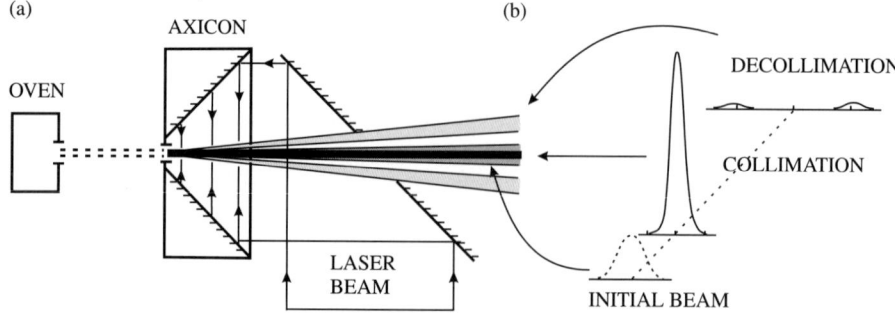

Fig. 5.10 Collimation and decollimation of an atomic beam by laser radiation: (a) schematic of a radiative atomic-beam collimator; (b) atomic-beam profiles.

friction and momentum diffusion leads to the establishment of a stationary atomic-velocity distribution which determines the ultimate collimation angle,

$$\Delta\varphi_{\min} = \left(\frac{1}{v_0}\right)\left(\frac{\hbar\gamma}{M}\right)^{1/2}, \tag{5.20}$$

where M is the atomic mass and v_0 is the longitudinal atomic velocity. For a thermal atomic beam, the ultimate collimation angle is of the order of 10^{-3}–10^{-4} rad.

This collimation experiment (Balykin et al. 1984) was conducted with a beam of sodium atoms. The right-hand side of Fig. 5.10 presents beam profiles prior to and after interaction with the laser field. The collimation process is very sensitive to the position of the laser radiation frequency. To effect collimation, the laser frequency was detuned toward the red side ($\omega - \omega_0 \simeq -\gamma$) for a longitudinal atomic velocity $v_0 \simeq 7.7 \times 10^4 \text{ cm s}^{-1}$. Comparison between the above beam profiles shows, first, a substantial increase (by a factor of 5) in the on-axis atomic-beam intensity and, second, a considerable narrowing (collimation) of the beam. Changing over to a positive detuning caused a material broadening (decollimation) of the beam. The on-axis beam intensity in that case was observed to decrease by more than three orders of magnitude.

5.4.3 3D cooling of atoms. Diffusional confinement

In 3D-cooling scheme, atoms are irradiated by three pairs of counterpropagating, red-detuned laser waves. In the simplest model of a two-level atom and at weak optical saturation, the radiation force on the atom, averaged over the wavelength of the laser field in the case of three-dimensional standing-wave irradiation, can be represented as a sum of three forces (eqn 5.10):

$$\mathbf{F} = \sum_{i=x,y,z} \hbar \mathbf{k}_i \gamma G \left\{ \frac{1}{1 + (\Delta - k_i v_i)^2/\gamma^2} - \frac{1}{1 + (\Delta + k_i v_i)^2/\gamma^2} \right\}, \tag{5.21}$$

where the $\mathbf{k}_i$ are the wave vectors of the laser waves propagating in the positive direction of the axes $i = x, y, z$ and v_i is the projection of the atomic velocity on the axis i. For a red detuning, $\Delta < 0$, the force given by eqn (5.21) in a linear approximation in the atomic velocity is reduced to a friction force $\mathbf{F} = -\beta \mathbf{v}$, where the friction coefficient β is defined by eqn (5.14). In the same model of a two-level atom and in linear approximation in the saturation parameter, the momentum diffusion coefficient averaged over the laser wavelength and taken at zero velocity is

$$D = 2\hbar^2 k^2 \gamma \frac{G}{1 + (\Delta^2/\gamma^2)}. \tag{5.22}$$

The stationary solution of the Fokker–Planck equation, which includes the friction force $\mathbf{F} = -\beta \mathbf{v}$, and the momentum diffusion coefficient (eqn 5.22), is a 3D Gaussian distribution

$$W(v) = \frac{1}{(\sqrt{\pi}u)^3} \exp\left(\frac{-v^2}{u^2}\right). \tag{5.23}$$

The half-width of the velocity distribution in eqn (5.23),

$$u = \sqrt{2k_B T_D/M}, \tag{5.24}$$

is defined by the Doppler temperature (eqn 5.15).

The above estimation of the atomic temperature is in good agreement with experimental observations in cases when the dipole interaction between the atoms and the laser field can be described by a two-level model. The first experiment on the 3D Doppler cooling of atoms was done with sodium vapor (Chu et al. 1985). A beam of sodium atoms was preliminarily slowed by a counterpropagating laser beam. Thereafter the slow atomic beam was directed into the intersection region of three orthogonal 1D standing laser waves producing a 3D standing wave. The minimal temperature $T_{\min}$ of the cold sodium atoms recorded in the experiment was $240\,\mu K$, which agreed well with the temperature T_D predicted by eqn (5.16) for two-level atoms.

The cooling of atoms down to their Doppler-cooling limit T_D is not accompanied by trapping of them. This is due to the fact that the kinetic energy kT_D of the cooled atoms is commensurable with the depth of the potential wells produced by the gradient force (eqn 5.11) (Balykin et al. 1988a). In that case, however, there takes place a diffusive confinement of the atoms, which perform a random walk because of the repeated photon reemission events. During the time $1/\gamma$ that elapses between successive photon reemission events (at $G \simeq 1$), a cooled atom moves in space for a distance $\Delta \ell \simeq u/\gamma$, equal to a few micrometers, where u is the velocity of the cooled atoms, defined by eqn (5.24). It takes approximately a second for a cooled atom to leave a region about 1 mm across by diffusion. This regime of diffusive motion of cooled atoms under conditions of viscous friction proportional to their velocity has been called "optical molasses" by analogy with the motion of macroparticles in a viscous liquid. A more detailed investigation into the velocity of atoms leaving an "optical molasses" region led to the discovery of the effect of sub-Doppler cooling for allowed transitions of atoms.

5.4.4 Cooling using forbidden transitions

At first glance, the Doppler cooling limit (eqn 5.15) seems a serious obstacle to the attainment of the recoil temperature limit (eqn 5.16). This has even led to the establishment of a hierarchy in the temperature classification, from the Doppler cooling limit to subrecoil cooling. In actual fact, Doppler cooling is a fairly universal mechanism allowing subrecoil temperatures to be reached, provided that the laser-cooling strategy is properly chosen. This fact was successfully demonstrated in experiments on the laser cooling and trapping of atoms of alkaline earth metals (Sr, Yb, and ^{40}Ca) using both allowed and spin-forbidden transitions (the second route to ultracold atoms in Fig. 5.4). Figure 5.11 presents a schematic diagram of such cooling for the ^{88}Sr atom (Katori et al. 1999), where ^{88}Sr atoms, precooled using the allowed $^1S_0 \to {}^1P_1$ transition at 461 nm, were further cooled in a magnetooptical trap using the spin-forbidden transition $^1S_0 \to {}^3P_1$ with the 689 nm intercombination line. Doppler cooling using the weak transition produced an atomic ensemble with a density of over $10^{12}\,\mathrm{cm}^{-3}$ and a minimum temperature of 400 nK (the recoil limit T_{rec} is 440 nK). The same strategy

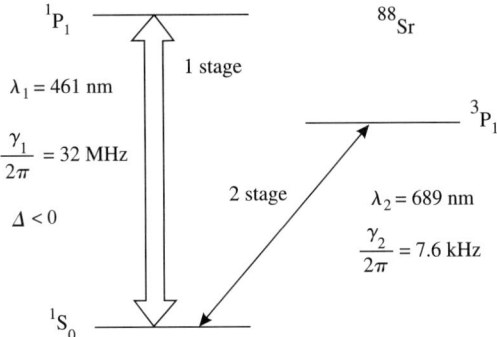

Fig. 5.11 Energy-level diagram of ^{88}Sr related to two-stage laser cooling using allowed and forbidden transitions.

was used to cool atoms of six Yb isotopes to a temperature of $20\,\mu\text{K}$ (Kuwamoto et al. 1999) and ^{40}Ca atoms to a temperature of $6\,\mu\text{K}$ (Binnewies et al. 2001). What is important is that Doppler cooling using such forbidden transitions can result in temperatures much lower than the recoil limit (Wallis and Ertmer 1989).

5.5 Laser polarization gradient cooling below the Doppler limit

In experiments on the 3D Doppler cooling of Na atoms (Lett et al. 1988), it was found, by way of direct measurement of the velocity of the cooled atoms leaving the cooling region, that the temperature of the atoms could in actual fact be almost ten times lower the Doppler limit $T_D \simeq 240\,\mu\text{K}$. Moreover, there were observed a number of differences between the actual optical molasses and its two-level model, specifically in the relationships between the atomic temperature and the various parameters of the molasses. First, it was demonstrated that the minimum atomic temperature was reached at a frequency detuning Δ a few times larger than γ. Secondly, the temperature of the atoms in the molasses was observed to depend sharply on the polarization of its component beams and on the strength of the additional stationary magnetic field. Lett et al. explained these differences by the inapplicability, in the case of optical molasses, of the two-level atomic model to a real atomic transition with levels that possessed a magnetic structure. This required that the Doppler-cooling model for idealized two-level atoms needed to be elaborated substantially to take account of the actual multi-sublevel structure of both the 3S and the 3P level of the Na atom.

The actual picture of the interaction between standing laser light waves and Na atom (and also other atoms) proved much more interesting. The interaction model developed by Dalibard and Cohen-Tannouji (1989) and Ungar et al. (1989) allowed for the effects of optical pumping of the Zeeman and HFS sublevels by polarized light, as well as the light-induced shift of the energy levels (see Chapter 4). Specifically, it was necessary to take into consideration the periodic spatial dependence of these effects in standing light waves. This formed the basis for the discovery of a mechanism of deep, sub-Doppler cooling of atoms that was termed "polarization gradient cooling." The effect of polarization gradient cooling can be produced by various combinations of types of polarization of the counterrunning light waves forming the standing

84 *Laser control of atoms and molecules*

wave. The largest effect is attained in the following two cases: (1) with orthogonal polarizations of the counterrunning waves (the linear ⊥ linear configuration), and (2) with opposite circular polarizations of the counterrunning waves (the σ^+–σ^- configuration). An excellent description of the mechanisms of laser polarization gradient cooling that made it possible to overcome the Doppler-cooling limit can be found in brief reviews by Cohen-Tannoudji and Phillips (1990) and Cohen-Tannoudji (1992). A detailed theoretical analysis of these mechanisms for various two-level atoms can be found in the review by Chang and Minogin (2002).

Consider an Na atom placed in the field of two counterpropagating light waves having the same frequency ω, but polarized linearly at right angles to each other. Figure 5.12(a) shows the Zeeman sublevels of the $3^2S_{1/2}$–$3^2P_{3/2}$ transition in Na, which correspond to different eigenvalues of the total angular-momentum projection on the propagation axis of the laser waves. When the Na atom is being excited by circularly polarized light, it undergoes optical pumping of one of the Zeeman levels of the ground state, depending on the sign of the circular polarization (see Chapter 4).

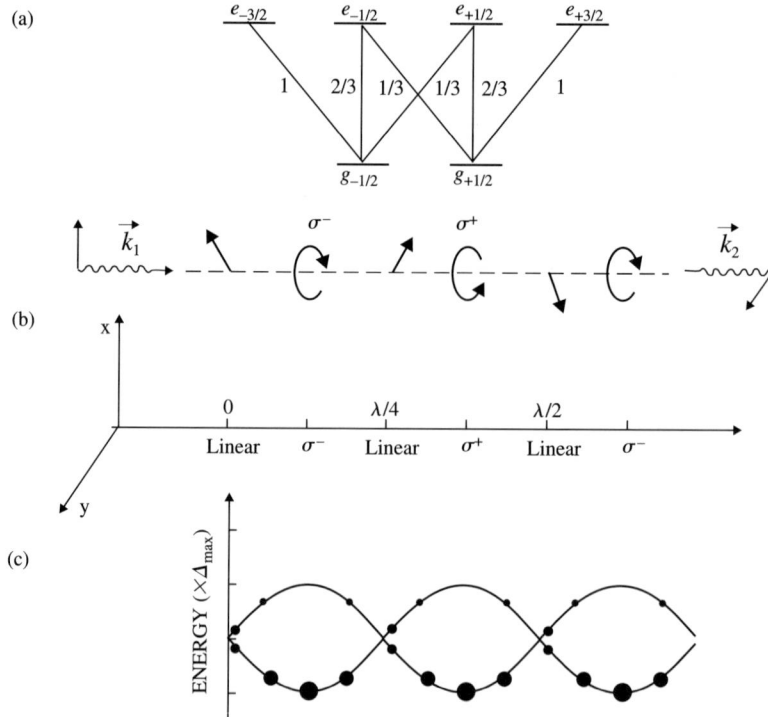

Fig. 5.12 Explanation of atomic cooling based on the effects of the polarization gradient and light-induced level shifts for a multilevel atom (a), with orthogonal linear polarizations of the counterpropagating laser beams, which produces a light field with a periodically changing circular polarization (b). The light shifts and the populations of the two ground-state sublevels vary periodically with the atom's position (c). (Reprinted with courtesy and permission of the Optical Society of America from Dalibard and Cohen-Tannoudji 1989.)

What is important is that the time τ_p of such optical pumping is many orders of magnitude longer than the spontaneous-decay time $1/\gamma$ of the excited state. Two counterpropagating laser beams with orthogonal linear polarizations form a standing light wave whose local polarization changes every eighth of a wavelength from the linear to the circular type (Fig. 5.12(b)). Depending on the sign of the local circular polarization, there takes place, thanks to optical pumping, an accumulation of atoms either in the sublevel $g_{-1/2}$ (for σ^-) or in the sublevel $g_{+1/2}$ (for σ^+), and thus a spatially periodic modulation of the circular polarization of the Zeeman sublevels of the ground state. The shift of the Zeeman sublevels also depends on the sign of the circular polarization: the σ^+ wave shifts only the $g_{+1/2}$ sublevels, whereas its σ^- counterpart shifts only the $g_{-1/2}$ ones. So, the light-shifted energies and populations of the two ground-state sublevels of the atom vary with the local polarization of the standing light wave, and hence with the atom's position, as shown in Fig. 5.12(c), where the sublevel populations are proportional to the size of the filled circles. The Stark-level shift, according to eqn (4.2) at large frequency detunings ($\Delta \gg \gamma$), is $\Delta E \sim \hbar \Omega_0^2/\Delta$, where Ω_0 is the Rabi frequency.

Let the atom move at a small angle to the wavefront of the light, so that its velocity projection is

$$0 < v_z \ll \frac{\gamma}{k}. \tag{5.25}$$

Consider for the sake of definiteness an atom starting to move from the point $z = \lambda/8$ (Fig. 5.12). Such an atom will cover a distance of $\lambda/4$ in a time shorter than τ_p, and so it will climb from the potential minimum to the maximum, while residing at one and the same sublevel, as illustrated in Fig. 5.13. The probability that optical pumping will cause the atom to move from one sublevel to the other here becomes higher. But the spontaneous-decay time is much shorter than the optical-pumping

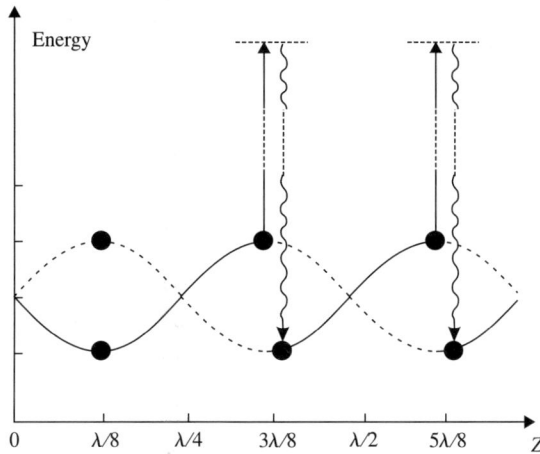

Fig. 5.13 Sisyphus mechanism of cooling for an atom moving with a low velocity in the OZ direction in a lin $\perp$ lin configuration of laser beams. (Reprinted with courtesy and permission of the Optical Society of America from Dalibard and Cohen-Tannoudji 1989.)

time, that is, $1/\gamma \ll \tau_p$, and therefore a spontaneous transition of the atom to the other sublevel takes place while it travels a distance of $\Delta z \ll \lambda/4$. As a result, the atom absorbs long-wave photons and emits short-wave ones. The energy difference between these photons is equal to the level shift and so is proportional to the light field intensity.

Thus, during each optical-pumping cycle with a period τ_p, the atom loses the energy related to the potential depth ΔE. The cooling force acting on a moving atom can be estimated from the relation

$$F = \frac{dU(z)}{dt} \frac{1}{v}, \qquad (5.26)$$

where $U(z)$ is the light-shifted energy of one of the ground-state sublevels of the atom. Taking into account the relation $dU/d \simeq \Delta E/\tau_p$ and the fact that optimal cooling take place at a velocity $v \simeq 1/k\tau_p$, eqn (5.26) assumes the form

$$F \simeq \Delta E \tau_p k^2 v. \qquad (5.27)$$

The pumping time is of the order of $\tau_p \simeq 1/n_2 \gamma \simeq \Delta^2/\gamma \Omega^2$, where n_2 is the relative population of the excited level, which is determined by eqn (2.76) in the weak-saturation regime $(G \ll 1)$. Substituting the relations for τ_p and ΔE into eqn (5.27), we have

$$F \simeq \hbar k^2 \frac{\Delta}{\gamma} v. \qquad (5.28)$$

From this expression we can see that the friction coefficient depends only on the detuning of the laser field. Obviously, the friction coefficient under conditions of polarization cooling, $\beta = \hbar k^2 \Delta/\gamma$, can be considerably higher than the maximum friction coefficient in the Doppler-cooling regime, $\beta \simeq \hbar k^2$, which is approximately determined by eqn (5.14) for $G \simeq 1$. In a similar manner, it is possible to estimate the diffusion coefficient

$$D = \frac{\langle \Delta p^2 \rangle}{dt} \simeq (F\tau_p)^2 \frac{1}{\tau_p} \simeq \frac{\hbar^2 k^2 \Omega^2}{\gamma}. \qquad (5.29)$$

From (eqn 5.28) and (5.29), there follows a relation for the ultimate atomic temperature,

$$T \simeq \frac{\hbar \Omega^2}{|\Delta| k_B}. \qquad (5.30)$$

The above expression shows that the atomic temperature drops as the strength of the light field is reduced and its frequency detuning is increased. However, there is a limit to such a reduction. It follows from the cooling mechanism examined above that energy is being taken out of the atom in quanta of the order of $\Delta E = \hbar \gamma^2/\Delta$, the atom making a transition between the ground-state sublevels. This transition is accompanied by the decay of the excited atomic state, imparting to the atom an energy R equal to the photon recoil energy. This heating, inevitable in polarization cooling, limits the minimum atomic temperature $T_{\min}$ to approximately $T_{\rm rec}$. A detailed theoretical analysis of this atomic cooling mechanism has been made by Dalibard and Cohen-Tannoudji (1989) and Ungar et al. (1989).

A spatial variation of the Stark level shift is not a necessary condition for achieving sub-Doppler temperatures. Sub-Doppler cooling of multilevel atoms can be effected in a field composed of two counterpropagating circularly polarized light waves of opposite polarizations (Dalibard and Cohen-Tannoudji 1989). At each point of space, the radiation is linearly polarized. However, the polarization direction rotates as the atom moves along the wave propagation axis. As in the case of the ordinary two-level molasses, the origin of the friction force is associated with an imbalance in the rates with which photons moving in opposite directions are absorbed. But the presence of several lower atomic levels brings about a cardinal change in the relationship between the friction coefficient and the radiation intensity. Whereas the friction coefficient for a two-level atom drops linearly with decreasing radiation intensity, that for a multilevel atom is independent of the intensity. On the other hand, the diffusion coefficient in both of these two cases decreases in a linear fashion. Therefore, when the intensities are low enough, atoms can be cooled below the $\hbar\gamma$ limit, because the cooling temperature is determined by the ratio between the diffusion and friction coefficients.

Sub-Doppler polarization gradient cooling for various atoms and various configurations of the polarized laser light fields has been investigated in detail using magnetooptical atomic traps, which are considered in Chapter 6, dealing with the cooling of atoms in atomic traps of various types.

5.6 Cooling below the recoil limit

The methods of cooling atoms below the recoil energy are closely associated with the stochastic nature of the atom–radiation interaction. The possibility of using the stochastic behavior of atoms in a laser field with a view to deep cooling of them was shown by Pritchard (1983). The idea of the method is as follows. In the closed cycle constituted by the stimulated absorption of a photon by an atom and the subsequent spontaneous emission of a photon, there exists a finite probability that the atom will transit to a translational state whose energy $E_{\rm at}$ is lower (as desired) than the recoil energy; that is, $E_{\rm at} \ll E_{\rm rec}$. If there exists some *selective* mechanism that provides for the repetition of the cycle only for fast atoms ($E_{\rm at} > E_{\rm rec}$), then, on completion of a sufficient number of cycles, a substantial fraction of the fast atoms will be cooled down to a temperature below the recoil energy. This method has received the name "phase-space optical pumping."

To date, two specific phase-space optical-pumping schemes have been demonstrated. One uses a two-photon Raman transition and the other is based on the velocity-selective coherent population-trapping (VSCPT) effect.

5.6.1 Raman cooling

Raman cooling uses a two-photon transition between two hyperfine-structure levels in the ground state of the atom and provides coherent displacement of parts of the velocity distribution toward zero velocity. Such cooling was successfully demonstrated in an experiment by Kasevich and Chu (1992). Consider a three-level Λ-atom in which $|1\rangle$ and $|2\rangle$ are the lower energy levels and $|e\rangle$ is the upper level, placed in the field

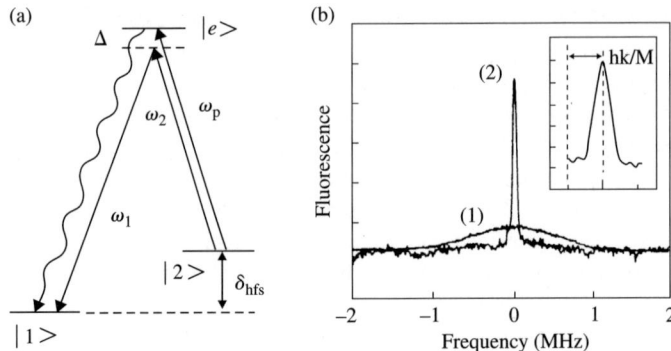

Fig. 5.14 (a) Energy-level diagram for Raman cooling based on a two-photon transition between two hyperfine-structure components of the ground atomic state. (b) Atomic velocity distribution before (1) and after (2) Raman cooling. The insert shows the central velocity peak and the velocity width relative to the recoil velocity. (Reprinted from Kasevich and Chu 1992 with courtesy and permission of the American Physical Society.)

of a light pulse formed by two-counterpropagating light waves differing in frequency (Fig. 5.14(a)). If the detuning of the frequencies of the waves from the corresponding transition frequencies satisfies the two-photon resonance condition, the three-level atom is reduced to a two-level one. The role of the frequency detuning in this case is played by the detuning Δ from the two-photon resonance. The effective Rabi frequency has the form

$$\Omega_{\text{eff}} = \frac{\Omega_{1e}\Omega_{e2}}{2\Delta}, \qquad (5.31)$$

where Ω_{1e} and Ω_{e2} are the Rabi frequencies of the corresponding one-photon transitions, and Δ is the detuning of the frequency of the first field from the one-photon resonance frequency.

If the duration of the laser pulse corresponds to a π-pulse, an atom originally residing in level 1 will completely move to level 2, the atomic momentum changing by an amount $2\hbar k$, where $\hbar k$ is the photon momentum. By virtue of the high Doppler sensitivity of the Raman transition, such a jump in momentum space proves possible only for atoms whose velocities fall within a narrow velocity interval ($\delta v \sim \Gamma_{21}/k$) in the vicinity of the velocity at which the two-photon resonance occurs. Here Γ_{21} stands for the width of the forbidden transition between levels 2 and 1. For this reason, the Raman atomic-excitation scheme allows one to selectively act upon various velocity groups of atoms. By using suitable sequences of Raman π-pulses of various intensities, frequencies, and directions, one can change the velocity distribution of atoms almost arbitrarily. Kasevich and Chu (1992) used a sequence of pulses which shifted different portions of the atomic velocity distribution, in consecutive order, toward zero velocity. The resultant ultranarrow peak in the velocity distribution corresponded to a laser-cooling temperature of ~ 100 nK ($0.1 T_{rec}$) (Fig. 5.14(b)).

5.6.2 Velocity-selective coherent population trapping of ultracold atoms

The coherent population-trapping effect is manifested in various domains of atomic and laser physics (see Section 4.4.2). Let a quantum mechanical system described by a nonperturbed operator $\hat{H}_0$ be placed in an external field. Its interaction with the field is described by an operator $\hat{V}$. In this case, the temporal evolution of the state $|\Psi\rangle$ of the system in the interaction representation is described by the Schrödinger equation as

$$i\hbar \frac{\partial}{\partial t} |\Psi\rangle = \exp(i\hbar^{-1}\hat{H}_0 t)\hat{V}\exp(-i\hbar\hat{H}_0 t)|\Psi\rangle. \tag{5.32}$$

In some special cases, the Schrödinger equation may have a solution $|\Psi_{\mathrm{nc}}\rangle$ satisfying the equation

$$\exp(i\hbar^{-1}\hat{H}_0 t)\hat{V}\exp(-i\hbar^{-1}\hat{H}_0 t)|\Psi_{\mathrm{nc}}\rangle = 0. \tag{5.33}$$

Equation (5.33) means that a quantum mechanical system in the state $|\Psi_{\mathrm{nc}}\rangle$ does not "feel" the external field. Such a noncoupled (nc) state possesses a series of remarkable properties. Assume than the noncoupled state $|\Psi_{\mathrm{nc}}\rangle = |\Psi_{\mathrm{nc}}(t)\rangle$ at the initial instant of time is a superposition of the states of the nonperturbed system described by the Hamiltonian $\hat{H}_0$. At all subsequent instants of time, the state $|\Psi_{\mathrm{nc}}(t)\rangle$ remains a linear superposition of the same nonperturbed states, despite the presence of the external field. This noncoupled state is customarily referred to as a 'trapped' state, and the phenomenon itself is called coherent population trapping. If the state $|\Psi_{\mathrm{nc}}(t)\rangle$ is a superposition of the ground states of the atom, an accumulation of atoms in the noncoupled state $|\Psi_{\mathrm{nc}}(t)\rangle$ can occur as a result of spontaneous atomic transitions from the excited atomic states. When the translational atomic motion is considered quantum mechanically, the noncoupled state can exist at a specific atomic momentum (or velocity). In this case, the effect is referred to as the velocity-selective coherent population-trapping (VSCPT) effect.

VSCPT has been discussed in connection with the control of atomic motion (see, for example, the review by Arimondo 1996). Aspect et al. (1989) treated the problem of coherent population trapping in a system consisting of a three-level Λ-atom and an electromagnetic field in a σ^+–σ^- configuration (Fig. 5.15(a)). The lower states of the atom were taken to be the $m = \pm 1$ components of a Zeeman triplet, and the upper state to be the $m = 0$ component of some other triplet. The translational motion of the atom was described in quantum mechanical terms. The coherent population-trapping states of such an atom turned out to be localized in two parallel planes $p_z = \pm \hbar k$ in momentum space, the z-axis being the field propagation direction. Coherent population-trapping states exist only for atoms with $v_z = 0$. If $v_z \neq 0$, the interference between the two amplitudes for the probability of transition from the lower sublevels to the excited states is destructive. As a result, at $v_z \neq 0$, an atom in any one of the sublevels can absorb a photon and thus be raised to an excited state. This circumstance is reflected in the name of the phenomenon, velocity-selective coherent

90 Laser control of atoms and molecules

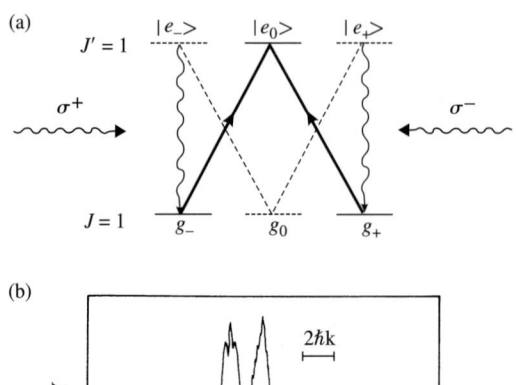

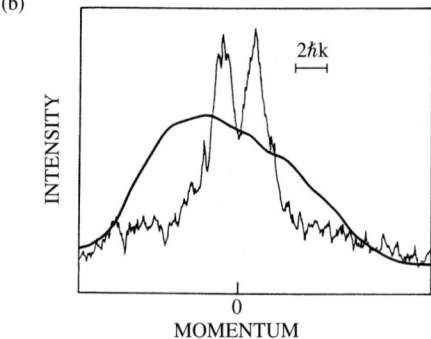

Fig. 5.15 (a) Three-level Λ-scheme describing the interaction of σ^+ and σ^- circularly polarized laser waves with an atomic transition $J = 1 - J' = 1$. The dashed and wavy lines show the ways by which the atoms are pumped into the Λ-system after a few cycles of absorption and spontaneous emission. (b) Atomic momentum distribution following a velocity-selective coherent population-trapping process. (Reprinted from Aspect et al. 1988 with courtesy and permission of the American Physical Society.)

population trapping. Owing to its spontaneous decays, the atom performs a Brownian motion in momentum space. Once it finds itself in the state $|\Psi_{\mathrm{nc}}(t)\rangle$, the atom remains there at all subsequent instants of time. Such a narrowing of the atomic velocity distribution can be interpreted as one-dimensional ultradeep cooling. Such cooling was demonstrated in an experiment by Aspect et al. (1988). These authors realized one-dimensional cooling of ^{4}He atoms in a metastable state. The widths of the peaks of the velocity distribution corresponded to a temperature $T \sim 2\,\mu\mathrm{K}$, which is half the recoil energy $T_{\mathrm{rec}} \approx 4\,\mu\mathrm{K}$ for the helium atom (Fig. 5.15(b)).

5.6.3 Other methods

The two methods of demonstrative character considered above should be supplemented with the classical Doppler-cooling method for forbidden transitions with a very small radiative broadening γ. In the first experiments of this kind (Katori et al. 1999) with an ensemble of ^{88}Sr atoms, a 3D-cooling temperature that was close to a single effective recoil was obtained, that is, the alternative cooling pathway shown on the left of Fig. 5.4 was realized. Moreover, in the case of polychromatic excitation, there is even a possibility of reaching a temperature below the recoil limit (Wallis and Ertmer 1989).

However, it was the "evaporative cooling" method that proved the most effective 3D subrecoil cooling technique. This method makes it possible to obtain 3D-cooled atoms by means of collisions between ultracold atoms in an ensemble of optically trapped atoms. This method has proved very efficient in producing Bose–Einstein condensation in ensembles of laser-cooled and laser-trapped atoms (Chapter 8).

6
Laser trapping of atoms

That the laser cooling and trapping of neutral atoms are interrelated has already been said in the introduction to Chapter 5. We shall, therefore, now consider various techniques whereby cold atoms can be trapped. All the known techniques for trapping laser-cooled neutral atoms can be classified according to a few basic methods. These basic methods are *optical trapping* using the forces of the electric dipole interaction between atoms and laser fields, *magnetic trapping* based on the use of the forces of the magnetic dipole interaction, mixed *magneto-optical trapping* using simultaneous interaction between atoms and magnetic and laser fields, and mixed *gravitooptical* and *gravito-magnetic* trapping. Traps for neutral atoms can have a wide variety of geometries and dimensions: (a) macroscopic traps with a size of $a \gg \lambda$, (b) microscopic traps with a size of $a \simeq \lambda/2$, (c) various combinations of macroscopic and microscopic traps (one-, two-, and three-dimensional) in the case of optical lattices, and finally (d) nanoscopic traps with a size of $a \ll \lambda/2$.

The trapping of cold neutral atoms is a powerful tool in experimental atomic physics that has made it possible to conduct many fundamental experiments, such as Bose–Einstein condensation and the production of Fermi-degenerate quantum gases. It is therefore one of the most vivid demonstrations of the capabilities inherent in the laser control of atoms.

6.1 Optical trapping

The motion of an atom in a spatially inhomogeneous laser field at generally governed by the dipole gradient force, the radiation pressure force, and momentum diffusion. For an atom moving slowly in a far-detuned laser field, the optical excitation is low. As a result, the radiation pressure force originating from the absorption of the laser light and the heating caused by the momentum diffusion are small. Accordingly, the motion of a cold atom in a far-detuned inhomogeneous laser field for not too long an interaction time is basically governed by the dipole gradient force. The minima of the potential produced by the dipole gradient force in a far-detuned laser field can thus be used for *optical trapping* of cold atoms over time intervals limited by the heating due to momentum diffusion.

At a low atomic velocity parallel to the laser beam, that is, $|v| \ll |\Delta|/k$, the gradient force depends only on the position of the atom. Accordingly, for atoms moving slowly along the laser beam; the gradient force can, in the lowest approximation, be treated as a velocity-independent potential force. In that case, one can put $\mathbf{k} \cdot \mathbf{v} = 0$ in eqn (5.11) and introduce a gradient force potential, putting its value equal to zero at infinity (Gordon and Ashkin 1980),

$$U_{gr}(\mathbf{r}) = \int_{\mathbf{r}}^{\infty} \mathbf{F}_{gr}(v=0) d\mathbf{r} = \frac{1}{2}\hbar\Delta \ln\left(1 + \frac{G(\mathbf{r})}{1+\Delta^2/\gamma^2}\right). \quad (6.1)$$

For a Gaussian laser beam with its intensity maximum on the symmetry axis and a negative (red-side) detuning $\Delta < 0$, eqn (6.1) defines the potential well for slowly moving atoms. In contrast, for a laser beam with its intensity minimum near the axis, for example, one produced by the TEM_{01}^* laser mode, a potential well is formed for a positive (blue-side) detuning.

The relationship between the dipole force and the sign of the frequency detuning Δ can easily be understood in terms of the light shift of the atomic energy levels (Section 3.2.2), and hence of the atomic potential energy (Fig. 6.1). The optically induced shift of the ground state exactly corresponds to the dipole potential for a two-level atom (eqn 6.1). Th excited state shows an opposite shift. In the interesting case of low saturation, the atom spends most of its time in the ground state, and we can interpret the *light-shifted ground state* as the relevant potential for the motion of atoms.

6.1.1 Trapping in laser beams

The most simple optical trap for cold atoms consists of a single focused beam formed by the TEM_{00} Gaussian mode. For a Gaussian beam propagating along the z-axis and with its focus at the origin of coordinates $(x, y, z = 0)$, the beam intensity varies as $I(\mathbf{r}) = I(0)(w_0/w)^2 \exp(-(x^2+y^2)/w^2)$, where w_0 is the beam radius, which depends on the longitudinal coordinate z; $w = w_0\sqrt{1+(\lambda z/2\pi w_0^2)^2}$; w_0 is the waist radius of the laser beam; and λ is the laser wavelength. In that case, for a red detuning potential, eqn (6.1) is reduced to a three-dimensional potential well. The depth U_0 of the potential well produced by the Gaussian beam is the same in all directions.

We can consider two specific detunings that are typical of real experiments (see Fig. 6.2). In a red-detuned dipole trap ($\Delta < 0$), the atoms are trapped in an intensity maximum with $U_0 < 0$, and the trap depth $U = |U_0|$ is usually large compared with the thermal energy of cold atoms $k_B T$. In a blue-detuned trap ($\Delta > 0$), a potential

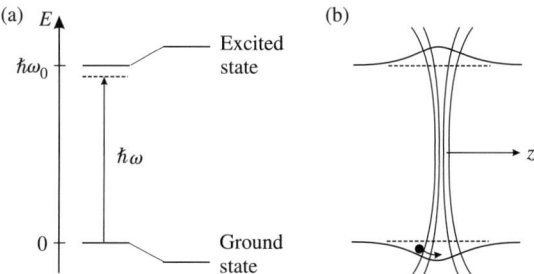

Fig. 6.1 Light shifts for a two-level atom: (a) red-detuned light ($\Delta < 0$) shifts the ground state down and the excited state up by the same amount; (b) a spatially inhomogeneous field such as a Gaussian laser beam produces a ground-state potential well, in which an atom can be trapped.

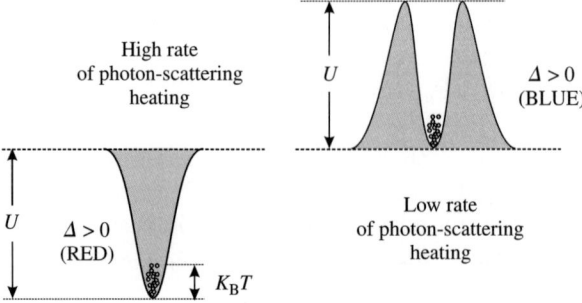

Fig. 6.2 Illustration of dipole traps with red and blue detuning. In the first case, a simple Gaussian laser beam is assumed. In the second case, a Laguerre–Gaussian LG$_{01}$ "doughnut" mode is chosen that provides the same potential depth and the same curvature in the trap center (note that the latter case requires e^2 times more laser power or a smaller detuning). (Reprinted from Grimm et al. 2000 with courtesy and permission of Academic Press.)

minimum corresponds to an intensity minimum, which in the ideal case means zero intensity. In this case $U_0 = 0$, and the potential depth U is determined by the height of the repulsive walls surrounding the center of the trap.

The properties of optical dipole traps depend largely on the magnitude of the detuning. In the *dipole far-off-resonance trap* (FORT), the frequency detuning is assumed to be substantially larger the homogeneous width, but the laser frequency is still close to that of the optical atomic transition. The motion of an atom in an off-resonance dipole trap, is well described by the rotating-wave approximation (Section 2.4.2). When the detuning is chosen to be comparable to the optical transition frequency, an optical dipole trap is often called a *quasi-electrostatic trap*. The properties of a quasi-electrostatic trap differ greatly from those of a far-off-resonance dipole trap, since the rotating-wave approximation can no longer be used to describe the motion of an atom in a quasi-electrostatic trap. Both types of traps produce a nearly conservative potential well for atoms, but incorporate inevitable heating due to momentum diffusion originating from the photon recoil. In a single-beam FORT, the detuning is assumed to satisfy the conditions $|\Delta| \gg \gamma, \Omega$, where $\Omega(\mathbf{r}) = dE_0(\mathbf{r})/\hbar$ is the Rabi frequency. Under this condition, the effect of the radiation pressure force (eqn 5.10) can be neglected compared with that of the gradient force (eqn 5.11). Accordingly, the potential well of the trap at a red detuning is defined by eqn (6.1), namely

$$U_{\text{gr}}(\mathbf{r}) = -\hbar \frac{\Omega^2(\mathbf{r})}{|\Delta|} \qquad (6.2)$$

The depth of the potential well in eqn (6.2) is $U_{\text{gr}}(0) = U_0 = \hbar \Omega^2(0)/|\Delta|$. Owing to the condition on the value of the detuning, the depth of the potential well (eqn 6.2) is always small compared with the value of the energy defined by the Rabi frequency, $U_0 \ll \hbar \Omega(0)$. Under typical experimental conditions, the potential-well depth U_0/k_B does not exceed 10 mK.

The lifetime of atoms in a single-beam far-off-resonance dipole trap is typically defined by the heating due to momentum diffusion and collisions with the background gas. The characteristic time of escape an atom from a far-off-resonance dipole trap caused by diffusive heating, τ_{diff}, can be estimated from the potential-well depth U_0 and the momentum diffusion coefficients (Minogin 1980) at the focus of the beam,

$$D_{ii} \approx D(0) = \hbar^2 k^2 \gamma \frac{\Omega^2(0)}{\Delta^2}, \quad (6.3)$$

$$\tau_{\text{diff}} \approx \frac{2MU_0}{D(0)} \approx \omega_r^{-1} \frac{|\Delta|}{\gamma}. \quad (6.4)$$

The characteristic value of the inverse recoil frequency $\tau_r = \omega_r^{-1}$ for atoms of medium mass M is of the order of 10^{-5} s. For a large detuning, the trapping time τ_{diff} can, accordingly, be much longer than the time τ_r. It should be noted that by increasing simultaneously the laser field intensity and the frequency detuning, one can lengthen the trapping time τ_{diff} while keeping constant the potential-well depth. The actual time of atom storage in a far-off-resonance dipole trap is determined by collisions with the residual gas.

The first far-off-resonance dipole trap used 220 mW of laser power at a detuning of 130 GHz to trap sodium atoms (Chu et al. 1986). The potential-well depth for sodium atoms was 10 mK, and the trapping time was limited to a few milliseconds because of the fast diffusive heating of the atoms at a relatively small frequency detuning. In a far-off-resonance dipole trap formed by two intersecting beams from Nd:YAG lasers ($\lambda = 1.06$ μm), the lifetime of sodium atoms amounted to a few seconds. The main disadvantage of the optical trapping of atoms in Gaussian laser modes is the fast diffusive heating. Hollow laser beams have their intensity minima on the axis, and atoms trapped by blue-detuned laser fields feature a low diffusive heating rate (Yang et al. 1986).

In a quasi-electrostatical optical trap (Letokhov 1968), the detuning of the laser beam is comparable to the optical frequencies of the atom. In this case, the trapping potential can be found from the general expression for the high-frequency Stark shift in an off-resonance light field. For the atomic ground state, the energy shift is

$$U(\mathbf{r}) = \Delta E_g = -\tfrac{1}{2}\alpha E_0^2(\mathbf{r}), \quad (6.5)$$

where α is the polarizability of the atom in the ground state. When the frequency of the light wave is well below that of the lowest atomic dipole transition (as is the case for a CO_2 laser trap), the polarizability α is always positive and close to the static polarizability α_s.

The most important advantage of the quasi-electrostatic trap is an extremely low diffusion-associated heating of atoms and, accordingly, a long lifetime of atoms. In this trap, the total photon scattering rate is generally defined by the Rayleigh and Raman scattering. Rayleigh scattering leaves an atom in its original state, while Raman scattering leaves an atom in a different hyperfine-structure sublevel. Typically, the cross section for the Raman scattering is much smaller than the cross section for the Rayleigh scattering. O'Hara et al. (1999) demonstrated an ultrastable CO_2 laser trap that provided tight confinement of atoms with negligible optical scattering and minimal laser-noise-induced heating (Fig. 6.3). By this technique, ^{6}Li atoms were

96 Laser control of atoms and molecules

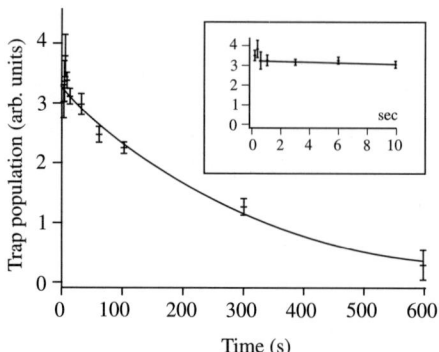

Fig. 6.3 Number of trapped atoms versus time for an ultrastable CO_2 laser trap. The solid line is a single exponential fit, $N(t) = A \exp(-t/\tau)$, with $\tau = 297$ s. (Reprinted from O'Hara et al. 1999 with courtesy and permission of the American Physical Society.)

stored in a 0.4 mK deep well with a lifetime of 300 s, which is consistent with a background pressure of 10^{-11} Torr. This is the longest storage time ever achieved with an all-optical trap, comparable to the best magnetic trap.

Note, finally, that the basic distinction between the far-off-resonance dipole trap and the quasi-electrostatic trap consists in the value of the diffusion heating. Since the population of the excited atomic state in the quasi-electrostatic trap is extremely small, the diffusion-associated heating of atoms in this trap is much smaller than in the far-off-resonance dipole trap.

The quasi-electrostatic trap can also be used for trapping neutral molecules. The main problem here is the production of cold molecules. At present, the process of photoassociation of laser-cooled atoms is the only known way to produce a sample of cold molecules. Takekoshi et al. (1998) have reported the first observation of optical trapping of cold Cs_2 molecules in a quasi-electrostatic trap (see Chapter 8).

6.1.2 Trapping in standing laser waves. Optical lattices

Standing light waves of various configurations (1D, 2D, and 3D standing waves) are of interest because, first, they restrict the motion of atoms to areas much smaller than λ (the Doppler-free Lamb–Dicke regime) and, second, they produce periodic optical potentials (optical lattices) in which the motion of atoms is similar to that of electrons in a crystal lattice. This was understood when the first suggestions were made about the trapping of atoms in standing waves (Letokhov 1968, 1973c). The situation, however, is much more complicated, for it is necessary to take account of the fluctuation heating of the atoms in the same light field. Specifically, for a two-level atom in a standing light wave, the minimum average kinetic energy is determined by eqn (5.15). This energy coincides in order of magnitude with the potential-well depth.

Figure 6.4 shows two experimental geometries used to observe the one-dimensional localization (channeling) of atoms in a standing (plane or spherical) light wave. An experiment using the geometry of Fig. 6.4(a) was performed by Salomon et al. (1987). In a direct experiment (Balykin et al. 1988b), a beam of sodium atoms was directed at a grazing angle with respect to curved potential wells (Fig. 6.4(b)). Fast atoms

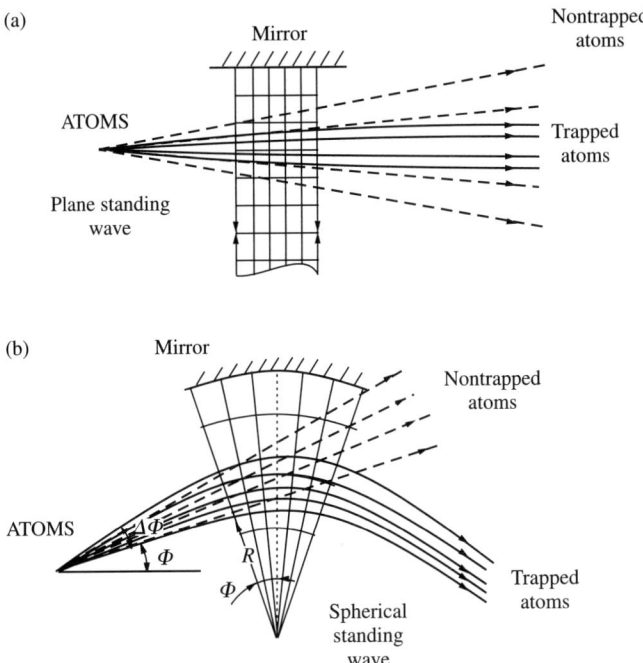

Fig. 6.4 One-dimensional localization (channeling) of atoms in a plane standing wave (a) and a spherical standing wave (b). (From Balykin et al. 1988b.)

that had a high transverse velocity traversed the laser field without changing their direction of motion in any perceptible way. In contrast, cold atoms were constrained in the potential wells of the spherical standing laser wave and deviated as a result through an angle governed by the magnitude of the dipole interaction between the atoms and the laser field. Deflection of slow atoms was observed to occur as a result of their channeling both at the nodes of the standing laser wave in the case of blue detuning and at the antinodes in the case of red detuning. Trajectories of motion of channeled and nonchanneled atoms are shown in Fig. 6.5, which explains the splitting of atomic beam in the experiments.

Another experimental proof of the localization of cold atoms at the minima of a periodic optical potential was obtained by recording the resonance fluorescence spectra of cesium atoms trapped in three-dimensional optical molasses (Westbrook et al. 1990) and rubidium atoms in a one-dimensional optical potential (Jessen et al. 1992) The resonance fluorescence spectrum of a motionless two-level atom consists of the well-known Mollow triplet, which includes a central peak at the laser frequency ω_1 and two side components displaced to the red and blue sides by an amount equal to the Rabi frequency (Mollow 1969). For a two-level atom oscillating in a potential well at a frequency lower than the Rabi frequency, each component of the Mollow triplet is split into side components corresponding to changes in the vibrational state of the atom. If the ratio between the oscillation amplitude of the atom in the potential well and the radiation wavelength (the Lamb–Dicke factor) is small, each component of the

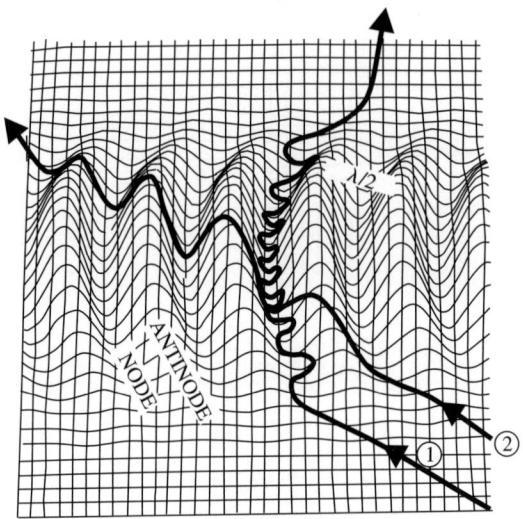

Fig. 6.5 Trajectories of localized (1) and nonlocalized (2) atoms in the potential of a standing spherical light wave.

Mollow triplet contains only the first side components (Fig. 6.6). The ratio between the intensities of the side peaks was used to determine the vibrational temperature of the atoms and to estimate that the localization region of the atoms was of size $\Delta z \approx \lambda/15$.

The maximum densities of atoms in optical lattices are typically limited to values of the order of 10^{11}–10^{12} cm^{-3}, mainly because of absorption and scattering of the laser light by the atoms. An increase of atomic density in optical lattices still remains an important problem, whose solution would allow one to approach the observation of quantum statistical effects and the possible production of dense atomic crystals. In this connection, attempts to trap atoms in the intensity minima of a laser field seem very promising (Grynberg and Courtois 1994). Such *dark optical lattices* can be produced by the optical pumping of atoms in a magnetic field to a magnetic sublevel that is not excited by the laser light. The first dark optical lattice was realized with ^{87}Rb atoms placed in a weak magnetic field of the order of 10 G (Hemmerich *et al.* 1995). A promising approach to decreasing the influence of the scattered and absorbed laser light is the use of far-off-resonance optical lattices. Friebel *et al.* (1998) reported on the trapping of Rb atoms in an extremely far-detuned 1D optical lattice produced by a standing wave from a CO_2 laser. This optical lattice was characterized by a long coherence time because of the low scattering rate. Friebel *et al.* assumed that by filling such a lattice with one atom per lattice site, a quantum bit (qubit) of information could be stored in each atom (e.g. in two ground-state magnetic sublevels) and individually addressed with a focused laser beam.

6.1.3 Trapping in optical-waveguide modes. Atom waveguides

The atomic-waveguide scheme that was first demonstrated in practice was that shown in Fig. 6.7, where atoms move inside the intensity maximum of the optical mode EH_{11} propagating through a hollow optical waveguide (Ol'shanii *et al.* 1993). The radial

(a)

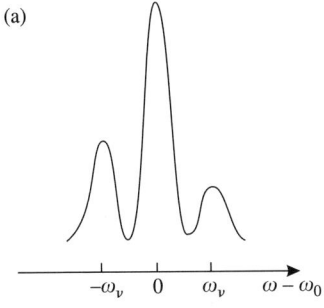

(b)

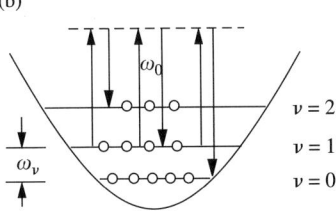

Fig. 6.6 (a) Resonance fluorescence of atoms trapped in periodic potential wells. The spectrum is shown near the central component of the Mollow triplet. (b) The quantum transitions of an atom between the vibrational states responsible for the side components. (Reprinted from Jessen *et al.* 1992 with courtesy and permission of the American Physical Society.)

distribution of the electric field of the EH_{11} mode, in the cylindrical coordinates z, ρ, φ, has the form

$$\mathbf{E} = \tfrac{1}{2}\mathbf{e}E_0 J_0(\chi\rho) + \text{c.c.}, \tag{6.6}$$

where J_0 is the Bessel function of order zero, $\mathbf{e}$ is the unit polarization vector, and χ is the complex propagation constant that characterizes the damping of the radiation in the shell of the fiber. In the case of a large negative detuning, the optical field in eqn (6.6) produces an attractive potential for atoms:

$$U = -\hbar\frac{\Omega_0^2}{|\Delta|}J_0^2(\chi\rho), \tag{6.7}$$

where Ω_0 is the Rabi frequency. This scheme was realized for rubidium atoms (Renn *et al.* 1995) (Fig. 6.7(a)). A glass hollow-core fiber 3 cm long and 40 μm in internal diameter was used in the experiment. Laser light was launched into the hollow region of the glass fiber. A beam of rubidium atoms propagated in the dipole potential, whose depth corresponded to an effective atomic temperature of 70 mK. This magnitude of the potential allowed atoms with transverse velocities of up to 40 cm/s to be transported through the waveguide. A direct experimental proof of the propagation of the atoms in the optical potential was obtained by measuring the atomic flux at the exit of the waveguide as a function of the sign of the detuning (Fig. 6.7(b)). A disadvantage of a waveguide based on the intensity maximum of an optical-waveguide mode is the diffusion-associated heating of the atoms. For this

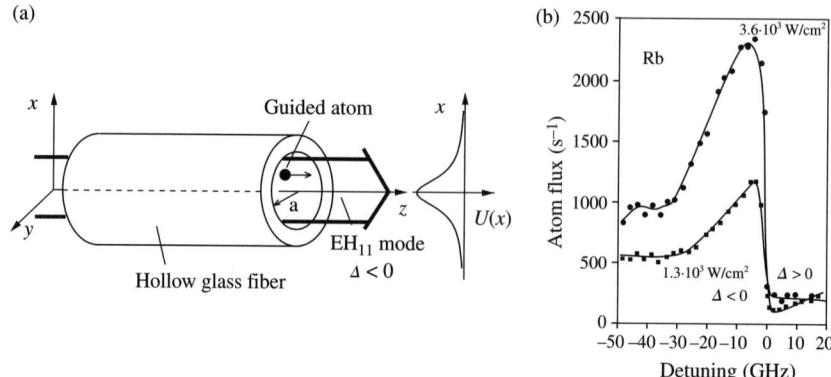

Fig. 6.7 (a) Scheme of a waveguide based on an evanescent wave EH_{11} propagating in a dielectric waveguide, and the optical potential for atoms; (b) experimental realization of a hollow fiber for guiding atoms in the case of red-detuning of the laser frequency. Thermal rubidium atoms are guided through 3 cm of hollow fiber with an internal diameter of 40 μm. The guided-atom flux is shown as a function of the detuning Δ for two EH_{11} mode laser intensities. (Reprinted from Renn et al. 1995 with courtesy and permission of the American Physical Society.)

reason, atom waveguides in which atoms propagate in optical-waveguide modes that have minima near the axis are much more attractive (Savage et al. 1993; Renn et al. 1996). At present, there exist various methods to generate such modes. A discussion of these methods can be found in review of atom waveguides (Balykin 1999).

The most important parameter of any atom waveguide is the value of the atomic flux that can be injected into and propagate inside the waveguide. Since the injected atoms typically have a relatively broad velocity distribution, one can easily see that in the case of a single-mode atom waveguide, only very small amount of atoms can be injected into the atom waveguide. To overcome this difficulty, an atom waveguide named the "atom hornfiber" has been proposed (Subbotin et al. 1997). An atom hornfiber is a hollow, tapered, curved waveguide "coated" inside by a blue-detuned evanescent wave. The waveguide has the shape of a curved horn with a large inlet opening, around the size of the atomic cloud released from a magnetooptical trop, which is a typical pumping source for atom waveguides. The outlet opening diameter of the horn corresponds to a single-mode atom waveguide. The key point of such a cooling hornfiber is the reflective cooling effect (Ovchinnikov et al. 1995), which will be discussed below (Section 6.4.1) in connection with the gravito-optical trapping of atoms.

6.2 Magnetic trapping

The magnetic trapping of cold atoms is especially effective in combination with optical fields that allow one to control the motion of the atoms (to cool and reflect them, and so on). But let us consider first the purely magnetic trapping of atoms and molecules.

All static magnetic traps for atoms use nonuniform stationary magnetic fields. In a nonuniform magnetic field $\mathbf{B} = \mathbf{B}(\mathbf{r})$, an atom with a permanent magnetic moment $\boldsymbol{\mu}$ has a magnetic dipole interaction energy

$$U = -\boldsymbol{\mu} \cdot \mathbf{B} = -\mu_l B, \tag{6.8}$$

where μ_l is the projection of the magnetic moment $\boldsymbol{\mu}$ onto the field direction. Accordingly, an atom in a field $\mathbf{B}(\mathbf{r})$ is acted on by a magnetic dipole force

$$\mathbf{F} = \nabla(\boldsymbol{\mu} \cdot \mathbf{B}) = \mu_l \nabla B. \tag{6.9}$$

Since the Maxwell equations do not allow a maximum of a static magnetic field in free space, the force in eqn (6.9) can be used to trap atoms only in a minimum of a static magnetic field. The force in eqn (6.9) can hold an atom near a minimum of a static magnetic field if the direction of the magnetic moment $\boldsymbol{\mu}$ is opposite to that of the magnetic field, that is, $\mu_l < 0$.

The most popular magnetic trap is the spherical quadrupole trap, proposed by Paul (1990). In this trap, two opposite circular currents produce a static magnetic field in the form of a spherical quadrupole (Fig. 6.8(a)). The gradient of the magnetic field in a quadrupole trap varies with direction. For this reason, the magnetic dipole force in the field is neither central nor harmonic. For a typical magnitude of the magnetic moment of about a Bohr magneton, that is, $\mu \approx \mu_B$, and a moderate magnetic field strength at the edges of the trap, $B = 100$ G, a quadrupole magnetic trap can hold atoms with a temperature of the order of 10 mK. The first experiment on trapping of cold atoms in a static magnetic field was done with a quadrupole magnetic trap (Migdal et al. 1985). The trap was loaded with ^{23}Na atoms, preliminarily cooled to a temperature of 17 mK by a counterpropagating laser beam. A microscopic quadrupole magnetic trap for atoms was realized with a combination of permanent magnets, coils, and

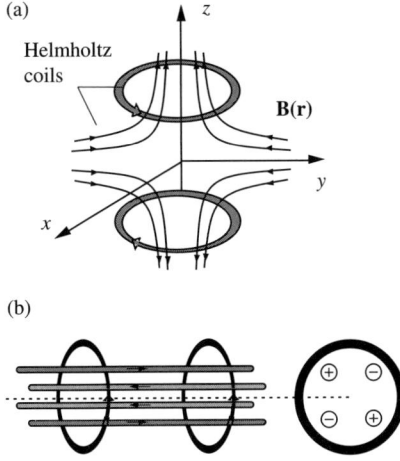

Fig. 6.8 Magnetostatic traps for cold neutral atoms: (a) quadrupole trap; (b) the Ioffe trap.

ferromagnetic pole pieces (Vuletic et al. 1998). The attainable magnetic field gradient of 3×10^5 G/cm imply a spatial extension of the ground state of a trapped atom much smaller than the wavelength of the optical transitions. The field gradient could be varied over a wide range, allowing efficient loading of 4×10^5 lithium atoms from a shallow potential by adiabatic transport and compression. During the compression, a 275-fold density increase was observed.

The principal shortcoming of magnetic traps that have a zero magnetic-field minimum (Fig. 6.9) is the presence of a channel for the atoms to leave the zero-field region. When a moving atom traverses the zero-field region sufficiently fast, its magnetic moment cannot adiabatically follow the rapidly changing magnetic-field direction. As a result, the change of the mutual orientation of the atom magnetic moment and the magnetic field causes the atom to leave the trap. Atom transitions from trapped to untrapped states (nonadiabatic Majorana transitions or spin flips) have a noticeable probability, even for very cold atoms (Bergeman et al. 1987). In a quadrupole magnetic trap, this escape mechanism usually limits the lifetime of cold atoms in the central region to a few seconds. An important example of a static magnetic trap without a zero-magnetic-field region is the Ioffe trap, first developed for use in plasma confinement schemes for nuclear fusion. In this trap, the magnetic field of two equal circular currents provides the axial confinement (Fig. 6.8(b)). The Ioffe trap has been used for trapping atoms at ultralow temperatures of the order of 100–10 nK, attained by the evaporative-cooling technique (Ketterle and van Druten 1996). A variety of the Ioffe trap is a trap proposed by Hansch and coworkers (Esslinger et al. 1998). This trap uses two circular currents, as in the case of the quadrupole trap, and a third circular current normal to the first two and located at a certain distance from the quadrupole configuration. A smooth activation of the third circular current transforms a magnetic trap of the first kind into one of the second kind. Atoms in such a trap are initially caught in the quadrupole field and then transferred to the trap of the second kind by activating the third circular current.

One direct way to close the channel whereby atoms escape from the central region of a quadrupole magnetic trap is to displace the potential (Fig. 6.9(a)) in the symmetry

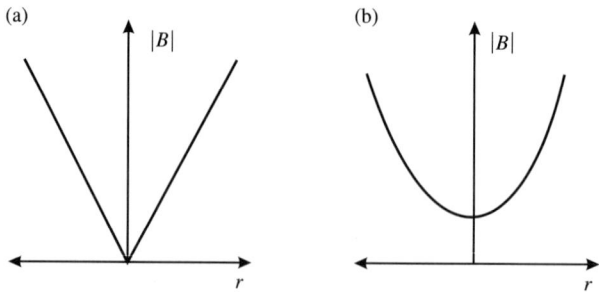

Fig. 6.9 The spatial dependence of field strength for the two configurations of a magnetostatic trap: (a) the quadrupole trap and (b) the Ioffe trap. For atoms with a linear Zeeman shift, such as alkali metal atoms, the trapping potential is proportional to the low magnetic field strength $|B|$.

plane of the trap and effect a rapid circling of it in this plane, that is, to convert the static potential into an orbiting potential (Fig. 6.9(b)). Thanks to the fact that the minimum of the orbiting potential moves about the central point of the quadrupole trap, the time-averaged orbiting potential proves to be other than zero at the center of trap, which closes the channel whereby atoms are lost on account of Majorana transitions (Petrich et al. 1995). One more way to "close" the central zero-magnetic-field region in a quadrupole magnetic trap is to irradiate this region with a blue-detuned laser beam. In this method, the atoms are acted on by the gradient force, which pushes atoms out of the central part of the trap. Davis et al. (1995) used as an "optical plug" an argon laser beam focused into a spot 30 μm across at the center of a quadrupole magnetic trap. The potential barrier produced by the laser beam was 350 μK, which was sufficient for "closing" the region where Majorana transitions were possible. The potential produced by this quadrupole magnetic trap with an optical plug has two minima displaced from the center of the trap by a distance of 50 μm. This trap with an optical plug was successfully used for the experimental observation of Bose–Einstein condensation in a gas of sodium atoms.

6.3 Magnetooptical trapping

The idea of magnetooptical trapping of atoms can be understood by considering the idealized one-dimensional scheme shown in Fig. 6.10. In this scheme, the atoms are placed in a weak magnetic field $\mathbf{B} = B\mathbf{e}_z$ which increases linearly in the positive direction of the z-axis, that is, $B(z) = az$, and is equal to zero at the central point $z = 0$. For simplicity, the atoms are assumed to have two electronic states: the ground state with an energy E_g^0 and a total angular momentum equal to zero, that is, $F_g = 0$, and an excited electronic state with an energy E_e^1 and a total angular momentum $F_e = 1$. The excited electronic state is split in the magnetic field into three Zeeman magnetic sublevels $m_e = 0, \pm 1$. The energies $E_e^{\pm 1} = \hbar\omega_0 \pm \mu B(z)$ of the two extreme magnetic sublevels $m_e = \pm 1$ depend on the atomic coordinate z, where ω_0 is the atomic transition frequency in zero magnetic field, that is, at the point $z = 0$, and μ is the projection of the magnetic moment $\vec{\mu}$ onto the field direction. The atoms are irradiated by two circularly polarized (σ^+ and σ^-) laser beams propagating in the directions $\pm z$. The frequency ω of the laser beams is assumed to be red-shifted with respect to the frequency of the unperturbed atomic transition, that is, $\omega < \omega_0$.

In this scheme, the atom is acted on by the radiation pressure force $\mathbf{F} = F\mathbf{e}_z$ caused by one-photon transitions between the ground state and the two upper-state magnetic sublevels $m_e = \pm 1$. For a motionless atom, the rate of excitation of the atom to the upper-state sublevels depends on the coordinate of the atom. When the coordinate z is positive, the atom is excited with a higher probability to the magnetic sublevel $m_e = -1$ and with a lower probability to the sublevel $m_e = 1$, and when the coordinate z is negative, it is excited with a higher probability to the sublevel $m_e = 1$ and with a lower probability to the sublevel $m_e = -1$. As a result, the direction of the radiation pressure force on a motionless atom depends on the sign of the coordinate z. When the atom coordinate is negative, that is, $z < 0$, it interacts mainly with the σ^+-polarized radiation, and experiences a force in the positive direction of the z-axis. In contrast, at a positive atom coordinate, that is, $z > 0$, the atom interacts mainly with the σ^--polarized laser light and is subject to a force in the negative direction of the z-axis. The

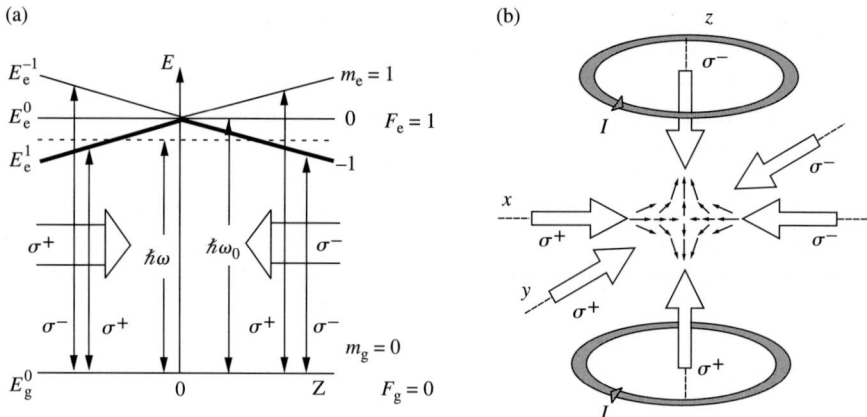

Fig. 6.10 Schemes of one-dimensional (a) and three-dimensional (b) magnetooptical trapping of atoms by two counterpropagating laser beams (a) or three pairs of counterpropagating laser beams (b) with circular polarizations σ^+ and σ^-.

radiation pressure force in the scheme of Fig. 6.10(a) is thus always directed opposite to the displacement of the atom. Accordingly, the radiation pressure force produces a potential well for atoms located at the central point $z=0$. For an atom moving near the center of the trap, the efficiency of excitation of the atom by red-detuned laser light depends strongly on the direction of the velocity of the atom. When the atom moves in the positive direction of the Oz axis it is more effectively excited by the σ^--polarized laser light. In contrast, an atom moving in the negative direction is mostly excited by the σ^+-polarized laser light. For these reasons, the radiation pressure force is directed opposite to the atomic velocity and thus produces cooling of atoms.

The one-dimensional magnetooptical trapping scheme considered above can easily be generalized to three dimensions. The most frequently used three-dimensional scheme of the magnetooptical trap (MOT) in practice includes a magnetic field in the form of a spherical quadrupole and a laser field produced by three orthogonal pairs of counterpropagating laser beams with polarizations σ^+ (Fig. 6.10(b)). In this basic configuration of the MOT, the radiation pressure force is responsible both for the formation of a three-dimensional potential well for atoms and for three-dimensional cooling of atoms near the center of the trap.

The simple model of the MOT describes the trapping and cooling of atoms by the radiation pressure force arising from one-photon processes. Accordingly, the model describes only the Doppler cooling of atoms inside the MOT. In this model, the minimum temperature is of the order of the Doppler-limit temperature, $T_{\min} = T_D = \hbar\gamma/k_B$. This minimum temperature accordingly defines both the minimum velocity width and the minimum spatial size of the trapped atomic cloud. The theory of the magnetooptical trapping of atoms in the case of an interaction scheme featuring two-photon or higher-order multiphoton processes differs substantially from that considered above, for multiphoton processes sharply change the friction force and momentum diffusion tensor.

The dynamics of atoms in any real three-dimensional MOT is rather complicated. Two factors substantially complicate the motion of atoms in a three-dimensional trap. First, in the atoms actually used, the lower and upper states are hyperfine-structure levels including several magnetic sublevels. For this reason, the atom–laser-field interaction always proceeds by a complex multilevel scheme. Second, the polarization direction and intensity of the laser field in the scheme of Fig. 6.10(b) vary on the scale of the optical wavelength. This circumstance, as in the case of optical lattices, causes a small-scale modulation of the magnetic-sublevel populations and causes coherences between the magnetic sublevels. These effects may give rise to additional friction forces due to nonadiabatic time evolution of the magnetic-sublevel populations and atomic coherences for a moving atom (Ungar *et al.* 1989; Weiss *et al.* 1989; Dalibard and Cohen-Tannoudji 1989).

The first magnetooptical trapping scheme was experimentally tested with sodium atoms (Raab *et al.* 1987). Since the ground state $3S_{1/2}$ of the sodium atom is split into two hyperfine-structure levels with total angular momenta $F = 1, 2$, the atoms were excited using the dipole transition $3S_{1/2} \to 3P_{3/2}$ by bimonochromatic laser light. This two-frequency excitation avoided optical pumping of the atoms to one of the two lower hyperfine-structure levels and thus provided for long-term atom–laser-light interaction. In the first experiment, about 10^7 atoms were confined in the MOT for about 2 minutes at a temperature below 1 mK. The trapping time of the atoms was mainly limited by collisions with the residual-gas particles.

The principal features of the MOT are its ability to simultaneously cool and trap atoms, a fairly large potential-well depth, a relatively large capture velocity, and a weak sensitivity to disturbances in the directions of the laser beams and imperfection in their polarization. The MOT is capable of effective operation not only in the standard geometry, but also with the use of only four laser beams in a tetrahedral geometry (Shimizu *et al.* 1991). An important advantage of the MOT over purely magnetic traps is that it uses very weak magnetic fields, approximately 100 times weaker. In the MOT, the magnetic field is used only to produce a small Zeeman shift of the magnetic sublevels, whereas pure magnetic traps require a substantial magnetic-field strength to produce a magnetic dipole force. No less an important advantage of the MOT is the possibility of injecting atoms into it both from an atomic beam and from an atomic gas (Monroe *et al.* 1990). In typical magnetooptical traps, the temperature of the atomic cloud ranges between 1 mK and 10 μK, and the density of atoms between 10^8 and 10^{11} cm^{-3}. Attaining high atomic densities in an MOT is mainly limited by collisions between the trapped atoms and residual-gas particles, the trapping of radiation, collisions between atoms in the lower states with the optically excited atoms, and the escape of atoms from the trapping region associated with above-barrier momentum diffusion.

A material increase in the atomic density has been attained in what is known as the dark MOT, wherein the localized atoms are optically pumped to a hyperfine-structure sublevel at which they are off resonance with the main localizing laser field. This method was successfully used to localize sodium atoms by Ketterle *et al.* (1993). These authors demonstrated a substantial improvement of the trap parameters in comparison with the standard MOT: the atomic density amounted to 10^{12} atoms/cm^3 with a large total number of localized atoms (over 10^{10}). In recent years, the MOT has become the most popular trap for cold atoms. To date, there is a very impressive list of atom isotopes that have been successfully localized in MOT's. This includes a long list of stable and radioactive isotopes of many atoms (see, for example, the review by Balykin *et al.* 2000).

6.4 Gravitooptical and near-field traps

Gravitooptical and near-field traps are based on a combined use of electromagnetic and gravitational forces.

6.4.1 Gravitooptical traps and cavities

Conceptually, the simplest gravitooptical atom trap and cavity consists of a single concave atomic mirror arranged horizontally (Fig. 6.11). In this geometry, the role of the second mirror is played by the gravitational field. In this cavity, the curvature of the mirror is small, and the vertical atomic-motion spectrum with quantum numbers n that are not very small (the quasiclassical approximation) is determined by the well-known quantum mechanical problem of the bouncing of a particle on an absolutely elastic plane in a gravitational field (see, for example, Flugge 1971). The corresponding spectrum is defined by the eigenvalues of the Airy functions,

$$\varepsilon_n = Mgl_g \left[\frac{3\pi}{2} \left(n - \frac{1}{4} \right) \right]^{2/3}, \quad (6.10)$$

where $l_g = (\hbar^2/2gM^2)^{1/3}$ is the characteristic gravitational length governed by the mass of the particle. For atoms of medium mass, the characteristic gravitational length is of the order of a micron, and the characteristic values of the energy given by eqn (6.10) correspond to effective temperatures of the order of 10 nK.

In such a cavity, the transverse (horizontal) size of the atomic mode can be expressed in terms of the distance L from the surface of the mirror to the classical turning point, which determines the length of the cavity. For estimation purposes, the gradient force potential can be taken to be stepped near the surface of the mirror. The shape of the surface of the mirror in the simplest approximation can he treated as a paraboloid of revolution,

$$z = \frac{x^2 + y^2}{2R}, \quad (6.11)$$

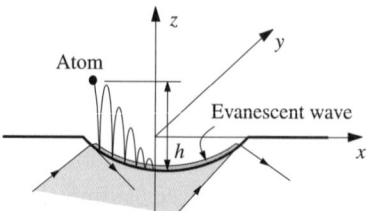

Fig. 6.11 Gravitooptical atom trap–cavity based on a single concave atom mirror oriented horizontally. (Reprinted from Aminoff *et al.* 1993 with courtesy and permission of the American Physical Society.)

where R is the radius of curvature of the mirror at its center. Assuming that the atom mirror has a parabolic shape, the transverse size of the atom mode in the vicinity of the mirror can be estimated as

$$\delta x_{\rm M} \cong \left(\frac{2l_g^3 R^2}{R-2L} \right)^{1/4}, \qquad (6.12)$$

and at the upper point of the classical trajectory at a distance L from the mirror as

$$\delta x_{\rm S} \cong (2l_g^3 (R-2L))^{1/4}. \qquad (6.13)$$

The distance to the upper point of the classical trajectory is determined by the atomic velocity in the neighborhood of the mirror, $L = v^2/2g$, and can be associated with the energy of longitudinal motion given by eqn (6.10). With the typical gravitational length being, as indicated above, of the order of a micron, the size of the atom mode is a few tens of microns.

A gravitooptical trapping scheme was investigated experimentally for cesium atoms (Aminoff et al. 1993). The atoms were preliminarily localized and cooled in an MOT. When the MOT was switched off, the atoms fell freely onto an atom mirror from a height of 3 mm. The atoms were observed to execute about ten bounces. In each reflection event, around 40% of the atoms were lost as a result of (a) photon scattering during reflection, (b) background gas collisions, and (c) residual misalignment of the mirror with respect to the vertical axis.

Note that the intensity of the evanescent wave in an atom mirror can be increased by two or three orders of magnitude on account of excitation of surface plasmons produced by introducing a thin metal layer into the dielectric–vacuum interface (Esslinger et al. 1993). Another method to intensify the evanescent wave is to introduce a dielectric film of high refractive index, which produces a dielectric optical fiber for the laser radiation. The repeated reflection of the laser light from the dielectric–vacuum and dielectric–dielectric interfaces substantially increases the intensity of the evanescent wave (Kaiser et al. 1994).

The reflecting surface formed by an evanescent wave is a very rich structural component for creating atomic traps and waveguides of varying geometry. In the past few years, such light surfaces have formed the basis for several types of half-open traps and waveguides. Further development of the simplest gravitooptical trapping scheme has resulted in vertically arranged pyramidal and conical traps (Dowling and Gea-Banacloche 1996; Ovchinnikov et al. 1995) and also atom gravitational cavities based on hollow optical fibers (Harris and Savage 1995). Figure 6.12 presents a schematic diagram of a gravitooptical surface trap (GOST), where atoms are cooled as a result of inelastic reflection from an evanescent wave (Söding et al. 1995). With this scheme, Cs atoms have been trapped in a horizontal plane by the gradient force produced by a hollow blue-detuned laser beam (Ovchinnikov et al. 1997). The trap implemented for cesium atoms embedded a cooling mechanism associated with the optical pumping of the atoms between the hyperfine-structure states (Fig. 6.12(b)) As inelastic reflection takes place when an atom enters the evanescent wave in the lower ground state and

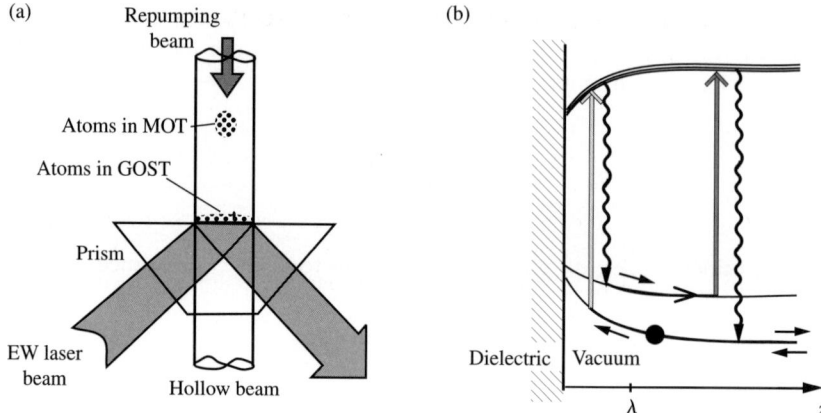

Fig. 6.12 (a) Scheme of a gravitooptical trap based on the inelastic reflection of atoms from an evanescent laser wave (EW, evanescent wave). (b) The atomic transitions between hyperfine-structure states responsible for cooling atoms in the trap. (Adapted from Söding et al. 1995.)

by the scattering of a photon during the reflection process, the atom is pumped into a less repulsive state. The filled circle in Fig. 6.12(b) shows a Cs atom as it approaches the dielectric surface in the lower hyperfine-structure state; it then scatters a photon, leaves the evanescent wave in the upper ground state, and is finally pumped back into the lower state.

6.4.2 Optical trapping of atoms in the near field

The optical near field is an interesting configuration of a light field confined to a small space with a size of the wavelength of light. It seems attractive to use the capabilities of the optical near field to utilize the potential of the gradient force, which is sensitive to the gradient of the light field. In the near field, intensity inhomogeneities in a light wave penetrating through an aperture (hole) whose size is small in comparison with the optical wavelength λ occur on a scale much smaller than λ (Bouwkamp 1954). These light field gradients can be used to trap ultracold atoms (Klimov and Letokhov 1995). There is the potential possibility of implementing, on the basis of this effect, a 2D lattice whose period is not associated with that of the standing light wave.

Other proposals for near-field traps were examined by Ohtsu (1998), where near-field effects in the vicinity of nanofiber traps, including combination with van der Waals forces, were discussed. All these proposals are undoubtedly of interest in the rapidly developing area of nanofabrication on the single-atom (or single-molecule) level in the spirit of the ideas put forward by Feynman (1992), which will be discussed in Chapter 7.

6.5 Optical trapping of cold atoms—new tools for atomic physics

The methods of trapping cold atoms, considered in Chapter 5 and the present chapter in a very brief and retrospective fashion, have become a very powerful tool in experimental physics. They have led to the development of atom optics, the observation and investigation of dilute quantum gases (Bose–Einstein condensation, atom lasers, Fermi-degenerate quantum gases, and ultracold molecules), and probably many other discoveries in the physics of ultracold atoms. These will be discussed in Chapters 7 and 8. But it would be expedient to consider at the end of this chapter a few examples of applications that lie beyond the mainstream, but are of physical interest.

6.5.1 Laser-trapping spectroscopy

The idea of Doppler-free ultrasensitive single-atom laser-trapped spectroscopy (Letokhov 1975b) seemed a distant goal in the 1970s. Nowadays, the method is commonly employed in many spectroscopic experiments. Two impressive applications are the following. First, one may note the use of cold trapped atoms in ultrahigh-resolution spectroscopy. A striking example is the observation of a two-photon transition 1S → 2S in cold hydrogen atoms localized in a magnetic trap (Cesar et al. 1996). Second, we should mention the first experiments with rare atoms. Lu et al. (1997) realized efficient collection of ^{221}Fr ($t_{1/2} = 4.9$ min) in a vapor cell MOT. These authors measured the energies and the hyperfine structure of the $7^2P_{5/2}$ and $7^2P_{3/2}$ states with 900 trapped atoms, with a signal-to-noise ratio of about 60:1 during 1 s. Rowe et al. (1999) measured the ground-state hyperfine structure of laser-trapped radioactive ^{21}Na ($t_{1/2} = 22$ s) collected in an MOT on line at the cyclotron at the Lawrence Berkeley National Laboratory. These experiments show the good prospects for laser-trapping spectroscopy of very rare atomic samples.

6.5.2 Nuclear physics

Cold, trapped radioactive atoms open up new experimental opportunities in nuclear physics. Trapped radioactive atoms can be used in experiments on the fundamental symmetries, including experiments on nuclear β-decay, atomic parity nonconservation, and the search for parity-violating and time-reversal-violating electric dipole moments. The first successful experiments on the trapping of radioactive atoms were performed with the isotope ^{21}Na (Lu et al. 1994). It is expected that further activity in this direction will be concentrated on efforts to undertake meaningful measurements with trapped radioactive species.

Recently, experiments on nuclear decay have started to use MOTs as a source of cold, well-localized atoms. The low-energy recoiling nuclei can escape from the MOT and be detected in coincidence with β-decays to reconstruct information about the properties of the particles coming from the nuclear reactions. An example of such an experiment is a beta–neutrino correlation measurement on laser-trapped ^{38m}K and ^{37}K (Behr et al. 1997). In the experiment, ions of ^{38}K and ^{37}K produced at the on-line isotope separators at TRIUMF in Canada were converted to neutral potassium atoms by stopping in a Zn foil. Next, the rethermalized low-energy atoms were captured by one MOT and, finally, transferred by a laser push beam and magnetooptical funnels into a second MOT, free of the large number of nonlocalized atoms. The overall capture efficiency into the second MOT was 7×10^{-4}. This experiment has detected several hundred thousand recoil–β^+ coincidences, sufficient for further use, with the goal of testing the Standard Model.

Electromagnetic trapping of atoms is also considered to be the most likely scheme for the production of atomic antihydrogen and its spectroscopic investigation. The crucial point here is the laser cooling of the hydrogen atoms. The only possible optical transition for the effective laser cooling of hydrogen is the 1S–2P transition, with a vacuum ultraviolet wavelength of 121.6 nm. Ultrahigh-resolution laser spectroscopy of a few trapped antihydrogen atoms provides a unique opportunity to compare the spectra of hydrogen and antihydrogen and thus realize a stringent test of the fundamental CPT symmetry (Bluhm et al. 1999), and also to compare the gravitational forces on matter and antimatter.

6.5.3 Ultrasensitive isotope trace analysis

An important property of the MOT is the ability to catch atoms whose optical frequencies are shifted from the laser frequency by only a few natural linewidths. This property has been applied for ultrasensitive isotope trace analysis. Chen et al. (1999) developed the technique in order to detect a counted number of atoms of the radioactive isotopes ^{85}Kr and ^{81}Kr, with abundances 10^{-11} and 10^{-13} relative to the stable isotope ^{83}Kr. The technique was called atom trap trace analysis (ATTA). At present, only the technique of accelerator mass spectrometry (AMS) has a detection sensitivity comparable to that of ATTA. Unlike the AMS technique based on a high-power cyclotron, the ATTA technique is much simpler and does not require a special operational environment. In the experiments by Chen et al. (1999), krypton gas was injected into a DC discharge volume, where the atoms were excited to a metastable level. 2D transverse laser cooling was used to collimate the atomic beam, and the Zeeman slowing technique was used to load the atoms into the MOT. With the specific laser frequency chosen for trapping the ^{81}Kr or ^{85}Kr isotope, only the chosen isotope could be trapped by the MOT. The experiment was able to detect a single trapped atom of an isotope, which remained in the MOT for about a second.

6.5.4 Optical-lattice atomic clocks

Laser-cooled slow Cs atoms can improve the accuracy of microwave Cs atom clocks based on spatially separated Ramsey oscillating fields, owing to a longer atom–field interaction time and a reduction of the quadratic Doppler shift (Letokhov and Minogin 1981b). Practical realization of this approach in a Cs atomic fountain allowed researchers to probe the microwave absorption of Cs atoms with a resolution of just 1 Hz (Clairon et al. 1995). With such a narrow resonance, Cs fountain clocks are stable to about one part in 10^{15}. Thanks to this development, atomic clocks are now so good that time and frequency can be measured more precisely than any other physical quantity. Unfortunately, this stability is possible only by averaging the signal over a period of about a day, which makes it difficult to use fountain microwave clocks in real time at this level of accuracy. In contrast, an optical-frequency clock with a frequency of 10^{14}–10^{15} Hz and a stability of one part in 10^{15} needs just a few seconds, rather than a day, for precise real-time measurements of time and frequency. This was one of the main motivations for the search for various approaches to optical frequency standards (Basov and Letokhov 1968).

A number of nice experiments with ultrastable, ultranarrow optical resonances have been performed with molecules, atoms, and ions since the time of the first proposals (Hall et al. 2001; Hollberg et al. 2001). Particularly, a trapped ion can be laser-cooled to the zero point of its motion, which suppresses Doppler effects and provides long interaction with the probing optical field. Excellent performance has been demonstrated with optical transitions in a variety of single-ion systems, including Hg^+, Yb^+, Sr^+, and In^+ (Diddams et al. 2004). Although a single trapped ion could be used in a highly accurate atomic clock, this would not be perfect, because the use of only one ion in the trap provides a limited signal-to-noise ratio for the atomic absorption signal. A possible alternative is to use forbidden transition in a cloud containing a large amount of ultracold atoms, confined in an "optical lattice" (imaged on the cover of this book). The first such optical-lattice clock (Takamoto et al. 2005) used Sr atoms with a two-frequency cooling scheme (using an allowed transition at 461 nm and a forbidden transition at 689 nm) (Fig. 5.11), with a cooling limit (eqn 5.15) $T = 180$ nK, which is lower than the recoil limit (eqn 5.16) of 450 nK. The advantages of optical-lattice clocks are the following: first, a very long interaction time and very small linewidth; second, confinement of the atoms to the Lamb–Dicke regime with $\Delta x < \lambda$ (no first-order Doppler effect); and, third, the large number of atoms (about 10^6) which provides a high signal-to-noise ratio. Optical-lattice clocks have opened up the most promising route toward more accurate and stable frequency standards, with a potential accuracy of 10^{-18}.

Such astonishing progress in obtaining extremely narrow optical resonances of cold atoms, together with the development of the femtosecond frequency synthesizer for the counting of optical cycles (Udem et al. 2002), will cause technology to have a serious impact on fundamental science. This will include, for example, fundamental tests of general relativity, measurements of the physical constants to improve global-positioning-system measurements, and better tracking of deep-space probes.

6.5.5 Optical trapping and cavity QED

New and interesting applications of the MOT have recently been demonstrated by Kimble and coworkers. His group have realized the trapping of single cold atoms in a cavity (Ye et al. 1999). This opened up an opportunity for the deterministic control of atom–photon interactions quantum by quantum. The interesting results reported by Kimble's group (Doherty et al. 2000) include results on the trapping of single atoms with single photons under conditions of cavity quantum electrodynamics (cavity QED). It has been a long-standing ambition in the field of cavity quantum electrodynamics to trap single atoms inside high-Q cavities in a regime of strong coupling. In this regime, the Rabi frequency Ω_0 for a single quantum of excitation exceeds the decay rates for the atom and the cavity mode. A critical aspect of this research is the development of techniques for atom localization that are compatible with strong coupling and do not interfere with cavity QED interactions, as is required for schemes proposed for quantum computation and communication via cavity QED. Within this setting, the experiments done by the Caltech Quantum Optics Group constitute important steps forward in enabling diverse investigations in quantum information science.

Cavity QED has led to many new effects, including the realization of a quantum phase gate (Turchette et al. 1995), the creation of Fock states of the radiation field (Varcoe et al. 2000), and a demonstration of quantum nondemolition detection of single photons (Nogues et al. 1999). Ye et al. (1999) have noted that all serious schemes for quantum computation and communication via cavity QED rely on developing techniques for atom *confinement*. This explains the importance of the experiments on the trapping of single atoms under conditions of cavity QED.

The list of applications of the methods of electromagnetic trapping of cold atoms and molecules will certainly grow fast and expand into different fields of science. In the coming years we should see many productive "marriages" of trapping techniques with many advanced fields of science and technology.

7
Atom optics

The methods developed for manipulating atomic motion with laser light, briefly described in Chapters 5 and 6, have led naturally to the creation of atom optics.

7.1 Introduction. Matter waves

The term *atom optics* is due to the natural analogy with *light optics*, or the optics of photons. Light optics is based on the two fundamental principles: (a) the wave properties of light and (b) the electromagnetic interaction between the light field and matter or, in other words, between light and bound charged particles (electrons or ions) in a medium. Owing to this interaction, the light field can be reflected by the medium or diffracted by it, or else light can propagate through the medium with some velocity other than the velocity of light in a vacuum, and so on (Born and Wolf, 1984).

According to de Broglie's idea, wavelike properties are associated with any particle of matter, and the de Broglie wavelength is defined by the fundamental relation

$$\lambda_{\mathrm{dB}} = \frac{h}{p} = \frac{h}{Mv}, \tag{7.1}$$

where $h = 2\pi\hbar$, and p, M, and v are the momentum, mass, and velocity, respectively, of the particle. The wave properties of particles were verified in experiments on the diffraction of electrons and used in the first analogue of light optics for particles—*electron optics*. Electron optics is based on (a) the wave properties of electrons and (b) the electromagnetic interaction between a moving electronic charge and electric and magnetic fields of an appropriate configuration. The most familiar application of electron optics is electron microscopy (Ruska 1980). Another analogue of light optics is *neutron optics*, which is based again on (a) the wave properties of ultracold neutrons and (b) the interaction between neutrons and atomic nuclei, which can be described by means of what is known as the optical potential. As distinct from electron optics, we deal here with more massive particles (ultracold neutrons), whose wave properties are manifested at low temperatures.

The next natural objects are neutral atoms or molecules. The wave properties of atoms and molecules, and various types of their interaction with matter and electromagnetic fields (from static to optical) make it possible to implement *atom* and *molecular optics*. It is precisely the great variety of methods for exerting an effect on an atom (or molecule) possessing a static electrical or magnetic moment, a quadrupole moment, or optical-resonance transitions (or a high-frequency dipole moment) that form the basis for several possible ways to realize atomic (and molecular) optics.

The known methods to implement atom optics (atomic-optical effects) can be classed into the following three categories:

(1) methods based on the interaction between atoms and matter;
(2) methods based on the interaction between atoms that have a magnetic or electric dipole moment and a static electric or magnetic field of a suitable configuration;
(3) methods based on resonant (or quasi-resonant) interaction between an atom and a laser field.

The first experiments on atom optics, realized by methods (1) and (2), were successfully conducted almost a century ago by Stern and Gerlach. The advent of the laser allowed the possibility to demonstrate atom optics based on the atom–light interaction. This was done on the basis of techniques for the manipulation of atomic motion with lasers. Numerous experiments were conducted to demonstrate the basic effects of the atom optics of matter waves: their reflection, refraction, focusing, interference, etc. A detailed analysis of these effects can be found in the books by Balykin and Letokhov (1995), Berman (1997), and Meystre (2001), and in the review by Adams et al. (1994). The analogy between light optics and atom optics is fairly deep. To illustrate, one can introduce a "refractive index" describing the propagation of matter waves in gases, which has been studied in interferometric experiments (Schmeidmayer et al. 1995).

The main effects of laser-controlled atomic-beam optics are as follows: (1) collimation of atomic beams, (2) reflection, (3) focusing, (4) guiding in hollow fibers, and (5) interferometry. The collimation of an atomic beam by making use of 2D laser cooling has already been considered in Section 5.4.2, and the laser guiding of atoms in a hollow optical fiber in Section 6.1.3. Therefore, we shall briefly consider the atomic reflection (atomic mirrors), atomic-beam focusing, and interferometry effects. In the latter case, laser radiation is used to produce the atomic beams necessary for the observation of the interference of atoms.

7.2 Reflection of atoms by light

The atom mirror is the key element of matter-wave optics. An electromagnetic mirror for neutral atoms was suggested by Cook and Hill (1982). The idea was to use the radiation force of an evanescent laser wave outside a dielectric surface to repel slow atoms. This evanescent-wave atomic mirror was realized experimentally by Balykin et al. (1988a).

7.2.1 The laser field and atomic potential in a evanescent wave

Let us first consider the scheme suggested by Cook and Hill (1982). When a plane traveling light wave is totally reflected internally at the surface of a dielectric in a vacuum, a evanescent wave is generated on the surface (Fig. 7.1). By application of Fresnel's reflection formulas (Born and Wolf 1984), the intensity of evanescent wave is given by

$$I(x) = qI_0 \exp\left(-\frac{x}{x_0}\right), \qquad (7.2)$$

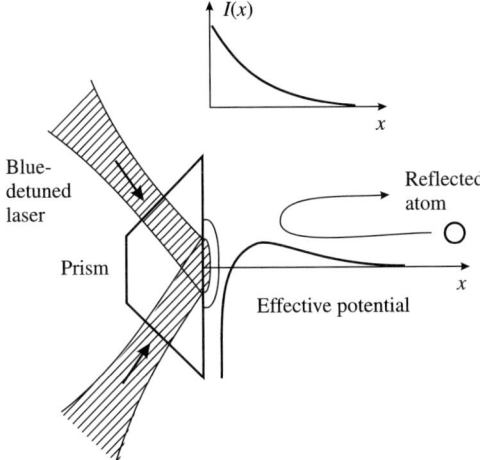

Fig. 7.1 Principle of atom reflection from an evanescent laser wave with a repulsive optical potential at a distance $\sim \lambda$, which is far away from short-distance van der Waals attractive potential.

where

$$q = \frac{16n^3}{(n^2 - 1)(n+1)^2 \cos^2 \phi}, \qquad (7.3)$$

$$x_0 = \frac{\lambda}{4\pi \left(n^2 \sin^2 \phi - 1\right)^{1/2}}, \qquad (7.4)$$

where I_0 is the intensity of the laser radiation that is incident on the surface of the dielectric in a vacuum, n is the refractive index of the dielectric, ϕ is the angle of incidence of the laser beam, and x is the distance from the dielectric surface. The evanescent wave propagates parallel to the surface; in the direction normal to the surface, this wave decays at a distance of $x_0 \cong \lambda/2$, where λ is the wavelength of the light wave. The characteristic depth of penetration of the field into the vacuum is of the order of the wavelength.

If we put an atom in such an evanescent wave, the atom will experience a radiation force due to momentum transfer from the wave. For a two-level atom, the radiation force has a component parallel to the surface (the light pressure force or radiation pressure force F_{rp}) and a component normal to the surface (the gradient force F_{gr}). By choosing the parameters of the laser field (the detuning, intensity, and beam size), it is possible to realize an interaction of the atom with the field where $F_{\text{rp}} \ll F_{\text{gr}}$ and $F \approx F_{\text{gr}}$: in this case the radiation pressure force will be a potential force. The maximum value of this force can be large enough to reflect the atom. It follows from eqn (5.11) that for a size of laser field $\rho = \lambda$ and a detuning of the order of the Rabi frequency $\Delta \approx \Omega_0$, the gradient force is

$$F_{gr} \approx \frac{\hbar \Delta}{4I}\nabla I \cong \frac{\hbar \Omega_0}{\lambda} \approx F_{rp}\left(\frac{\Omega_0}{\gamma}\right). \tag{7.5}$$

As can be seen from eqn (7.5), the gradient force is directed along the gradient of the laser field and the sign of this force is determined by the sign of the detuning. At a positive detuning, the gradient force expels an atom from the laser field. These properties of the gradient force can be used for reflection of atoms from a laser field. The maximum value of the gradient force is reached at a detuning of the order of the Rabi frequency. At such a detuning, the light pressure force is considerably smaller than the gradient force and can be neglected in the interaction of the atom with an evanescent wave. It is this surface wave that can serve as an atomic mirror for an atom running into it: with a positive detuning, the gradient force expels atoms out of the field. It can be shown (Cook and Hill 1982), from consideration of the motion of atoms under the action of the gradient force eqn (7.5) (and when the amplitude of the light changes adiabatically slowly in comparison with the relaxation of the internal atomic motion), that the angle of reflection of the atom is equal to the angle of incidence, which means that we have here a specular reflection of the atom.

The reflection of an atom from a "mirror" may also be considered as the result of its being expelled out of a potential field, whose energy, according to eqn (6.1), is

$$U(x) = \frac{1}{2}\hbar(\Delta - kv_x)\ln\left\{1 + \frac{G(x)}{1 + (\Delta - kv_x)^2/\gamma^2}\right\}. \tag{7.6}$$

If the perpendicular atomic-velocity component v_x is large enough ($v_x > v_{\max} = [2U(0)/M]^{1/2}$, where M is the atomic mass), the atoms reach the surface. In this case, they either adhere to or are reflected from the surface. Usually, the reflection of atoms by the surface is diffusive. Specular reflection can occur when the de Broglie wavelength of the atom divided by the grazing angle is larger then the local surface roughness (Ramsey 1956; Anderson et al. 1986). The atomic mirror can effectively reflect atoms whose maximum velocity component perpendicular to the mirror surface is given by

$$v_{\max}^2 = \frac{2U(0)}{M}. \tag{7.7}$$

For typical atomic parameters (e.g. for a sodium atom), $M = 4 \times 10^{-23}g$, $v = 6 \times 10^4$ cm/s and $\gamma/2\pi = 5$ MHz, and for laser parameters $k = (2\pi/\lambda) = 10^5$ cm^{-1}, $G(0) = 10^5$, and a laser detuning $\Delta = 2.6$ GHz, we have $v_{\max} = 430$ cm/s. Accordingly, for thermal atomic velocities, the maximum grazing angle is 7×10^{-3} rad.

In the reflection of atoms by an evanescent wave, the amplitude of the light field does not necessarily change adiabatically slowly in comparison with the relaxation of the internal atomic motion. In this case eqn (7.6) for the light gradient (dipole) force acting on the atom is only the zeroth-order term in an expansion of the force in powers of the inverse interaction time (Ol'shanii et al. 1992). The next term in the expansion gives rise to a dissipative part in the gradient force. Such nonadiabaticity can happen if the time of interaction of the atom with the field is comparable to γ^{-1}. In this case the specular character of the reflection of atoms can be disturbed.

7.2.2 Quantum-state-selective reflection of atoms

One of the remarkable properties of an atomic mirror is its ability to reflect atoms in a certain quantum state. That the atomic mirror is quantum-state-selective follows from the character of the relationship between the gradient force and the detuning. When the detuning is positive, the gradient force repels an atom from the surface, and thus specular reflection takes place. With negative detuning, the force attracts an atom to the surface, and so diffusive reflection is observed.

Let an atom (or molecule) have several sublevels in the ground state. For an atom, there may be, for example, fine and hyperfine ground-state sublevels (and for a molecule, vibration–rotational sublevels of the ground state). Atoms and molecules in a sublevel for which the transition frequency to an excited state is lower than the laser frequency are reflected from an atomic mirror. Thus, if a beam of atoms or molecules distributed among several ground-state sublevels is incident upon the mirror, the reflected beam will contain only atoms in a single chosen quantum state.

Quantum-state-selective reflection has been studied with sodium atoms (Balykin et al. 1988b). The ground state of the sodium atom, $3S_{1/2}$, is split, because of the hyperfine interaction, into sublevels, one with a quantum number $F = 2$ and the other with $F = 1$, the distance between which is 1772 MHz. Owing to the statistical weights of these sublevels, 62.5% of sodium atoms in a thermal beam are in the sublevel with $F = 2$ and 37.5% in that with $F = 1$. If the laser frequency ν is selected so that $\nu_1 < \nu < \nu_2$ (ν_1 and ν_2 being the frequencies of the transitions from the sublevels with $F = 1$ and $F = 2$ to the excited state $3P_{3/2}$), the reflected beam will contain only the atoms in the quantum state with $F = 2$. The atomic mirror used was a parallel-face plate of fused quartz 0.4 mm thick and 25 mm long, into which the laser beam entered through a beveled side (Fig. 7.2). Multiple total internal reflection of the laser beam was used to increase the surface area of the atomic mirror. The power of the laser beam was 650 mW, and its diameter was 0.4 mm.

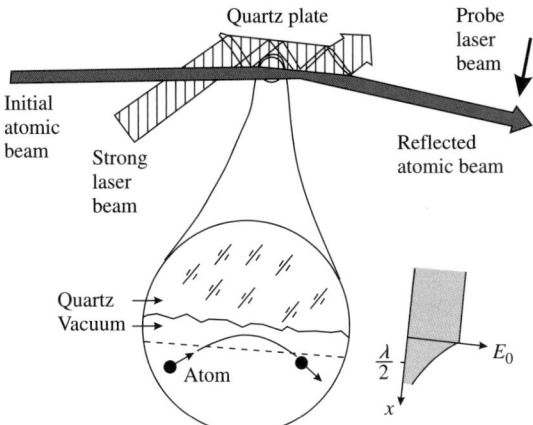

Fig. 7.2 Experimental observation of the reflection of an Na atom thermal beam at grazing incidence. (From Balykin et al. 1988a.)

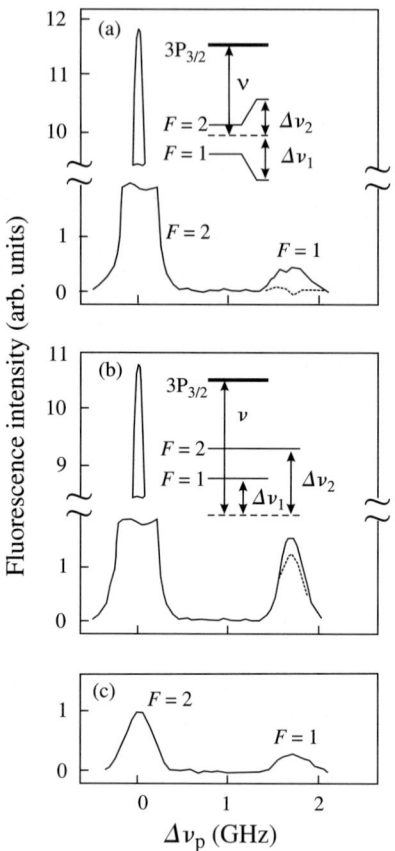

Fig. 7.3 Demonstration of quantum-state-selective reflection of Na atoms by an evanescent laser wave with proper detuning of the frequency $\Delta\nu$ relative to the hyperfine-structure components of the D$_2$ line. (From Balykin et al. 1988a.)

Figure 7.3 shows quantum-state-selective reflection of an atomic beam from an atomic mirror. When the detuning of the frequency of the laser field forming the atomic mirror is positive relative to the transition $F = 2$, $3S_{1/2}$–$3P_{3/2}$ and negative relative to the transition $F = 1$, $3S_{1/2}$–$3P_{3/2}$, the absorption spectrum of the reflected atomic beam contains atoms only in the sublevel with $F = 2$ (Fig. 7.3(a)). The presence of a small number of atoms in the sublevel $F = 1$ is explained by the scattering of sodium atoms by the residual-gas molecules. Subtraction of the signal due to scattering of atoms (dashed line in Fig. 7.3(a)) shows that there are no reflected atoms in the sublevel $F = 1$. At a positive detuning of an intense laser field relative to both transitions, the reflected beam contains atoms in both the sublevels (Fig. 7.3(b)). The absorption spectrum of scattered atoms (for a laser power $P_{\text{las}} = 0$) contains atoms in both sublevels (Fig. 7.3(c)). The reflection selectivity of the atomic mirror, which represents the ratio of the reflection coefficients of the atoms in the sublevels $F = 2$ and $F = 1$, was also measured. The maximum experimentally observed reflection selectivity was about 100 and was determined by the noise of the registration system.

In another experiment (Kasevich *et al.* 1990), the normal-incidence reflection of atoms was observed by dropping a sample of laser-cooled atoms on an evanescent wave. The sample of cold atoms was prepared in the following way. Na atoms were loaded into a magnetooptical trap by slowing down a thermal atomic beam with a counterpropagating frequency-chirped laser beam. The trap was formed by three mutually orthogonal pairs of counterpropagating circularly polarized laser beams intersecting in the zero-field region of a spherical quadrupole magnetic field. After loading of atoms for 0.5 s, the field coils were turned off, leaving a small residual magnetic field in order to cool the atoms in optical molasses. The final stage of cooling was done by gradual extinction of the light. The final temperature of the atoms (a cloud of 3 mm diameter) was $T \approx 25\,\mu\mathrm{K}$. The evanescent wave was created by total internal reflection of laser beam of $8.9\,\mathrm{W/cm^2}$ intensity at a detuning of 400 MHz. The atoms were dropped from a height of 2 cm, and two bounces were registered in the initial trapping region. After the first bounce, about 0.3% of the atoms were observed in the trapping region. The number of atoms after the second bounce was 10% of those observed after the first bounce. The main loss of atoms was due to the ballistic expansion of the sample of atoms, which had a certain initial spatial and velocity extent.

An atomic mirror could be a device of practical importance in applications such as a recombiner in an atom interferometer, for deep focusing (in the form of a concave mirror), and in storage of atoms in an atomic cavity. In all of these potential applications it is essential that the atom mirror be an "ideal mirror" in that (a) the reflection is specular reflection, and (b) the phase shift introduced by reflection can be determined ("coherent" reflection).

There are several processes in the reflection of atoms by light which could lead to nonideal properties of the mirror. First, spontaneous emission of photons by an atom during its interaction with light, and second, spatial variation of the laser intensity due to inhomogeneity of the laser beam and roughness of the dielectric surface. The effect of diffuse atom reflection by a rough mirror has been studied in detail (Henkel *et al.* 1997). Atom reflection is diffuse rather than specular if the surface roughness is comparable to the wavelength of the incident atom. The diffuse atom reflection from an evanescent wave above a dielectric surface may be used as a probe of the surface on the scale of the atomic wavelength. This method can be considered as matter waves probing a dielectric surface.

Quantum-state-selective reflection can be expected in the case of molecules as well. This will offer the unique possibility of preparing and spectroscopically analyzing beams of molecules in a single chosen vibrational–rotational state which can be changed by varying the laser frequency.

7.2.3 Atom reflection by surface plasmon waves

Surface plasmon waves are surface electromagnetic modes that travel along a metal–dielectric interface as bound nonradiative waves with their field amplitude decaying exponentially perpendicular to the interface (Raether 1988). Surface plasmons are usually excited by coupling them to an evanescent wave at a dielectric surface. A plasmon wave atom mirror can be formed on a glass surface with a thin deposited metallic layer (with a typical thickness of 70 nm). A laser field is totally internally reflected from the glass surface. By varying the angle of incidence of the light beam, the wave vector of the evanescent wave can be varied. When the wave vector coincides with the surface plasmon vector, most of the laser photons are converted into surface plasmons. An attractive feature of a plasmon wave as an atom mirror is the large field enhancement of the initial laser beam intensity. The field enhancement ratio ϑ is determined by the ratio between the maximum intensity of the evanescent field with plasmon excitation and the field intensity of the evanescent wave on the bare dielectric surface. The maximum electric-field enhancement is (Raether 1988)

$$\vartheta_{\max} = \left(\frac{1}{\varepsilon_2}\right)\left(\frac{2|\varepsilon_1'|^2}{\varepsilon_1''}\right)\frac{a}{(1+|\varepsilon_1'|)}, \tag{7.8}$$

where $\varepsilon_1 = \varepsilon_1' + i\varepsilon_1''$, ε_2, and ε_0 are the dielectric constant of the metal, the vacuum, and the dielectric, and $a^2 = |\varepsilon_1'|(\varepsilon_0 - 1) - \varepsilon_0$. If the dielectric is quartz ($\varepsilon_0 = 2.2$), the metal layer is Ag($\varepsilon_1' = -17.5, \varepsilon_1'' = 0.5$), and the wavelength λ is 700 nm, then the enhancement factor can be of the order of 100. The resonance condition for plasmon excitation depends on the thickness of the metal layer, which results in a thickness dependence of the enhancement of the plasmon wave.

A simple evanescent wave allows one to create a high homogeneous field along the surface of an "atomic mirror", without any limitation on the intensity of the field in the evanescent wave. The size of an atomic mirror is usually restricted by the laser intensity required for reflection. The main motivation for using a plasmon wave as an atomic mirror is the possibility of reaching a high field intensity with a low laser power. There are several drawbacks to this scheme of using plasmon waves as an atom mirror. The metal layer strongly absorbs light, which leads to destruction of the layer and this process limits the intensity. The intrinsic roughness of the interface produces a strong variation of the local field intensity of the plasmon wave and couples out some of the intensity to the vacuum side. Both effects lead to a broadening of the atomic beam during reflection.

Several groups have reported experimental realizations of an atomic mirror which were based upon the dipole force exerted by a surface plasmon wave. Specular reflection of a supersonic beam of metastable argon (Seifert et al. 1994) and metastable neon (Feron et al. 1993) has been demonstrated using a plasmon wave at a metal–vacuum interface. The enhancement of the laser field by plasmons increased the maximum reflected angle by a factor of 2 to 3.

7.3 Laser focusing of an atomic beam

The principal element in any sort of optics is a lens. It is therefore essential to create laser field configurations capable of focusing neutral atomic beams. There are at present two possibilities for focusing an atomic beam by means of laser light: by using the gradient force or by using the light pressure force. In the former case, it is possible to effect a theoretically ideal focusing of the atomic beam, and in the latter case the degree of focusing will be far from perfect because of the substantial contribution from fluctuation motion of the atoms, that is, it will be similar to the focusing of light by a lens with scattering. The first experiments were conducted by use of both methods.

7.3.1 Atom focusing by the gradient force

Let us first find the condition that the radiation force must satisfy in a laser lens. The lenses used in light and electron optics satisfy the following condition: a divergent concentric beam is transformed by means of the lens into a convergent concentric beam. This requirement means that the deviation angle $\delta\phi$ of an atom from its initial propagation direction in the paraxial-optics approximation should be proportional to its displacement $\delta\rho$ from the beam axis, that is, $\delta\phi = -\alpha\,\delta\rho$, where α is a constant. For this reason, the change in the transverse atomic-velocity component will be $\delta v_\perp = -\delta\phi v_\ell = -\alpha\delta\rho v_\ell$, where v_ℓ is the longitudinal atomic velocity. On the other hand, the change of the transverse atomic velocity is

$$\delta v_\perp = \frac{F t_{\text{int}}}{M} = \frac{FL}{M v_\ell}, \tag{7.9}$$

where F is the force exerted on an atom, M is the atomic mass, t_{int} is the interaction time, and L is the extent of the interaction region. From this expression, we obtain the relationship between the force and the displacement of the atom:

$$F = \frac{-\alpha M v_\ell^2 \rho}{L} = -\beta\rho. \tag{7.10}$$

Thus, we conclude that the force effecting the focusing of an atom in a beam should be proportional to the atomic displacement. This criterion must be satisfied for the force to produce a true "image" if this criterion is not met, the image is blurred, that is, there are aberrations.

A *particle-optical approach* can also be used for the treatment of focusing of atoms by laser light (McClelland and Scheinfein 1991). In this approach the atoms are treated as classical particles that move in the potential field of a laser beam. This method was originally developed for charged particle optics, for calculation of trajectories in a cylindrically symmetric potential field. The equation of motion can be derived from the Lagrangian $L = Mv^2/2 - U(\rho, z)$, where $U(\rho, z)$ is the potential energy eqn (6.1) and z is the axis of symmetry. In cylindrical coordinates, the radial equation of motion is

$$\frac{d^2\rho}{dt^2} + \left(\frac{1}{M}\right)\left(\frac{dU(\rho,z)}{d\rho}\right) = 0. \tag{7.11}$$

By making the assumption that the potential energy $U(\rho,z)$ is less than the kinetic energy of the atom E_0 ($U(\rho, z) \ll E_0$), and that $d\rho/dz \ll 1$ (both conditions are usually valid in real experimental situations), the equation simplifies to

$$\frac{d\rho}{dz^2} + \left(\frac{1}{2E_0}\right)\left(\frac{dU}{d\rho}\right) = 0, \quad z \sim v_z t. \tag{7.12}$$

The quadratic radial dependence of the potential is a necessary condition for the focusing of atoms. Expanding the real potential $U(\rho, z)$ and keeping only the lowest quadratic term in the expansion, we find

$$U(\rho, z) \cong k(z)\rho^2. \tag{7.13}$$

Equation (7.12) becomes

$$\frac{d^2\rho}{dz^2} + \frac{k(z)}{2E_0\rho} = 0, \qquad (7.14)$$

where $k(z)$ is determined by the parameters of the laser field. The higher-order terms in the expansion (7.13) give the spherical aberration. The chromatic aberration can be found by calculation of trajectories of atoms with different initial kinetic energies. The diffusive spontaneous and dipole aberrations can be treated by adding in eqn (7.14), and explicit expressions for all aberrations of the atom lens can be obtained.

The focusing of an atomic beam by means of the gradient force was demonstrated first at Bell Laboratories (Bjorkholm et al. 1978, 1980). In the scheme used there, the atomic lens was created by a CW dye laser, which was focused to 200 μm and superimposed upon an atomic beam of sodium. The laser power was 50 mW and the frequency detuning Δ was -2 GHz. The atomic beam propagated along and inside a narrow Gaussian laser beam. The laser frequency was tuned below the atomic transition frequency, so that the gradient force was directed toward the laser beam axis. The radial potential here is determined by eqn (6.1), with the saturation parameter

$$G(z) = \frac{I}{I_s}\left[\frac{\gamma^2}{\gamma^2 + \Delta^2}\right]\exp\left[\frac{-2\rho^2}{\rho_0^2(z)}\right]. \qquad (7.15)$$

In the experiment, the width of the atomic beam was compressed down to a spot diameter of 28 μm (Fig. 7.4). This minimum achievable spot diameter in the experiment was determined by the fluctuation of the momenta of the atoms due to spontaneous emission. In this experiment, every atom scattered a small number of photons because of optical pumping, so that the duration of the resonant interaction of each atom with the field was less than the flight time of the atoms through the laser beam. If the atoms interacted with the field all the time, the transverse motion of the atoms would be a periodic focusing and defocusing.

The idea that atoms can be channeled in a standing laser light wave (Letokhov 1968) contained important elements of atom optics, namely, the reflection, focusing,

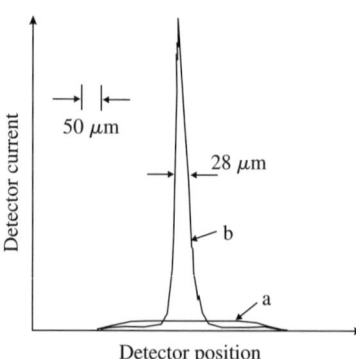

Fig. 7.4 Spatial profile of sodium atomic beam focused by the gradient force of a copropagating Gaussian laser beam. (Reprinted from Bjorkholm et al. 1980 with courtesy and permission of IEEE (USA).)

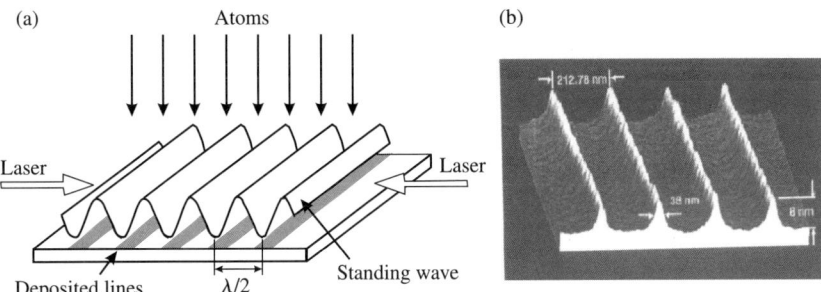

Fig. 7.5 (a) Schematic representation of focusing of a chromium atomic beam in a standing wave. (b) Image of Cr lines on Si substate. The lines are 50 nm wide, spaced by 212 nm. (Reprinted from McClelland *et al.* 1993 with courtesy and permission of AAAS (USA).)

and guiding of atoms by laser light. Therefore, the next natural light field configuration used to focus an atomic beam was precisely a standing light wave. Further development of this idea of using a standing light wave was performed in several successful experiments. In the first experiment (Timp *et al.* 1992), an optical standing wave was used as an array of cylindrical lenses (each period in the standing wave was a lens) to focus a perpendicular sodium atomic beam. The atomic beam was focused into a grating on a substrate with a period of $\lambda/2$. The next important experiment (McClelland *et al.* 1993) demonstrated the principle of laser focusing using deposition of chromium atoms on a surface. Figure 7.5(a) is a schematic illustration of this experiment. A collimated uniform atomic chromium beam is directed onto a silicon surface. Grazing along the surface is an optical standing wave formed by retroreflecting a laser beam onto itself. Each node of the standing wave acts as a cylindrical lens for the atoms. The atoms are thus focused into a series of lines with a spacing equal to half of the wavelength. One-dimensional optical molasses was setup in the atomic beam before it crossed the standing wave. In molasses region, the atoms had been cooled transversely to a temperature of 76 μK with an angular divergence of 0.3 mrad. This small angular divergence permitted a sharp focusing of the atoms. The chromium atoms were deposited on the silicon substrate and observed with either a scanning electron microscope or an atomic force microscope. Figure 7.5(b) shows an image of the Cr lines. The widths of the lines are about 50 nm. These pioneering experiments became the basis for nanofabrication by atom optics (see the review by Meschede and Metcalf 2003).

7.3.2 Focusing of atoms by the light pressure force

Another light field configuration for the focusing of an atomic beam is based on using a dissipative light pressure force, and was demonstrated by Balykin *et al.* (1988c). The atomic lens was formed by divergent Gaussian beams propagating pairwise in opposite directions along the x-axis perpendicular to the atomic beam (Fig. 7.6). The waists of these beams were situated at equal distances from the center of the atomic beam. The lasers were tuned to precise resonance with the atomic absorption frequency. Under

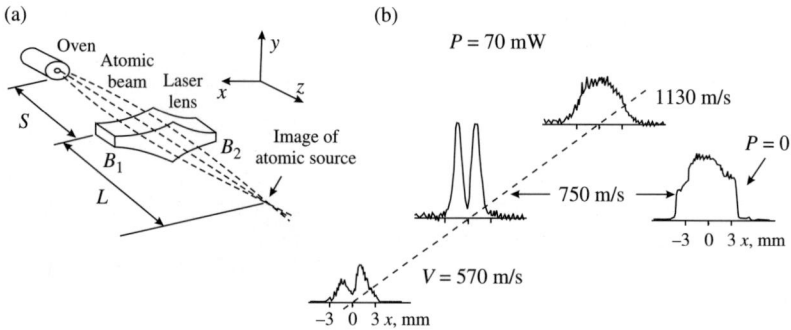

Fig. 7.6 Imaging of two atomic sources. (a) Simplified scheme of experiment. (b) Atomic-beam profiles in the image plane as a function of the longitudinal atomic velocity. On the right, the profile in the image plane without light is shown (From Balykin et al. 1988c.)

such conditions an atom moving away from the atomic-beam axis is acted upon a light pressure force that tends to bring it back to the beam axis. The effect of the gradient force in this case is insignificant. Such a configuration could be used as an atom lens with scattering, for a beam of neutral atoms. Such a lens is similar to a cylindrical optical lens made of scattering glass.

The atom lens, like its counterpart in geometrical optics, suffers from aberrations. The focal length is proportional to the square of the longitudinal atomic velocity. This leads to chromatic aberrations. Since an atomic beam has a Maxwell velocity distribution, atoms with different velocities will be focused at different points. Another source of aberration is the finite cross-sectional dimension of the atomic beam. A violation of the requirement that the atomic trajectories are paraxial ($\rho \ll 1$) results in additional terms that are not proportional to the displacement ρ from the z-axis. This causes spherical aberrations.

In the above experiment, the imaging of the source of an atomic beam was also observed. The atomic lens was formed by two divergent Gaussian laser beams whose waists were at a distance 2 mm from the symmetry axis of the lens. The atom–radiation interaction length was 10 mm. The beam of sodium atoms was formed by one exit hole for focusing of the atomic-beam, or by two exit holes separated by 2 mm symmetrically about the atomic beam axis for the imaging of the atomic source. To observe the focusing or imaging, the spatial distribution of the atomic density in the beam was registered by observing the fluorescent signal from a single-frequency probe laser beam as a function of the transverse atomic coordinate. The diameter of the probe laser beam was considerably less than the size of atomic beam in the image plane. This laser beam crossed the atomic beam in a certain region along the axis of the atomic beam. The probe laser frequency was tuned to resonance with the D_2 transitions of Na in resonance with the Doppler profile of the atomic beam so as to intercept atoms, with a definite velocity. Imaging by means of an atom lens of a source that has a more complex configuration, for example a source with two holes, is a more interesting task. Such an atomic source would correspond to a two-point source in light optics. Figure 7.6 demonstrates the image of two atomic beam point sources by a atomic lens. The density distribution of the atomic "image" consists of two peaks which correspond to the two holes in the atomic source.

An atom lens suffers from chromatic aberration. In this experiment, the effect of chromatic aberration on the formation of the source "image" and on the resolution of the atomic lens was measured. For this purpose, the beam profile was recorded after it was focused at a definite, fixed distance from the laser lens, and the longitudinal velocity of the atoms being registered was varied by altering the position of the probe laser frequency within the Doppler absorption profile of the atomic beam (Fig. 7.6). The original profile of the beam of atoms, with a velocity of 7.5×10^4 cm/s, is shown on the right side of the picture. The focused atomic-beam profiles, arranged in a line, from the two sources are shown on the left. The longitudinal velocity was varied from 5.7×10^4 cm/s to 11.3×10^4 cm/s. One can appreciate qualitatively that with a fixed laser power, there should be a group of atoms with the optimum longitudinal velocity, which are focused precisely into the registered region, the slower atoms being focused before and the faster ones after this region. That is to say, there should be some optimum velocity at which the "image" of the two sources will be defined most sharply. This can be seen exactly in Fig. 7.6. At velocities below 5.7×10^4 cm/s no satisfactory focusing is observed at all. As the longitudinal atomic velocity was increased, there appeared a two-peak structure corresponding to the focusing of the atomic beam from the two sources. The best resolution was observed at a velocity equal to 7.5×10^4 cm/s. As the atomic velocity increased further, the resolution of the laser lens decreased until, finally, the two-peak structure vanished.

7.3.3 Focusing of an atomic beam to the nanoscale

In all the experiments mentioned above, the spatial resolution was in the range of tens of micrometers to 100 nm. The resolution can be considerably improved by using the same idea of the gradient force but now with a different laser field configuration and a different atom–field interaction geometry, as was proposed by Balykin and Letokhov (1987). The new atomic objective lens is a focused TEM_{01}^* laser beam tuned above resonance (Fig. 7.7(a)). The atomic beam propagates along the lens axis, where the intensity of light and therefore the rate of spontaneous emission and momentum diffusion is minimum. This solves the destructive diffusion problem. If all the characteristic dimensions of the focusing field are large in comparison with the de Broglie wavelength of the atom, then, to estimate the width d of the focal spot at half-maximum, use can be made of the usual expression

$$d = \frac{0.61 \lambda_{\text{dB}}}{a}, \tag{7.16}$$

where is one-half of the angle at which the aperture is viewed from the focal point. Clearly, the size of the focal spot can range between 1 Å and 10 nm. This is of potential interest for nanoscience and nanotechnology.

The TEM_{01}^* optical-field mode is very attractive because the motion of an atom along its axis occurs in a low-intensity region, which allows one to achieve a small value of aberrations due to the fluctuations of the dipole force. The focusing of thermal atomic beams with the aid of this mode was considered by Balykin and Letokhov

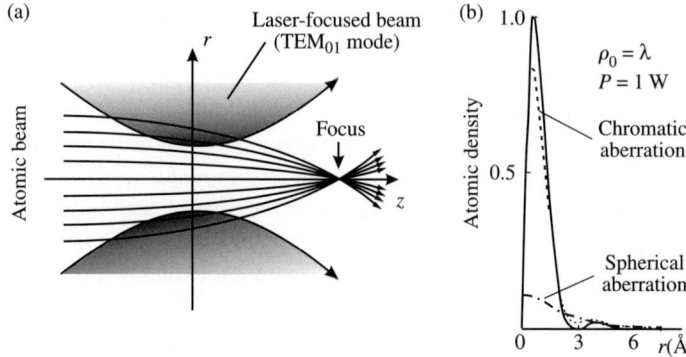

Fig. 7.7 (a) Configuration of laser TEM$_{01}^*$ mode with minimum intensity on optical axis with a copropagating focused atomic beam; (b) distribution of the atomic beam in the focal plane of an atom lens based on the TEM$_{01}^*$ laser mode field. The solid curve corresponds to the aberration-free case, the dashed curve shows the distribution with chromatic aberration, and the dashed and dotted curves represent the distribution with spherical aberration allowed for. The distribution calculated with diffusion aberration coincides with that for the aberration-free case. The minimum waist of the laser beam is $\rho_0 = \lambda$. (From Balykin and Letokhov 1987.)

(1987), McClelland and Scheinfein (1991), and Gallatin and Gould (1991). McClelland and Scheinfein found solutions of the classical equations for paraxial rays in the field of the gradient force and estimated on that basis the spherical, chromatic, diffusion, and diffraction aberrations. It follows from these estimates that the main contribution to the width of the focal spot is, as a rule, the contribution from the diffraction aberrations, the contribution from the dipole force fluctuations sometimes also being substantial.

To calculate the atomic density distribution in the focal plane, Kirchoff's diffraction theory was used by Balykin and Letokhov (1988). Figure 7.7(b) illustrates the atomic-beam distribution in the focal plane of the atom objective lens formed by a laser beam focused into a spot $\rho_0 = \lambda$. The solid curve corresponds to the aberration-free case (the diffraction-limited spot diameter), the dashed curve shows the distribution with chromatic aberration, and the dashed-and-dotted curve represents the distirbution with spherical aberration allowed for. The distribution calculated with diffusion aberration coincides with that for the aberration-free case. The curves were calculated for the following parameters of the laser and atomic beams: a laser power $P = 1$ W, a velocity of the atoms $v = 2.2 \times 10^5$ cm/s, ratio of the diameter of the atomic beam to the waist of the laser beam of 0.25, and a relative velocity spread $\delta v/v = 10^{-3}$. It can be seen from the figure that with aberration taken into account, the size of the atomic beam at the focal point does not differ very greatly from the diffraction-limited spot diameter.

For one to be able to implement precision control over an atomic beam and to focus it into nanometer-size regions, one should know how to calculate exactly its dynamics in various light fields differing greatly in geometry and for different values of the parameters, too. The dynamics of an atomic beam in the field of an electromagnetic wave is a complex matter, for it simultaneously involves the dynamics of the internal degrees of freedom of the atoms, the dynamics of their translational degrees of freedom, and radiation processes. However, far away from a resonance, the spontaneous decay rate becomes low and the fluctuations in the atomic momentum and dipole force can, as a first approximation, be neglected. As a result, the effect of the dipole force becomes predominant and the quantum mechanical dynamics of the atomic beam can be described by a Schrödinger equation wherein the potential of the optical gradient force plays the role of the potential:

$$\frac{\hbar^2 k^2}{2M}\Psi(\mathbf{r}) = \left[-\frac{\hbar^2}{2M}\Delta + U(\mathbf{r})\right]\Psi(\mathbf{r}). \qquad (7.17)$$

Here k is the wave vector of the incident atom at infinity, and $U(\mathbf{r})$ is the potential of the optical gradient force defined by eqn (6.1). By solving the Schrödinger equation (7.17) by some method or other, we obtain a spatial description of the atomic beam, including in its focal region (Klimov and Letokhov 2003).

In conclusion, let us list the requirements that must be met by the laser radiation and by the atomic beam to enable deep focusing of an atomic beam. The focusing potential field is produced by using the TEM_{01}^* laser mode, strongly focused to a size of the order of wavelength of the light. The radiation power needed to focus beams that have thermal velocities is several hundred milliwatts. The diffraction resolution of an atomic objective could be realized at an atomic-beam monochromaticy $\delta v/v = 10^{-3}$. Using the atomic-beam deep-focusing technique considered here, it is quite possible to conceive of an atomic-beam microscope similar to a reflection or transmission scanning electron microscope.

7.4 Diffraction of atoms

The diffraction of atoms by both light waves and material structures plays an important part in atom optics, especially in the development of atomic interferometers, coherent atom beam splitters, etc.

7.4.1 Diffraction by a standing light wave

Diffraction (deflection or scattering) of an atomic beam results when momentum is transferred from the light field to the atoms. If the light wave consists of a single plane wave, momentum is transferred to the atoms at the spontaneous rate γ. Such momentum transfer proceeds at the spontaneous rate because absorption followed by stimulated emission into the same light field mode involves no net transfer of momentum, while absorption followed by spontaneous emission transfers an average of one quantum of momentum for each spontaneous event. If the light field is composed of two or more plane waves, an atom can absorb a photon from one of the plane waves, and stimulated emission can cause that photon to be emitted into a different plane wave, with a resultant transfer of momentum at the rate of stimulated transitions, which may exceed the spontaneous rate by many orders of magnitude in an intense laser field. In a standing laser wave, the momentum transfer proceeds at the stimulated rate and provides fast transfer of momentum from both traveling waves of the standing wave to the atoms, that is, strong diffraction of atoms occurs in a laser standing wave.

128 *Laser control of atoms and molecules*

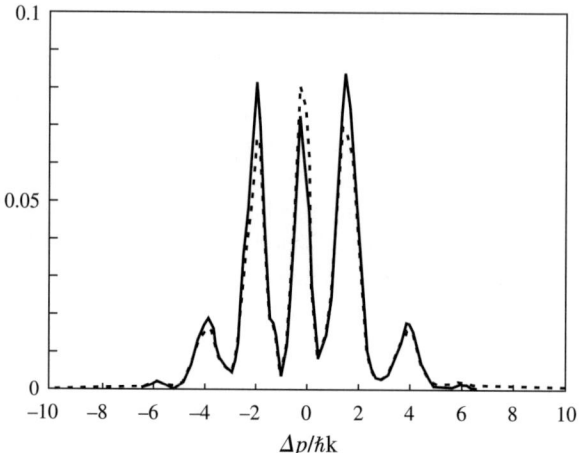

Fig. 7.8 Na Atomic diffraction patterns obtained for Na atoms with a standing laser wave, with a clear splitting of the diffraction peaks of $2\hbar k$ (theory-dashed line, experiment-solid line). (Reprinted with courtesy and permission of the American Physical Society from Gould et al. 1986)

Diffraction of atoms by light has received considerable interest with regard to its practical importance for atomic interferometers (for use in a coherent beam splitter) and its intrinsic features. Pritchard and coworkers have demonstrated the diffraction of sodium atoms at normal incidence by a transmission grating consisting of an optical standing wave (Gould et al. 1986). In the experiment, a collimated beam of sodium atoms was diffracted at angles corresponding to transfer of momentum from $-8\hbar k$ to $+8\hbar k$. Figure 7.8 presents the results of the experiment. The observed sharp quantization of the transferred momentum can be interpreted as the diffraction of the atomic de Broglie waves by the intensity grating of the standing laser wave, which diffracts at intervals of $2\hbar k$ as a result of its periodicity of $\lambda/2$.

The next very interesting step was the observation of diffraction of complex molecules by a structure made from light ($\lambda = 514$ nm, 27 W, Ar laser). Nairz et al. (2001) demonstrated that structures made from light can be used to coherently control the motion of complex molecules (C_{60} and C_{70}). They demonstrated that the principles of light gratings can be successfully carried over to fullerenes, which are internally in a thermodynamic mixed (noncoherent) state. They concluded that an optical grating possesses the scaling properties that the mass and polarizability have roughly the same scaling behavior, because they are both proportional to the volume of the object. In principle, light gratings may ultimately even be used for particles the size of which is comparable to the grating period.

7.4.2 Diffraction of atoms by an evanescent wave

It is possible to combine the reflection and diffraction of atoms by using a standing evanescent wave (Hajnal and Opat 1989). The required optical wave field can be produced by total internal reflection of a laser beam at the surface of a refractive medium and retroreflecting of the light back along its original path. The evanescent field decreases exponentially in the direction perpendicular to the surface and is

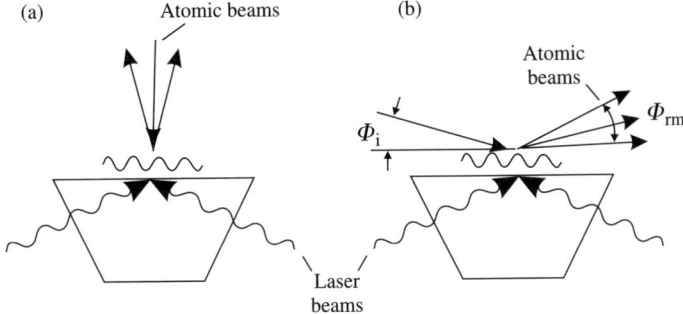

Fig. 7.9 Scheme of an evanescent-wave reflection grating: (a) at normal incidence; (b) at grazing incidence.

modulated sinusoidally along the surface (Fig. 7.9). The result of diffraction of atoms standing evanescent wave can be predicted by considering the incident atomic beam as a plane de Broglie wave incident upon the periodic structure formed by the standing evanescent wave. The diffraction angles Φ_{rm} of the reflected atomic beams are given by the grating equation:

$$\frac{\lambda}{2}(\sin\Phi_{rm} - \sin\Phi_i) = m\lambda_{dB}, \qquad (7.18)$$

where Φ_i is the incident glancing angle of the evanescent wave. At small glancing angles, the diffraction angle can be considerably larger (by a factor of 100) than that for the diffraction of atoms at normal incidence on a transmission grating consisting of a standing wave. For atoms with a thermal velocity of 10^5 cm/s and a glancing angle of 10^{-3} rad on a standing evanescent wave, the diffracted beams may be separated by angles of the order of 10^{-2}–10^{-3} rad. The theory of atomic diffraction for normal and grazing incidence is presented by Henkel et al. (1999).

Another way to consider diffraction of atoms by an evanescent wave is a photon picture (Baldwin et al. 1990), in which diffraction is considered to be a result of absorption and emission of photons, leading to a change of atomic momentum. The exponential profile of an evanescent wave in the direction perpendicular to the interface has contribution from waves of all directions. This permits an atom to acquire, through absorption–stimulated-emission processes of photons of these waves, a momentum in a direction perpendicular to the interface in the case of *specular reflection*. An atom entering the evanescent wave can also absorb a photon from either of the two counterpropagating waves in the vacuum–dielectric interface. The atom can then reemit that photon by a stimulated process back into the same wave. In this case there is zero net change of the momentum of the atom parallel to the interface. The atom can also reemit the photon in a direction opposite to the evanescent wave. In this case the momentum of the atom parallel to the interface will be changed by $2\hbar k$. The absorption and emission of photon pairs changes the momentum in the direction of the standing wave but not the kinetic energy of the atom, owing to energy conservation. This means that the modulus of the total atomic momentum remains unchanged but the momentum normal to the interface must be changed from that in specular reflection. This is a photon explanation of the appearance of the first *diffraction* order in the reflection of atoms by a standing evanescent wave. A change of the atomic momentum with a value $m \cdot 2\hbar k$ gives the mth order of atomic diffraction. The atom can also leave the evanescent wave in an excited state. In this case the atom acquires from the field an additional energy, which equals the difference between the atomic transition frequency and the photon energy. This kinetic energy changes the atomic momentum only in the direction perpendicular to the surface. In the direction parallel to the surface, the atomic momentum can only be changed by one unit of the photon momentum and hence the atom cannot absorb additional kinetic energy in this direction.

7.4.3 The atomic Talbot effect

The Talbot effect is the self-imaging of a periodic structure illuminated by quasi-monochromatic coherent light, which is well known in classical optics and has many applications (see the review by Patorski 1989). This effect is also well known in the field of electron optics, with many applications to electron microscopy (Heindenrich 1964). The history of this effect is very interesting. Talbot made his remarkable observation in 1836, which was explained by Rayleigh in 1881. Rayleigh showed that for a periodic grating illuminated by a plane wave, identical self-images of the grating are produced downstream at distances that are integer multiples of the following length:

$$L_{\text{Talbot}} = 2\frac{d^2}{\lambda}, \tag{7.19}$$

where d is the grating period, λ is the wavelength of the incident radiation, and L_{Talbot} is the Talbot length. Later work showed that identical self-images, laterally shifted by half a period, are also produced at distances midway between those derived by Rayleigh and that other images with smaller periods d/n ($n = 2, 3, 4, \ldots$) are produced at intermediate distances. The basic Talbot effect can be understood by considering the image formed at $(1/2)L_{\text{Talbot}}$, as shown in Fig. 7.10 (Chapman et al. 1995).

The Talbot effect plays an important role in atomic interferometry (see Section 7.5). Also, this effect provides an excellent example of near-field atom optics because self-imaging of the grating take place in the near field, where the curvature of the atom wave fronts must be considered (Fresnel diffraction of atomic waves).

7.5 Atom interferometry

Interference of matter waves has played a central role in the creation of quantum physics. It is enough to refer to de Broglie's prediction of electron waves, the evidence of which was demonstrated by Davisson and Germer's experiment. The concept of matter waves triggered an amazing chain of discoveries: the Bohr–Sommerfeld quantization rule, the Schrödinger equation, etc. In this case interference of electron matter waves takes place inside microsystems such as atoms and molecules. Today, the concept of matter wave interference extends to a macroscopic scale in artificial laboratory systems using atoms, molecules, and even biological molecules. In today's experiments, the

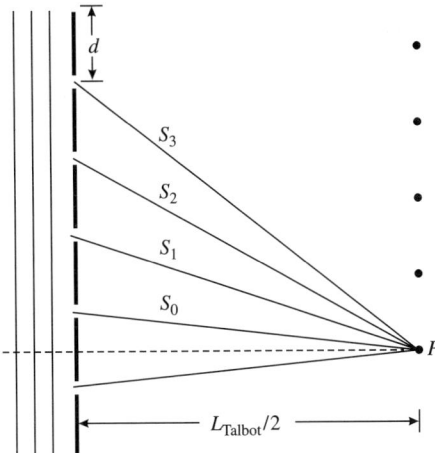

Fig. 7.10 A simple model illustrating the self-imaging of a grating illuminated by a plane wave. It is readily shown that the path lengths S_n from an opening in the grating to the point P are given by $S_n \approx S_0 + n(n+1)\lambda/2$. Hence, any two path lengths differ by an integer multiple of λ, resulting in an intensity maximum at P and, by symmetry, at the other points indicated. (Reprinted from Chapman et al. 1995 with courtesy and permission of the American Physical Society.)

matter wave is associated with the center-of-mass motion of massive particles, instead of the internal microscopic motion of particles with a complex structure.

Atom interferometry is in many respects similar to optical interferometry, but instead of light waves, whose behavior is described by Maxwell's equations, we are dealing here with the wave functions $\Psi(\mathbf{r},t)$ of atoms described by the Schrödinger equation. As distinct from optical interferometry, the main difficulty with atom interferometry is to design atomic-beam splitters and recombiners that are analogous to the semitransparent mirrors used in light optics. This problem, however, can be solved quite successfully by using the diffraction of atoms by a surface or a light wave. Atoms (or molecules) can be detected in a universal way, for example by way of their ionization by laser radiation.

We know of many types of optical interferometer (the simple double-slit Young interferometer, the Mach–Zehnder interferometer, the Fabry–Perot interferometer, the Talbot interferometer, etc.). A similar situation occurs in atom interferometry. Artificial laboratory devices exploit various types of structure for atom interferometry: both material bodies (slits and gratings) and nonmaterial light structures. All these atom interferometers will be considered very briefly; we refer readers for details to the book by Berman (1997) and reviews by Baudon et al. (1999), Kasevich (2002), and Chu (2002).

7.5.1 Matter-wave interferometry with material structures

A simple Young's double-slit atom interferometer was demonstrated by Carnal and Mlynek (1991). This experiment is schematically illustrated in Fig. 7.11. It used a

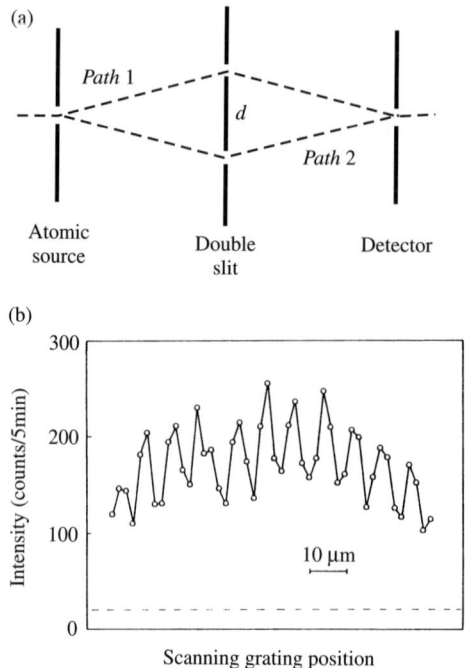

Fig. 7.11 (a) Scheme of a double-slit interferometer. The atoms can move along two spatially separated paths from the source to the detector. (b) Atomic density profile, monitored with an 8 μm grating in the detector plane, as a function of the lateral displacement of the grating. The dashed line is the detector background. The line connecting the experimental points is a guide to the eye. (Reprinted from Carnal and Mlynek 1991 with courtesy and permission of the American Physical Society.)

supersonic beam of helium atoms ($\lambda_{dB} = 1$ Å) and a slit separation of $d = 8\,\mu$m. The entrance slit served as a coherent atomic-wave-function splitter. It provided for spatial coherence between the two exit slits of the double-slit arrangement, so that these two slits, with spacing d, acted as two coherent atomic sources, which led to the formation of an interference pattern in the detection plane at a distance of L from the slits, interference fringes recurring with a period of $\Delta x = L\lambda_{dB}/d$. Similar experiments were also conducted with cold metastable neon atoms ($\lambda_{dB} \simeq 20$ nm) (Shimizu et al. 1992). A three-grating atom interferometer was demonstrated by Keith et al. (1991). This was characterized by the spatial separation of the interfering coherent atomic beams formed by diffraction of a beam of sodium atoms ($\lambda_{dB} = 0.016$ nm) incident at right angles on the first grating. In this experiment, use was made of microfabricated gratings with periods of 200 and 400 nm. Distinct atom interference fringes were observed, with a measured contrast of 13% and a period of 400 nm. Interference was detected by measuring the intensity of the atomic beam passing through the system upon transverse scanning of the central grating.

The matter-wave interferometry of large molecules is another a very interesting line of research, allowing one to investigate the wave properties of massive particles (fullerenes, biomolecules, and viruses); it originated with the experiments conducted by Zeilinger (Arndt et al. 2002). The Talbot–Lau interferometer is ideally suited to such experiments: this also consists of three successive gratings, but operates in the near field, or Fresnel zone, where the characteristic size of the diffraction pattern scales as the square root of both wavelength and distance. A Talbot–Lau interferometer can accept a spatially incoherent beam, which implies that no collimation is needed, and it works with a spatially extended detector. Therefore, with a Talbot–Lau interferometer, the atom count rates can exceed those obtained with a three-grating Mach–Zehnder interferometer by several orders of magnitude. Because of the scaling properties of near-field interferometers, the Talbot–Lau interferometer has actually been proposed for experiments using quantum microobjects up to the size of a virus (Arndt et al. 2002).

7.5.2 Atom interferometry with light waves

Several types of atom interferometers have been realized that use various light-field configurations to split and recombine atomic beams. The main requirement for the light field is that it should split the atomic wave by the maximum possible angle without multiple scattering of the beam. So far, researchers have failed to devise an ideal laser atomic-beam splitter that would be comparable with the semitransparent mirrors in optical interferometry.

The Bragg atom interferometer, using Bragg deflection of a collimated atom beam, was first discussed by Dubetskii et al. (1985), and the Bragg scattering of atoms from a standing light wave was first observed in work with sodium (Martin et al. 1988). In this interferometer, metastable Ne atoms underwent Bragg scattering up to the third order, giving a maximum of $6\hbar k$ transverse-momentum difference between the two arms of the interferometer. Such Bragg interferometers can be used with a cold beam of atoms to drastically increase its area and improve sensitivity in applications (atomic gyroscopes etc.). Various elegant atom interferometer schemes have been devised, including in particular the Ramsey–Borde interferometer (Borde 1989), which developed further the Ramsey method of spatially separated beams (Ramsey 1950), extended to the optical region by Chebotayev and coworkers (Baklanov et al. 1976b). It was suggested that the method of three spatially separated light fields could also be used for atom interferometry purposes (Chebotayev et al. 1985). In this interferometer, interference of not only internal but also external degrees of freedom was obtained. Finally, Borde (1989) put forward the concept of an atom interferometer using a four-zone geometry with spatially separated atomic paths. The spatial separation and recombination of the pathways was accomplished by use of the recoil effect.

Figure 7.12(a) shows a schematic arrangement of an atom wave interferometer using diffraction gratings made of light (Rasel et al. 1995). This atom interferometer is an exact "mirror image" of an interferometer for light, with the roles of atoms and photons interchanged. This interferometer directly demonstrates the coherence of the diffraction of atomic waves by standing light waves. The incident atoms are diffracted by the first standing light wave, which produces a coherent superposition of mainly zeroth- and first-order beams. These beams impinge on the second standing light wave, where each beam is coherently split. Finally, at the third standing light wave, each of the incident beams is once more coherently split and a number of emerging

(a)

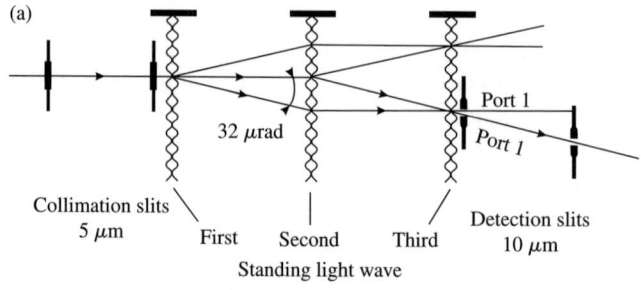

(b)

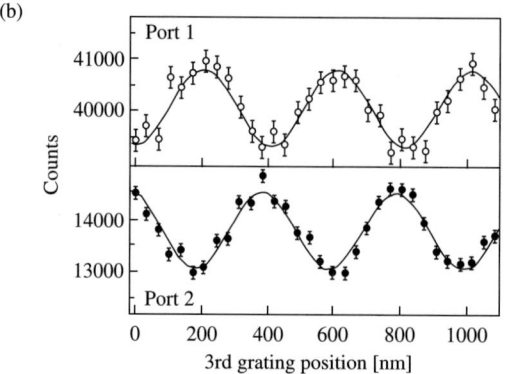

Fig. 7.12 (a) Schematic arrangement of interferometer setup (not to scale). The collimation slits for the incoming beam, the three standing light waves created by retroreflection at the mirrors, and the two final slits, one selecting a specific interferometer (thick lines) and the other selecting a specific output port, are shown. For reasons of presentation, the wavelength of the light beams has been greatly exaggerated. In the experiment, the atomic beam was wide enough to cover more than 12 light wave antinodes. (b) Measured atom interference pattern for both output ports of the interferometer. The complementary intensity variations in the two output ports observed is a consequence of particle number conservation. The solid line is a fitted sinusoid. (Reprinted from Rasel et al. 1995 with courtesy and permission of the American Physical Society.)

beams result, some of which are coherent superpositions of different paths through the interferometer with different relative phases. Rasel et al. used either one of the two skew-symmetrical interferometers formed by the zeroth and first diffraction orders at the first grating, the first diffraction orders at the second grating, and finally again the zeroth and first diffraction orders at the third grating. The interferences were detected by translating the third grating and observing the intensity alternately in the two outgoing beams of the selected interferometer in the far field. The two output ports of the Mach–Zehnder interferometer show complementary intensity oscillations (Fig. 7.12(b)).

Atom interferometers based on optical transitions have made possible ultrahigh-precision measurements of gravity (Peters et al. 2001), gravity gradients (Snadden et al. 1998), rotation (Gustavson et al. 1997), and the photon recoil effect for atoms

(Hensley et al. 2000). Cold-atom interferometry holds much promise, especially when use is made of a Bose–Einstein condensate, that is, a coherent collective of well-organized, very cold atoms in the ground state (see Chapter 8), for developing guided-atom interferometers (Kreutzmann et al. 2004), etc.

7.6 Atomic holography

Optical holography, which allows an optical wavefront to be reproduced by passing an optical beam through a hologram, is well known. A hologram can be produced by photographically recording two interfering optical beams (a reference beam and a beam reflected from the object of interest). The progress in the production of laser-cooled atomic beams has made it possible to obtain a holographic image by means of Ne atoms and a computer-generated binary hologram (Fujita et al. 1996). A laser-cooled ($50\,\mu$K) beam of metastable Ne* atoms was passed through a hologram produced on a 100 nm thick silicon nitride membrane. The hologram comprised numerous holes (typically 200 nm across) in the membrane, their positions being precisely specified by the computer, so that the diffracted Ne atoms generated the reconstructed pattern.

In principle, many atoms can be used for atomic-beam-holography purposes. It is possible to deposit the atoms directly on a substrate to produce a desired pattern, with a theoretical resolution of about 100 nm under typical conditions. Atomic-beam holography has considerable potential for the production of patterns with a nanometer-scale resolution. The present state of the art of atomic holography is rather primitive. However, it is a promising technique for atom manipulation in three-dimensional space, which could be used for control of the spatial phase and amplitude structure of atomic de Broglie waves in the future (Shimizu 2000).

7.7 Towards atom nanooptics

The prefix "nano" has in recent years been applied to one of the most rapidly developing avenues of investigation, namely, nanoscience and nanotechnology. Optics and photonics are not keeping away from this avenue; new research areas are emerging, near-field nanooptics for one. The development potential of nanooptics is great enough, especially as regards nanoscopy and optical imaging and diagnostics, but laser light makes it possible not only to passively investigate objects with a nanometer-scale spatial resolution, but also to modify matter on a nanoscale (Ohtsu 1998). And certainly the manipulation of free atoms in a nanoscale (Balykin et al. 1994) also pertains to the field of nanooptics that can be called "atom nanooptics." Atom nanooptics is aimed at forming ensembles or beams of neutral atoms with a characteristic size lying in the nanometer range. The possibility of focusing of an atomic beam to spot of a nanometer size has already been discussed in Section 7.3.3.

Atom optics based on the use of traveling and standing light waves suffers from a number of restrictions stemming from the spatially "nonlocalized" character of laser light fields. This "nonlocalized" nature of laser light fields makes atom-optical elements "nonlocalized." Hence the imperfections of these elements: aberrations of atom-optical lenses, low diffraction efficiency, limitations on the contrast of interference fringes in atom-optical interferometers, and so on. It is evident from general physical considerations that the use of spatially localized fields (and, accordingly, spatially localized atomic potentials) can offer new possibilities for constructing atom-optical elements.

136 *Laser control of atoms and molecules*

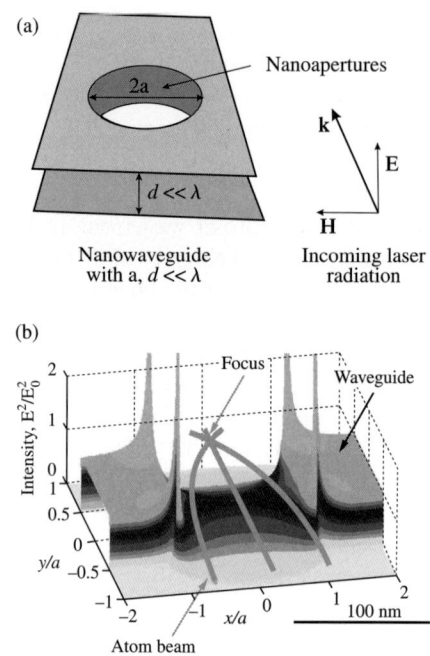

Fig. 7.13 (a) Formation of a "photon hole" by laser light with a vertically oriented electric-field component in a waveguide with coaxial holes at a distance $d \ll \lambda$; (b) light field intensity distribution near the aperture, and atom focusing by the "photon hole." (From Balykin *et al.* 2003.)

Optical nanofield configurations free from the above shortcomings have been suggested by Balykin *et al.* (2003). Figure 7.13(a) schematically illustrates the production of one version of an optical nanofield. Two plane conductive plates spaced by a distance of the order of or smaller than the wavelength of the light ($d \ll \lambda$) form a plane waveguide for laser radiation coupled into it from one side. If the electric-field-strength vector of the laser radiation is normal to the plane of the waveguide, the radiation can propagate through the waveguide, no matter what its thickness. Now let two small coaxial apertures be made in the conductive screens forming the waveguide. If the diameters of these apertures are smaller than the wavelength of the radiation coupled into the waveguide, the radiation will practically not exit through the apertures, but the light field near them will be strongly modified. To find the electromagnetic-field distribution in the vicinity of the apertures in the waveguide walls is an involved problem of electrodynamics. In the particular nanoaperture case ($a, d \ll \lambda$) under consideration, the problem has been solved in a quasi-stationary approximation. Figure 7.13(b) shows the light field intensity distribution near the apertures inside and outside of the waveguide in the case where the thickness of the waveguide is equal to the radius of the apertures. It can be seen from the figure that there is a minimum in the light field intensity in the direction normal to the plane of the waveguide. It is only natural to call such a field configuration a "photon hole." Its characteristic volume is determined by the size of the apertures and the thickness of the waveguide: $V \sim a^2 d \ll \lambda^3$. The sharp peaks in the field intensity near the aperture edges are due to the hypothetical infinite conductivity of the waveguide walls. In waveguides with walls of finite conductivity, the amplitude of the field intensity peaks will not be so strongly manifested.

When the detuning Δ of the laser radiation frequency relative to the atomic transition (resonance) frequency is positive, an atom in the laser light guide is drawn into the weak-field region of the guide. In the case of a "photon hole," the nanometer-size weak-field region is surrounded by the strong field inside the waveguide, and if the detuning of the light field frequency is positive, atoms flying through the apertures in the waveguide walls will be attracted to the axis of the system, that is, they will be focused (Fig. 7.13(b)). An attractive feature of the light field configurations being considered here is the possibility of making a large number of pairs of apertures in a waveguide and, correspondingly, the same number of nanolocalized light fields. Such field arrays would potentially allow the motion of many atomic beams to be controlled simultaneously.

8

From laser-cooled and trapped atoms to atomic and molecular quantum gases

The invention and development of laser methods for cooling and trapping atoms have led to the advent of the physics of ultracold matter. Low-temperature quantum physics has so far dealt with condensed media, wherein quantum phenomena (superfluidity and superconductivity) usually manifest themselves at temperatures of a few kelvin. Laser cooling and trapping of atoms have made it possible to successively attain millikelvin, microkelvin, and nanokelvin temperatures, at which the de Broglie wavelength of the atoms becomes commensurable with the interatomic separation, even in dilute gases. The interaction of atoms in ultracold dilute gases becomes essentially quantum mechanical. The physics of such quantum gases is more akin to the well-developed physics of condensed matter at low temperatures. It is in essence a boundary field between atomic, laser, and condensed-matter physics. Following the first, pioneering experiments on the Bose–Einstein condensation of alkali-metal atoms, this field has continued to develop at a very fast pace, which can be explained by the vast variety of particles (atoms, isotopes, and molecules) that can be laser cooled and trapped under degeneracy conditions. This is much more than physicists had at their disposal previously (^{3,4}He). Moreover, laser radiation and magnetic fields additionally make it possible to control not only the intrinsic quantum state of the interacting ultracold particle species, but also the interaction of the particles. All this makes the research in the field of quantum gases a great breakthrough, both in atomic–molecular physics and in many-body-interaction physics, which heretofore was the prerogative of low-temperature condensed-matter physics.

The present short Chapter has no intention of reviewing this rapidly growing domain of science, for this field has already been covered by Nobel lecture reviews (Cornell and Wieman 2002; Ketterle 2002) and the reviews by Leggett (2001) and Courteille et al. (2001), and will soon become the subject of many other reviews, textbooks, and monographs. The objective of this Chapter is, rather, to illustrate the new wide possibilities and challenges opened up by the invention of new methods for controlling the motion of atoms and molecules, namely, the laser cooling and trapping of particles. Therefore, the presentation of the material of this Chapter is of necessity illustrative and superficial.

8.1 Introduction

The variety of new physical effects occurring in dilute quantum gases is enormous. One of the most interesting of them is the quantum-statistical behavior of dense atomic gases made up of bosons, that is, atoms possessing an integer spin, or of fermions, that is, atoms possessing a half-integer spin. When bosonic atoms are cooled down to a very low temperature, the characteristic size of every atomic wave packet, determined by the de Broglie wavelength, becomes very large, so that the individual wave packets in a sufficiently dense gas overlap. Figure 8.1 presents scales of the kinetic energy E of atomic motion (12 orders of magnitude), in terms of the temperature $T = E/k_B$ and de Broglie wavelengths (for the mass of the ^{23}Na atom), together with the ranges covered by various fields of research and various methods of cooling and trapping particles, and also typical atomic energy scales. At nanokelvin temperatures the de Broglie wavelength equals 1 μm, and at an atomic density over 10^{15} cm^{-3} it becomes comparable to the interatomic separation. In this region of parameters, it is essential to consider the quantum mechanical interaction of atoms. The overlapping of wave functions in the case of integer-spin atoms gives rise to a constructive interference between the individual atomic wave functions and, accordingly, creates a macroscopic coherence in the atomic gas. Such a specific macroscopic state of an atomic gas is described by a single macroscopic wave function and is known as a Bose–Einstein condensate. The Bose–Einstein condensation of atomic alkali-metal gases was experimentally achieved in atom traps (Anderson *et al.* 1995; Davis *et al.* 1995; Bradley *et al.* 1995).

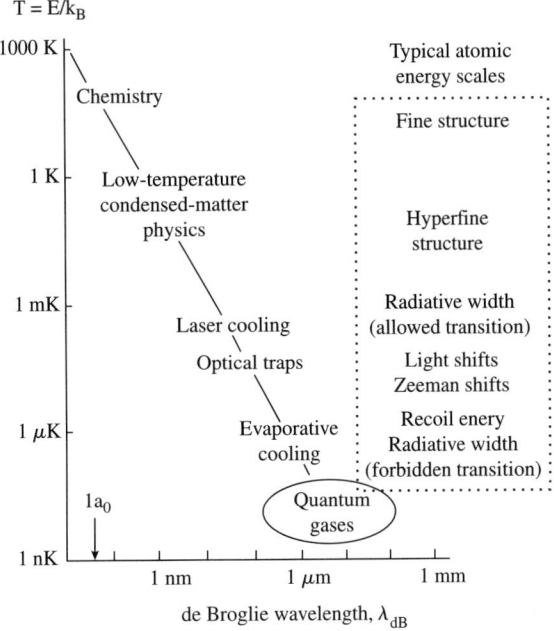

Fig. 8.1 Temperature of atomic motion versus the de Broglie wavelength λ_{dB}, with indication of various corresponding fields of science and cooling/trapping methods, and typical atomic energy scales. (Adapted and modified from Burnett *et al.* 2002.)

The experimental realization of Bose–Einstein condensation has shown that this phenomenon makes it possible to construct an atom laser by releasing a condensate from an atom trap (Mewes et al. 1997). An atom laser is a source of coherent collimated atomic de Broglie waves, which was nicely demonstrated in the observation of interference between two Bose–Einstein condensates (Andrews et al. 1997). Like a conventional optical laser emitting coherent light waves, an atom laser emits coherent waves of atomic matter. This means that atom lasers may have an impact on coherent atom optics, which may lead to new achievements in atom interferometry, holography, and microscopy. The development of coherent atom optics may in turn lead to new fundamental observations, including nonlinear and nonclassical effects in coherent-matter media.

The next very important step was the observation of Fermi–Dirac degeneracy in a trapped atomic gas of the fermionic atoms ^{40}K (DeMarco and Jin 1999). To illustrate the radically different behavior of bosons and fermions in quantum degeneracy conditions, Fig. 8.2 presents the distributions of these atomic species among energy states. The bosons tend to occupy the lowermost quantum state and, conversely, not more than one fermion can reside in one and the same quantum state. If the density of bosons in the lower state is high enough, they interact to form a collective state, called a Bose–Einstein condensate (BEC). The behavior of a Fermi-degenerate quantum gas is much more involved because of repulsive interactions, which is very important for the understanding of the striking phenomenon of "superconductivity" of low-temperature condensed matter.

Thereafter followed the fireworks of remarkable experiments on the production of BECs of molecules, Fermi–Dirac degeneracy in molecular gases, etc. Thus originated a new domain of physics—the physics of ultracold matter. At the beginning of the investigations into the laser cooling and trapping of atoms (1968–1982), described in

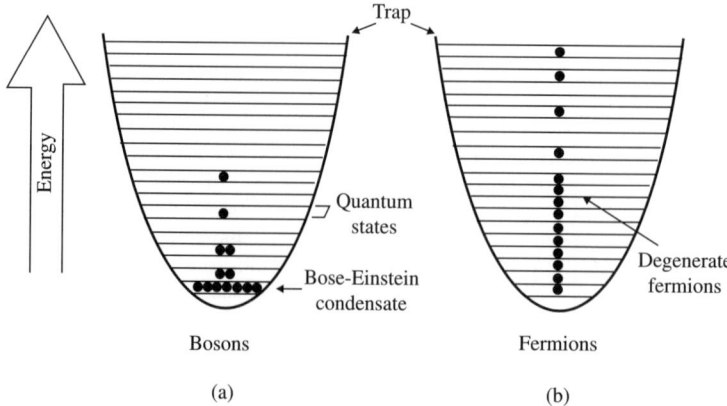

Fig. 8.2 A quantum degenerate gas of ultracold atoms reaches degeneracy when the matter waves of neighboring atoms overlap: (a) at absolute zero, gaseous bosonic atoms all end up in the lowest energy state; (b) fermions, in contrast, fill the states with one atom per state, and the energy of the highest filled state at $T = 0$ is the Fermi energy E_F.

Chapters 5 and 6, one could even hardly imagine such rapid progress of atomic and laser physics.

8.2 Bose–Einstein condensation of atomic gases

To observe the Bose–Einstein condensation of atomic gases was a real challenge to atomic physicists in the 1980s. The first systematic pioneering investigations along these lines were conducted with spin-polarized hydrogen in a strong magnetic trap with cooling by a He buffer gas (Silvera and Walvaren 1980). These studies had their predecessors (Stwalley and Nosanow 1976). The development of simpler and more effective techniques for the laser cooling and trapping of alkali-metal atoms using Bose–Einstein forestalled condensation experiments on the Bose–Einstein condensation of hydrogen. But taken together, they meant a crucial breakthrough in ultracold-quantum-gas physics.

8.2.1 Physical background

In an ideal atomic gas at a temperature T, the characteristic size of the atomic wave packet is governed by the de Broglie wavelength

$$\lambda_{\mathrm{dB}} = \left(\frac{2\pi\hbar^2}{Mk_{\mathrm{B}}T} \right)^{1/2}, \qquad (8.1)$$

where M is the atomic mass and k_{B} is the Boltzmann constant. At a gas density of n, the typical distance between atoms is about $n^{-1/3}$. Accordingly, quantum-statistical effects in an atomic gas become possible if the condition $\lambda_{\mathrm{dB}} \approx n^{-1/3}$ is satisfied, or if the gas temperature and density are related by

$$T_{\mathrm{c}} \approx \frac{2\pi\hbar^2 n^{2/3}}{k_{\mathrm{B}}M}. \qquad (8.2)$$

For example, when atoms are cooled down to a temperature of $T=1$ mK, quantum-statistical effects can manifest themselves at a density of $n=2\times 10^{18}$ cm^{-3}. At a temperature of $T=1$ μK, quantum-statistical effects become important at a density of $n=7\times 10^{13}$ cm^{-3}, and at an extremely low temperature of $T=1$ nK these effects become manifest at a very low density of $n=2\times 10^{9}$ cm^{-3}.

If the atoms in the gas have an integer spin, that is, if they are bosonic atoms, they are distributed among the quantum states in accordance with the Bose–Einstein distribution

$$f(E) = \frac{1}{\mathrm{e}^{(E-\mu)/k_{\mathrm{B}}T} - 1}, \qquad (8.3)$$

where μ is the chemical potential, which is the energy required to add an additional particle to the system. In the case of bosonic atoms, dramatic changes can occur in the atomic gas at the temperature defined by eqn (8.2). Above this transition temperature, the atoms are distributed among many quantum states and the gas behaves as a classical system. Below the transition temperature, the atoms mostly occupy the ground state, and the thermodynamic properties of the gas undergo dramatic changes. In particular, above the transition temperature the heat capacity of the atomic gas is contributed to by the population of many atomic states, whereas at temperatures below the transition value the heat capacity depends mainly on the ground-state population. In general, at the critical temperature T_c, the number of atoms in the ground state, N_0, is related to the total number of atoms, N, by the relation

$$N_0 = N\left[1 - \left(\frac{T}{T_c}\right)^{3/2}\right]. \tag{8.4}$$

Thus, at $T < T_c$ the number of atoms in the ground state is comparable to the total number of atoms in the gas.

It should be noted that in real experimental situations an atomic gas is always confined in a potential well with a certain wall steepness. In such practical situations, the value of the critical temperature T_c depends on the shape of the potential well. The steeper the potential well, the higher the critical temperature.

The expressions considered above are valid for the phase transition to the Bose–Einstein condensate in an ideal gas of noninteracting atomic species. In any real atomic gas, atomic interactions alter the properties of the transition to the BEC.

The interaction of ultracold atoms depends on the long-range atomic potentials and can conveniently be described in terms of the scattering length a. Since the energy of ultracold atoms is very low, they influence one another at distances tens of times as long as the short-range interaction length $r_{\text{sh.r.}}$ that is important for atoms at normal temperatures. The magnitude of $r_{\text{sh.r.}}$ is a few Bohr radii a_0, so that the interaction of ultracold atoms takes place under conditions where

$$r_{\text{sh.r.}} \ll a \ll \lambda_{\text{dB}}. \tag{8.5}$$

The scattering length a is a few tens of a_0 and depends on the details of the atomic potential. Very small changes in the atomic potential can give perceptible changes in the scattering length a and its sign. Positive scattering lengths correspond to repulsive interaction potentials between the atoms, whereas negative scattering lengths correspond to attractive atomic interaction potentials. The scattering of atoms whose de Broglie wavelengths are much greater than the size of the scattering object is of isotropic character (s-scattering). The low-energy elastic scattering cross section for identical bosonic atoms is $8\pi a^2$. The average energy of an atom interacting with similar atoms in a quantum gas is given by

$$E_{\text{int}} = \frac{4\pi a \hbar^2 n}{M}, \tag{8.6}$$

where n is the atomic density in the condensate and M is the mass of the atom. It is exactly this term that is included in the Schrödinger equation to describe the condensate wave function. The resultant equation has come to be known as the Gross–Pitaevskii equation (Dalfovo et al. 1999).

The scattering length differs between different atoms and between different internal spin states. For example, the scattering lengths of the ^{23}Na and ^{87}Rb atoms are $55a_0$ and $105a_0$, respectively, for the internal states that they assume following their Bose–Einstein condensation in a weak magnetic field (see below). For ^{7}Li, the length a is negative ($-27a_0$), so that the attractive interactions between ^{7}Li atoms prevent any stable existence of a BEC in an infinite homogeneous gas, except in a small-size trap. In this case, the quantum zero-point motion in the trap helps to keep the atoms apart, allowing a condensate to form.

Note also the principal difference between the structure of a BEC in a spatially homogeneous and an inhomogeneous atomic gas. In the case of a spatially homogeneous gas, the Bose–Einstein condensation occurs as a phase transition in momentum space. In that case, both the thermal fraction and the BEC fraction are homogeneously distributed in space, but have different momentum distributions. The thermal fraction is distributed over a momentum range from zero to the thermal momentum, while the BEC fraction is localized at zero momentum. In the case of a spatially inhomogeneous gas, the BEC fraction is localized near the central region, where the gas density is near its maximum value. In experimental observations, the appearance of a BEC is often detected as a sudden change in the gas density profile at the critical temperature by the time-of-flight imaging technique.

8.2.2 Observation of Bose–Einstein condensation in atomic gases

At the time the first experiments on the laser cooling of atoms were performed, it seemed that combining laser-cooling and optical-trapping techniques could be a direct way to produce Bose–Einstein condensates in atomic gases. Later on, it became obvious that the density of laser-cooled and optically trapped atomic gases is limited by rescattering effects. To overcome this difficulty, optical traps were replaced by magnetic traps, and the laser-cooling technique was supplemented by an evaporative-cooling technique that was initially proposed for cooling atomic hydrogen (Hess, 1986).

The evaporative cooling of atoms is based on a process of removing hot particles from an ensemble of particles (see the review by Ketterle and van Drutten (1996)). Assume that the particles are confined in an atom trap produced by a potential of finite depth. In such a case, sufficiently hot atoms always leave the trap. This process is accompanied by the rethermalization of the rest of the atoms as a result of their elastic collisions. In the course of rethermalization, the temperature of the atomic ensemble is reduced. Of course, if the depth of the potential well is fixed, the new, lower temperature of the ensemble is reached within a very short time interval, spanning only a few atomic collisions. However, if the trapping potential is lowered continuously, the rethermalization process continues until all the atoms have evaporated from the trap. This process, called forced evaporation, allows one to reach extremely low temperatures as a result of the reduction of the number of trapped atoms. In many experiments, a lowering of the atomic-ensemble temperature to nanokelvin values has been achieved upon evaporation of 99% of the atoms. The phase space density of the remaining 1% of the atoms is in that case typically increased by six orders of magnitude.

The evaporative-cooling process requires that the elastic collision cross section, governed by the elastic scattering length a, should be larger than the inelastic collision cross section. In the case of inelastic collisions, some energy is released or unconfined atoms appear, which is undesirable as far as evaporative cooling is concerned. By appropriately selecting the internal state of the atoms being cooled, one can make inelastic collisions occur with a lower probability and at distances shorter than the scattering length a. Technically, evaporative cooling can be performed in various traps and by various techniques. Evaporative cooling has been realized in various types of magnetic traps, including time-orbiting-potential (TOP) traps and optical traps. Evaporation was achieved by lowering the trapping potential or by making the atoms escape from the trap by means of a radio-frequency field (rf evaporation).

The development of laser-cooling techniques, magnetooptical traps (MOTs), and the evaporative-cooling technique opened the way to producing cold, dense atomic samples satisfying the Bose–Einstein condensation condition in eqn (8.2). The first successful experiments were performed with alkali-metal atoms, namely Rb (Anderson et al. 1995) and Li (Bradley et al. 1995), and also with Na atoms (Davis et al. 1995). Later on, the critical phase density was achieved with atomic hydrogen (Fried et al. 1998). This list is being constantly extended and now includes various isotopes, the metastable atomic state $^3S_1$ of He (Robert et al. 2001), two-electron atoms (Yb and Sr), etc. The extension of Bose–Einstein condensation to molecules has also proved possible. Not even a decade has yet elapsed since the initially very difficult experiments with BECs became standard in numerous laboratories throughout the world.

In an experiment by Anderson et al. (1995), about 10^7 rubidium atoms were trapped in a dark MOT at a pressure of about 10^{-11} Torr. Once loaded, the MOT was switched off and a quadrupole TOP trap turned on. The evaporative cooling of the atoms was initiated by reducing the potential of the TOP trap. This resulted in cooling of the atomic cloud of 4×10^6 atoms to a temperature of 90 μK and increasing the atomic density to 2×10^{10} cm^{-3}. After that, additional evaporative cooling was produced by means of a radio-frequency field acting at the edge of the potential. In the last stage the atomic cloud was rethermalized for 2 seconds, released from the trap, and ballistically expanded. The creation of a BEC was detected as a sudden change in the gas profile at the critical temperature. An example of such a dependence is shown in Fig. 8.3, which presents the peak density of an atomic gas trapped in a parabolic potential well. The expanded cloud was probed with a laser beam tuned to resonance with a strong cycling transition. Figure 8.4 shows the absorption pictures observed in the neighborhood of the phase transition. As can be seen from these pictures, the

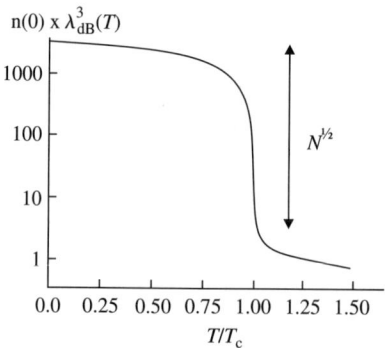

Fig. 8.3 Density of a rubidium atomic gas at the center of a parabolic trap as a function of temperature. The total number of atoms was $N = 10^6$, and the trap oscillation frequency was $\omega/2\pi = 16$ Hz. (Adapted from Courteille et al. 2001.)

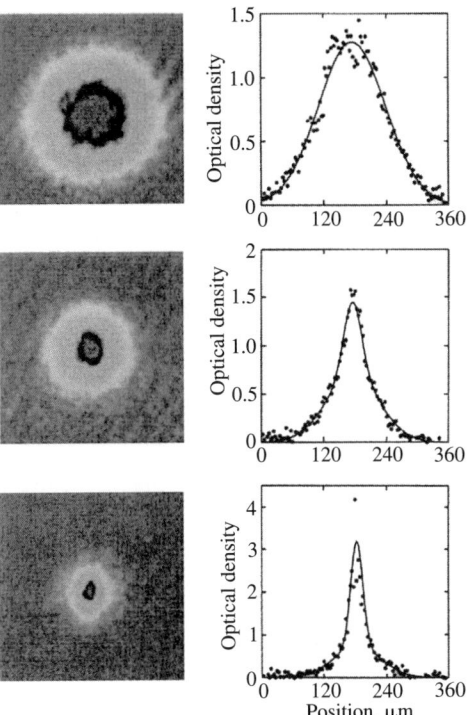

Fig. 8.4 Absorption pictures observed around the phase transition for a cloud of rubidium atoms. (Reprinted from Han *et al.* 1998 with courtesy and permission of the American Physical Society.)

manifestations of the BEC are a bimodal density distribution with an increased density at the center of the cloud, a sharp dependence on the frequency of the rf field, and an anisotropic shape of the central peak.

In an experiment by Bradley *et al.* (1995), a cloud of Li atoms was cooled in a permanent magnetic trap. Subsequent to evaporative cooling for 5 minutes, the cloud of some 10^5 Li atoms cooled down to a temperature of about 300 nK. In the experiment, about 1400 atoms were condensed. The phase transition to a BEC was detected by a near-resonant absorption imaging technique applied to the optically thick atomic cloud.

In an experiment by Davis *et al.* (1995), an MOT was loaded with sodium atoms. In the final stage, the atoms were stored in a quadrupole magnetic trap. To suppress Majorana spin transitions near the centre of the trap, a far-off-resonance laser beam was used to plug the hole. Subsequent to evaporative cooling, some 5×10^5 atoms cooled down to a temperature of around 2 μK at a density of 4×10^{14} cm^{-3}.

Following the successful experiments with alkali-metal atoms, a BEC was achieved in atomic hydrogen. In an experiment by Fried *et al.* (1998), hydrogen molecules were dissociated by a cryogenic discharge. The hydrogen atoms thus obtained were loaded into an Ioffe–Pritchard magnetic trap and stored in a cell whose walls were coated

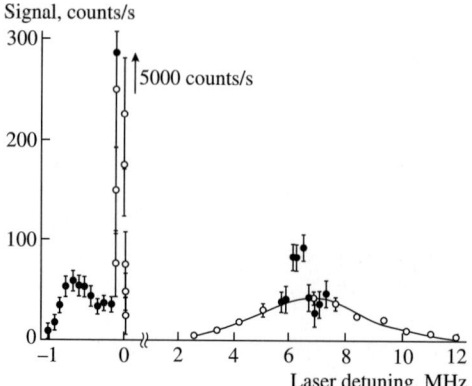

Fig. 8.5 Absorption spectrum in the two-photon 1S → 2S transition in atomic hydrogen. The appearance of the BEC fraction was followed via the wings in the Doppler-free and Doppler-sensitive peaks at negative detunings. (Reprinted with courtesy and permission of the American Physical Society from Fried *et al.* 1998.)

with liquid ^{4}He at a temperature of 120 mK. The atoms were cooled by collisions with the cold walls. Deep cooling of the atomic cloud was attained through rf evaporation. Bose–Einstein condensation was achieved at a temperature of 50 μK and a density of about 5×10^{15} cm^{-3}. Out of the total of 10^9 atoms, about 10% were condensed. The atomic cloud was probed by two-photon optical excitation of the atoms at the 1S → 2S transition. Fluorescence from the 2S state was observed by its decay via the 2P state. Basic information about the properties of the hydrogen BEC was obtained from the absorption spectrum shown in Fig. 8.5. The spectrum includes a broad Doppler line due to the absorption of photons from the same laser beam, and a narrow Doppler-free peak due to photon absorption from a counterpropagating laser beam.

The next important step was the realization of the Bose–Einstein condensation of ytterbium atoms. Bose–Einstein condensates of two-electron atoms such as ytterbium and alkaline-earth metal atoms are remarkable because of the capacity of such atoms for narrow intercombination transitions, the simple structure of their spinless ground state, and their metastable triplet states. The presence in such atoms of two coupled transitions, one allowed and the other forbidden, makes it possible to effect Doppler cooling first with the allowed and then with the forbidden transition. The Doppler-cooling limit eqn (5.3) for forbidden transitions lies in the submicrokelvin range. This makes it possible to reach the recoil limit temperature by the alternative method illustrated in Figs. 5.4 and 5.11. The first to be achieved was the laser cooling and trapping of Yb atoms (Honda *et al.* 1999), and after that the Bose–Einstein condensation of Yb atoms was realized by an all-optical method (Takasu *et al.* 2003). Following the evaporative cooling of high-density ^{174}Yb atoms by lowering the depth of a far-off-resonance dipole trap (FORT), the atomic cloud released from the trap was observed to undergo anisotropic expansion, which is characteristic of a BEC. The existence of abundant stable Yb isotopes (five bosons and two fermions) will allow one to study various interesting quantum-degenerate gases using Yb isotopic atoms and isotopic mixtures of them. The successful realization of the Bose–Einstein condensation of Yb atoms promises the production of BECs of other two-electron atoms.

8.2.3 The atom laser

The production of a BEC in an atomic gas can be considered as a realization of the atom laser. As in the conventional optical laser, which uses some pumping process to produce the active medium, in an atom laser coherent atomic matter—a Bose–Einstein condensate—is produced by pumping from a thermal atomic cloud cooled by evaporation. The storage of a BEC in an atom trap is also similar to the storage of a light field in a laser cavity. Again, similarly to the formation of a specific optical mode in a single-mode laser, in an atom laser the BEC is formed in the ground state of the trap. The release of the BEC from the trap is similar to the outcoupling of a light pulse from a conventional pulsed laser. The coherence length of an atom laser is approximately the same as its physical size.

One more similarity between the conventional laser and the atom laser can be understood in terms of stimulated processes. In the conventional laser, the photon field stimulates the excited atoms to emit additional photons into the laser mode. If the laser photon field contains n photons, the stimulated photon emission is proportional to $n + 1$. In the atom laser, a similar process of bosonic stimulation produces more atoms in the ground state of the trap, that is, in the condensed state. If the condensed matter already includes N atoms, the bosonic stimulation is proportional to a factor $N + 1$.

The basic difference between an atomic beam from a BEC (an atom laser) and an atomic beam from a thermal atomic oven is the same as the difference between a laser beam and a spectrally filtered and collimated light beam from a thermal light source. According to quantum theory (Glauber 1963), a laser field is a coherent state of light with minimal fluctuations in its amplitude and phase. In other words, the essence of an atomic beam from a BEC, as well as a laser beam, lies in its statistical properties.

Direct evidence of bosonic stimulation was obtained in experiments with sodium atoms evaporatively cooled to a temperature near the phase-transition value (Miesner et al. 1998). A cloud of about $2 \text{ s} \times 10^7$ atoms was cooled to 1.5 µK. By rapid variation of the frequency of the rf field used, a substantial number of hot atoms were removed from the atomic cloud to produce an oversaturated nonequilibrium cloud near the phase transition threshold. This resulted in the exponential growth of a BEC cloud within the thermal cloud, and the bosonic-stimulation effect thus became evident. The exponential growth of the BEC was clear evidence of the bosonic-stimulation process.

To operate an atom laser in a continuous-wave mode was not an easy problem, for this task could only be coped with by decreasing the losses and replenishing the condensate. This goal was achieved by Chikkatur et al. (2002), who produced condensates in two different vacuum chambers. The condensate periodically produced in one of the chambers was fed to the other chamber to merge with the condensate contained therein. This procedure provided a permanent reservoir of condensed atoms. The continuous release of the condensate from the former chamber thus provided for the continuous-wave operation of the atom laser.

It is worth noting that an atom laser can be realized not only with Bose–Einstein condensates stored in magnetic traps, but also with condensates stored in optical traps.

Such an all-optical atom laser has recently been successfully realized by Gennini et al. (2003). In that experiment, rubidium atoms were stored in a dipole trap produced by a single focused CO_2-laser beam. Using the evaporative-cooling technique to cool the atom cloud stored in the dipole trap, these authors produced a condensate of about 1.2×10^4 atoms distributed among the magnetic states $m_F = -1, 0, +1$. Thereupon, they applied an inhomogeneous magnetic field to remove from the condensate the atoms in the magnetic states with $m_F = -1, +1$. This allowed them to obtain a condensate of some 7000 atoms in the single magnetic state $m_F = 0$. By reducing the potential of the dipole trap, they finally obtained a well-collimated atom-laser beam.

The realization of Bose-Einstein condensation and the atom laser has opened up an opportunity to observe and use in experiments the coherent properties of atomic matter. Coherent atomic sources can be used for the purposes of atom interferometry and holography. A direct observation of interference in a BEC was made in an experiment by Andrews et al. (1997). A BEC cloud obtained from with 5×10^6 sodium atoms was cut by a blue-detuned laser beam into two parts. The two parts of the condensate were then released from the trap. In the course of their free expansion, the two parts of the BEC overlapped and produced an interference pattern, which was probed by absorption imaging. The interference fringes showed good contrast, thus pointing to the conservation of long-range order in the condensates.

8.3 Fermi-degenerate quantum atomic gases

The next important breakthrough in the physics of ultracold matter was made in an experiment on the production of an ultracold Fermi-degenerate atomic gas (DeMarco and Jin 1999). Fermions, such as electrons, photons, and neutrons, constitute the whole of the matter surrounding us. Fermi–Dirac statistics governs the structure of atoms, electrons in condensed matter, nuclei, etc. However, the particles in the matter accessible and inaccessible (astrophysical matter) to us have a high density and interact strongly with one another. Therefore, to obtain Fermi-degenerate matter in the form of a dilute quantum gas under controllable conditions seems very important in the physics of ultracold matter for many reasons. First, the physics of Fermi-degenerate gases is obviously richer exactly because of the interaction between the particles, which can be both repulsive and attractive. Secondly, two Fermi particles of half-integer spin can be made to form one Bose particle of integer spin, whereas the opposite is impossible. This forms the basis of such remarkable macrophysical quantum phenomena as superfluidity and superconductivity. The energy distribution of the Fermi particles obeys Fermi–Dirac statistics:

$$f(E) = \frac{1}{e^{(E-E_F)/k_B T} + 1}, \qquad (8.7)$$

where E_F is the Fermi energy, that is, the energy of the highest filled state. Pauli's exclusion principle allows Fermi particles to occupy a given quantum state only singly (Fig. 8.2(b)). This has a strong impact on the behavior of fermions in nature, including dilute quantum gases.

To obtain and observe an ultracold Fermi-degenerate gas proved a more difficult task than obtaining a BEC. This was due, first, to the fact that fermions do not undergo any sudden phase transition under ultralow-temperature conditions. Second, ultracold fermionic atoms have "strange" collisional properties because of Pauli's principle. In particular, two fermionic atoms in the same quantum state are not subject to the elastic collisions that are necessary to reach submicrokelvin temperatures by way of evaporative cooling. Therefore, it took the ingenious idea of using a two-component mixture of fermions in two spin states (and in subsequent work, a mixture of two different atoms or isotopes) in order to use the strategy of successive laser and evaporative-cooling with a view to reaching temperatures $T < T_F$, where $T_F = E_F/k_B$ is the Fermi temperature (typically below 1 μK).

In the first successful experiment on the observation of Fermi–Dirac degeneracy in an ultracold trapped atomic gas, the fermionic atoms, ^{40}K, were laser precooled, magnetically confined, and cooled by evaporation to temperatures T below 300 nK. The gas thus fell into the region of temperatures below the Fermi temperature T_F at which quantum-statistical effects became important. These effects are manifested in the momentum distribution and the total energy of the ultracold fermionic gas. Figure 8.6 shows the emergence of quantum degeneracy, as manifest in the energy of the trapped fermionic gas. A momentum analysis was used to find the energy of the gas from time-of-flight measurements of atoms released from the trap. The relative excess energy $\delta U/U_{cl} = U - U_{cl}$ is plotted against T/T_F, where U is the measured energy and $U_{cl} = 3Nk_BT$ is the energy of a classical gas with a total number N of trapped atoms at the same temperature. The excess energy measured at low T/T_F agrees well with the thermodynamic theory for a noninteracting Fermi-degenerate gas (the solid line in Fig. 8.6).

A Fermi-degenerate dilute gas is a fountain of quantum phenomena characteristic exactly of Fermi gases. Specifically, there can be observed the phenomenon known as "Pauli blocking." Pauli blocking is a consequence of the fact that identical fermions cannot occupy the same quantum state. Under quantum-degenerate conditions, all the lower quantum states are occupied. This limits the possibility for the interacting

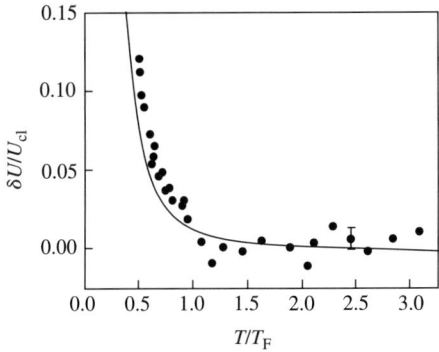

Fig. 8.6 Emergence of quantum degeneracy, as seen in the energy of a trapped Fermi gas (Reprinted with courtesy and permission of AAAS (USA) from DeMarco and Jin 1999.)

fermionic atoms to change their energy and leads to the "collective excitation" of the trapped gas (De Marco et al. 2001).

A Fermi-degenerate gas can coexist with a Bose–Einstein condensate. This was observed, for example, in experiments with an ultracold atomic mixture of ^{7}Li atoms (bosons) and ^{6}Li atoms (fermions) at a temperature of $T = 0.28\,\mu\text{K}$ and $T_\text{F} = 0.2T_\text{c}$, where T_F and T_c are the Fermi and the Bose–Einstein condensation critical temperature, respectively (Schreck et al. 2001).

The possibility of pairing fermionic atoms (Cooper pairing) to form bosonic atoms has far-reaching consequences, as far as studies into such phenomena as superfluidity and superconductivity under the controlled conditions of quantum degenerate gases are concerned (see Section 8.5).

8.4 Formation of ultracold molecules

Laser cooling of atoms, based on cyclic atom–light interaction, proved impossible to extend to molecules because of the huge number of vibrational–rotational states and, accordingly, radiative decay pathways that they have in their electronically excited states. But another approach was found and proved a success, namely, the photoassociation of translationally laser precooled atoms. This idea and the theory of the process were introduced by Thorsheim et al. (1987). At present, this process is the main method for producing ultracold molecules in traps. It is being widely used to obtain and investigate quantum molecular gases and also finds application in the spectroscopy of molecules near the dissociation limit (see the reviews by Lett et al. 1995, Bahns et al. 2000, and Masnou-Seeuws and Pillet 2001).

8.4.1 Laser-induced photoassociation of atoms

Photoassociation is process of formation of a molecule that is induced by laser light in the course of collision between two atoms. This process is the opposite of the more well-known photodissociation process, whereby a molecule under the effect of light separates into two fragments. The photoassociation process for the K_2 molecule is illustrated in Fig. 8.7. Free ground-state atoms collide under the action of the molecular potential for the ground-state atoms, which varies little at long distances, namely $V = -c_6/R^6$, where R is the internuclear separation. At room temperature the energy of the colliding atoms falls within the wide range $k_\text{B}T \simeq 200 \text{ cm}^{-1}$, that is, the initial state is smeared. At the same time, the upper excited electronic–vibrational state has very narrow radiative and Doppler widths. For this reason, the process of laser-induced photoassociation is extremely ineffective at room temperature. But for colliding ultracold atoms, the energy range $k_\text{B}T$ of the initial state is very small, and so all the colliding atoms can participate in photoassociation. If they absorb a photon while at a great distance from each other, as illustrated in Fig. 8.7, they can drift apart without forming a bound state. However, if in the course of their oscillation they fall within an interval R where the Franck–Condon factor for the radiative transition to the ground state is large, they then form, after spontaneously emitting a photon, a bound vibrationally excited state. By using laser radiation of appropriate frequency, one can excite a molecular state below the dissociation limit, that is, one can at once produce a bound electronically excited molecular state.

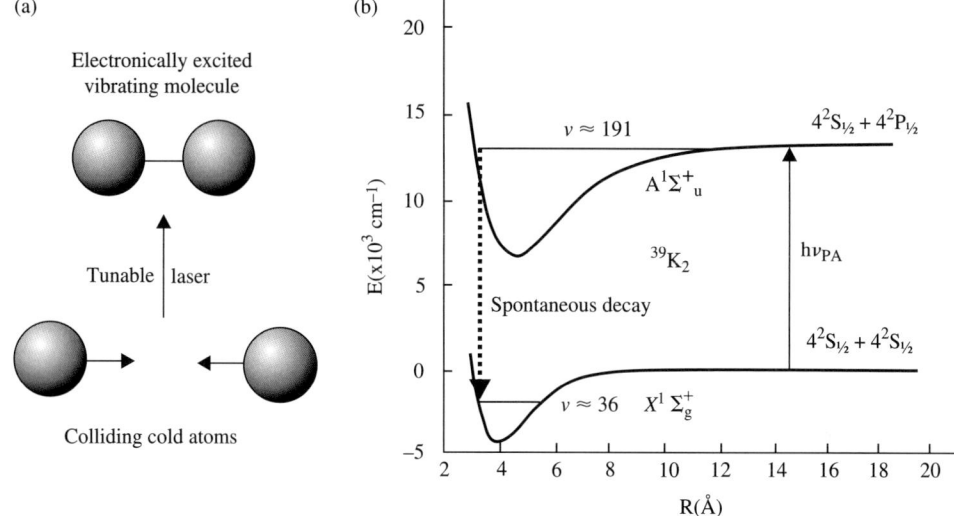

Fig. 8.7 Laser-induced photoassociation: (a) resonant excitation of the colliding cold atoms to an electronically excited state of a molecule; (b) single-photon photoassociation of ultracold ^{39}K$_2$ atoms yields translationally ultracold molecules in vibrationally excited levels of the electronic ground state. (Reprinted with courtesy and permission of the American Physical Society from Nikolov et al. 1999.)

An experiment by the scheme illustrated in Fig. 8.7 was carried out with ^{39}K$_2$ molecules by Nikolov et al. (1999) using a magnetooptical trap. In this experiment, the trapped laser-cooled potassium atoms were photoassociated to $v = 191$ of the $A^1\Sigma_n^+$ state:

$$K + K + h\nu_{PA} \rightarrow K_2^*(A^1\Sigma_u^+, v = 191). \tag{8.8}$$

The spontaneous decay of the excited state proceeds by two channels, one leading to a bound state and the other to a free state:

$$K_2^*(A^1\Sigma_u^+, v = 191) \diagdown^{K_2(X^1\Sigma_g^+) + \hbar\nu_{sp}}_{K + K}. \tag{8.9}$$

Most decays are bound–free transitions, but a small proportion (0.15%) are bound–bound ones, giving rise to a distribution of highly vibrationally excited levels, with Franck–Condon factors suggesting that the most populated vibrational state is $v = 36$.

The ultimate goal in the laser control of the photoassociation process is the production of molecules in the vibrational–rotational ground state ($v'' = 0$, $J'' = 0$). In this case, collisions between the trapped ultracold molecules produced through photoassociation will not result in the release of vibrational–rotational energy in vibrational–translational (V–T) or rotational–translational (R–T) energy transfer processes, and hence there will be no translational heating of the molecules and no escape of them from a trap whose optical trapping potential is not very high. To attain this goal, Band and Julienne (1995) suggested the two-color photoassociation scheme presented in Fig. 8.8(a). The first stage in this scheme involves the photoassociation of free

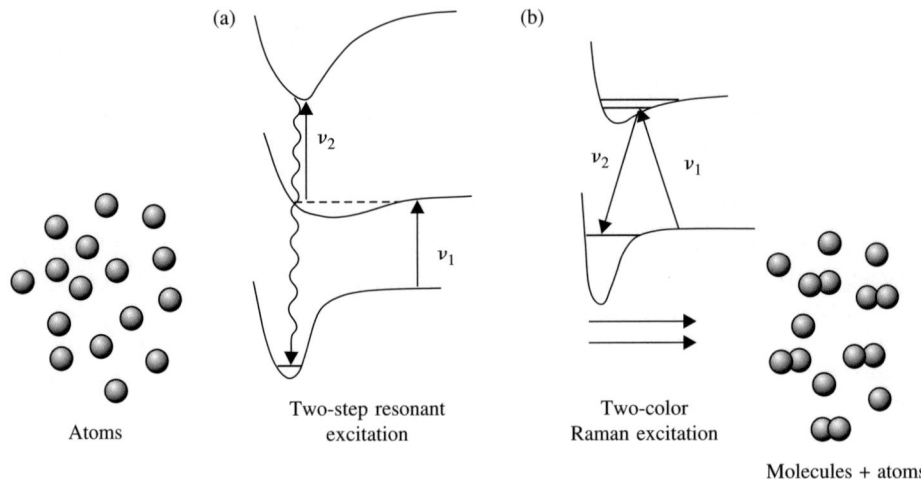

Fig. 8.8 Illustration of the two-color formation of ultracold molecules in the electronic ground state by way of (a) two-step resonant excitation and (b) Raman photoassociation.

atoms to produce molecules in a high-lying vibrational state with a relatively long radiative lifetime, which allows them to vibrate many times before undergoing spontaneous decay. The second stage is the absorption by the molecules of a laser photon $\hbar v_2$. The spontaneous decay of the resultant doubly excited molecules to their ground state is a bound–bound transition to $v'' = 0$ with a 10% emission probability. That can be realized by Raman excitation to the ground electronic state of the molecules (Fig. 8.8(b)).

The most interesting implementations and applications of laser-induced photoassociation of ultracold atoms have emerged in experiments with quantum gases (BECs and Fermi-degenerate gases). These experiments made it possible to obtain and investigate molecular quantum gases. They are briefly discussed in Section 8.5.

The photoassociation of ultracold atoms in a trap by means of a tunable laser makes possible the spectroscopy of the energy states of the molecules formed near the dissociation limit, with a very high spectral resolution. To this end, it is necessary to detect the electronically excited molecules being formed. Photoionization of the excited molecules with an additional laser has become the standard technique. Such detection methods were developed earlier for atoms and molecules at room temperatures (Letokhov 1987) (see Chapters 9 and 10), but thanks to their exceptional detection sensitivity (up to single atoms and molecules), they have proved fairly effective in experiments with a small number of excited ultracold molecules confined in traps. Photoassociation spectroscopy uses two approaches: (1) spectroscopy of the excited molecular states near the dissociation limit, with variation of the radiation frequency of the first laser, used for the purpose of photoassociation, and (2) spectroscopy of the ionized bound or free states, with variation of the radiation frequency of the second laser, used for the photoionization of the excited molecules.

In short, photoassociation spectroscopy of cold trapped atoms is a very powerful tool for exploring molecular energy levels heretofore inaccessible to conventional spectroscopy. This technique owes its advent to the development of two effective laser methods for controlling atoms and molecules, namely, laser-induced photoassociation and laser-induced photoionization.

8.4.2 External-field-controlled Feshbach resonances

The formation of an ultracold molecule upon collision between two ultracold atoms is extremely sensitive to the details of the atomic interaction potential, which is, in turn very sensitive to the internal (spin) states of the colliding atoms. The interaction of the colliding particles can be of a resonant character. This effect, in nuclear physics, was discovered by Feshbach. It proved very important in collisions between ultracold atoms because of its sensitivity to a magnetic field, which has an effect on the Zeeman shift of the resultant bound molecular state, and hence on the resonance with the scattering state. The possibility of the existence of magnetically tunable resonances in ultracold collisions was theoretically suggested for collisions between two H atoms (Stwalley 1976) and two Cs atoms (Tiesinga et al. 1993) and was observed in ^{23}Na (Inoye et al. 1998) and in ^{85}Rb Bose–Einstein condensates (Cornish et al. 2000).

The physics of the Feshbach resonance in the formation of ultracold molecules is illustrated in Fig. 8.9. Figure 8.9(a) presents the molecular potentials involved in the Feshbach resonance. The colliding atoms reside on the flat long-range part of the potential A. The energy of the excited level of the potential B is close to the energy of the colliding atoms, the energy difference being ΔE. However, the dissociation limit of the potential B is situated higher than that of the potential A. Therefore, if a weakly bound molecule is made to transit from the potential A to the potential B, it will find itself in a well-bound state. To this end, it is necessary to make the energy difference ΔE tend to zero ($\Delta E \to 0$), which will lead to a resonance-enhanced population of the bound state, and it is exactly this phenomenon that is referred to as the Feshbach resonance. In order to tune the system to resonance, it is necessary to shift the energies of the nearest levels (the dissociation limit of the channel A and the bound level of the molecular potential B) with respect to each other. Since the magnetic moments of these levels are generally different, this can be done by applying

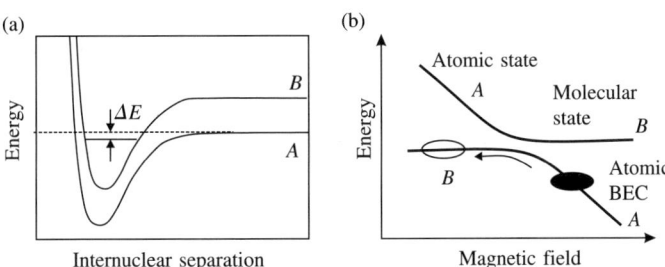

Fig. 8.9 Illustration of the physics of the magnetic Feshbach resonance: (a) molecular potentials involved in a Feshbach resonance; (b) formation of ultracold molecules B from ultracold atoms A using a magnetically tunable Feshbach resonance.

an external magnetic field to the system. In the vicinity of the Feshbach resonance, the elastic and inelastic scattering lengths change drastically. The Feshbach resonance was experimentally observed in a variety of alkali-metal atoms (Inoye et al. 1998; Courteille et al. 1998) and became a powerful method to control the formation of ultracold molecules from ultracold atoms, Bose–Einstein condensates in particular.

Since the magnitude and sign of the scattering length a are very important for the Bose–Einstein condensation of atomic gases, the possibility of controlling these parameters with the aid of an external laser field is of great interest. The idea of such near-resonance laser control of elastic atomic collisions was put forward by Fedichev et al. (1996). It is based on the following effect. A pair of atoms absorbs a photon and undergoes a transition to an electronically excited transient quasi-molecular state. After reemitting the photon, the atoms return to the initial electronic state at the same kinetic energy. As the interaction between the atoms in the excited state is much stronger than in the ground state, an external laser field of moderate intensity can help to significantly change the magnitude of the scattering length. The possibility of implementing laser control of the scattering length was demonstrated in experiments with ^{86}Rb atoms (Theis et al. 2004). In these experiments, a laser-induced Feshbach resonance was used to control the scattering length by varying the laser frequency and intensity. Figure 8.10 illustrates the variation of the scattering length a and the inelastic-collision rate coefficient K_{inel} as a function of the laser frequency detuning from the photoassociation resonance for typical parameters of the above experiments. This idea and its experimental implementation illustrate the great possibilities for the laser control of collisions between ultracold atoms and the formation of ultracold molecules.

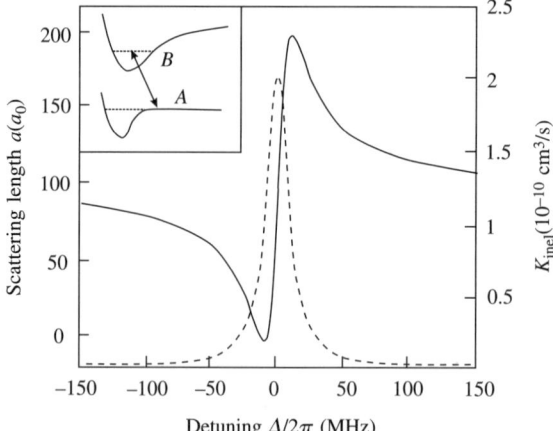

Fig. 8.10 Laser-induced Feshbach resonance: scattering length a (in terms of the Bohr radius a_0) (solid line) and inelastic-collision rate coefficient K_{inel} (dashed line) as a function of the detuning of the laser from the photoassociation resonance for typical parameters of ^{87}Rb. *Inset*: schematic diagram of the optical coupling of the scattering unbound state A with the excited bound state B. (Reprinted from Theis et al. 2004 with courtesy and permission of the American Physical Society.)

8.5 Molecular quantum gases

The production of laser-cooled ultracold molecular quantum gases was always considered a difficult and remote task, it being very difficult to effect multiple, cyclic interaction between molecules and resonant laser light. Nevertheless, a way out was found indirectly, by preparing ultracold molecules by way of making laser-cooled ultracold atoms associate using the Feshbach resonances in the interacting atoms. What is remarkable is that molecular quantum gases were obtained exactly in this way from both BECs and Fermi-degenerate gases.

8.5.1 Formation from a BEC of an atom gases

From the general quantum-statistical standpoint, Bose–Einstein condensation can be achieved with both atoms and molecules. However the realization of BEC with a molecular gas is a much more formidable task because of the difficulties involved in the deep cooling of molecules.

Nevertheless, this problem has been solved for gases of laser-cooled cesium molecules (Herbig *et al.* 2003) and ^{6}Li$_2$ molecules (Zwierlein *et al.* 2003). Both these experiments used the same strategy for obtaining a molecular Bose–Einstein condensate: the production of molecules from ultracold atoms in the presence of a magnetic field near a Feshbach resonance (Tiesinga *et al.* 1993; Inoye *et al.* 1998), where molecules can form without any heat release. Figure 8.9 illustrates the method of production of cold molecules using Feshbach resonances used in those experiments.

In the experiment by Herbig *et al.* (2003), the starting point was a BEC of about 5×10^4 Cs atoms in the hyperfine ground state with a total angular momentum $F = 3$ and a magnetic quantum number $m_F = 3$, trapped in an optical dipole trap. This state has a Feshbach resonance near 20 G, with an estimated width of 5 mG. In the experiment, about 3000 Cs$_2$ molecules were produced from the BEC by sweeping the magnetic field across the resonance from a higher field value at a constant rate of typically 50 G/s. The magnetic moment of the molecular state was determined to be 0.93 Bohr magnetons, whereas that of two atoms in the $(F = 3, m_F = 3)$ state is 1.5 Bohr magnetons. A measurement of the expansion energies of the molecular cloud gave energies of a few nK, which is evidence of a macroscopic molecular matter wave.

In the experiment by Zwierlein *et al.* (2003), lithium atoms were sympathetically cooled with sodium atoms in a magnetic trap and then transferred into an optical trap produced by a single far-off-resonance laser beam. Thereupon the lithium atoms were made to move to the lowest-lying hyperfine state by means of an adiabatic frequency sweep of an rf field with a frequency close to that of a hyperfine transition in the ^{6}Li atom. In the presence of a constant magnetic field, the cold lithium atoms formed weakly bound molecular states. As a result, the trap contained a mixture of lithium atoms and molecules. The lithium molecules had twice as high an atomic polarizability as the atoms, and so they experienced twice as deep a trap potential. The mixture of atoms and molecules was cooled by reducing the laser power and, accordingly, the depth of the trap potential. Thanks to the difference between the trap potential depths for atoms and molecules, it was mostly the atoms that were evaporated. This finally yielded a molecular gas at a temperature around 1 microkelvin. The molecules were detected by dissociating them by sweeping the magnetic field around the Feshbach resonance and then probing the resultant ballistically expanding atomic cloud. The onset of Bose–Einstein condensation was observed as the appearance of a bimodal spatial distribution of the expanded atomic gas.

8.5.2 Formation from Fermi quantum gases

The production of molecular Bose–Einstein condensates from atomic BECs, especially with the use of Feshbach magnetic resonances, made it possible to develop fine methods for controlling colliding ultracold bosonic atoms. This experience proved very valuable for the production of ultracold molecules and molecular BECs from the Fermi quantum atomic gases. It took only a few years for the "terra incognita" of the physics of ultracold atoms to turn into a flourishing field of progressing discoveries at the interface of atomic physics, laser physics, and "condensed" -matter physics.

The first step was the production of ultracold ^{40}K$_2$ molecules from a quantum-degenerate Fermi gas of ^{40}K atoms at a temperature below 150 nK (Regal et al. 2003). The low binding energy of the molecules was controlled by detuning the magnetic field away from the Feshbach resonance. Clear evidence of diatomic molecules was achieved through direct, radio-frequency spectroscopic detection of ^{40}K$_2$ molecules.

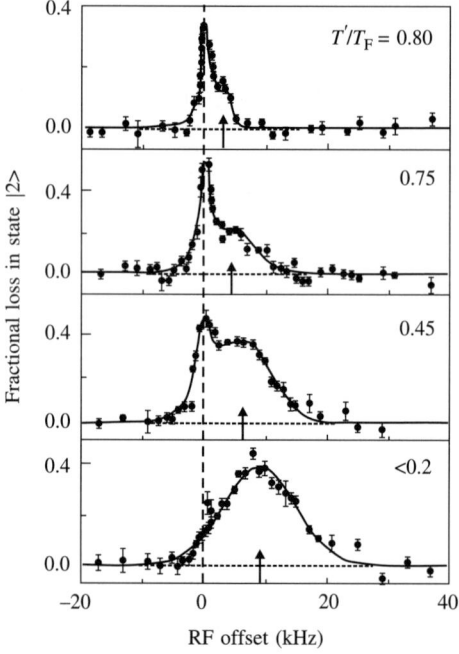

Fig. 8.11 Evidence of a pairing gap in a strongly interacting Fermi gas: radio-frequency excitation spectra of a pairing ultracold two-component gas of ^{6}Li at various temperatures T ($T_F = 2.5$ μK). The vertical dotted line marks the atomic transition, and the arrows indicate the effective pairing gap $\Delta\nu$. (Reprinted with courtesy and permission of AAAS (USA) from Chin et al. 2004.)

The next step was the conversion of an atomic Fermi gas into a long-lived molecular Bose–Einstein gas. This was achieved by several groups of investigators working with ^{6}Li (Strecker *et al.* 2003; Jochim *et al.* 2003) and ^{40}K (Greiner *et al.* 2003) atoms. This step required great experimental skill, for the ultracold molecules obtained were vibrationally excited and had a short lifetime (around 1 ms). Strecker and coworkers used a mixture of some 10^{10} bosons (^{7}Li) and 10^9 fermions (^{6}Li) in a magnetooptical trap, both species being cooled by evaporation. This "dual evaporation" scheme proved very efficient, resulting in a hundredfold increase in the number of trapped ^{6}Li atoms and an ultimate lowering of their relative temperature down to $T = 0.1 T_\mathrm{r} \simeq 1.4$ μK. Molecular condensates produced from the ^{6}Li fermionic atoms exhibited a remarkable stability (seconds) with respect to collisional decay.

The principal goal of the experiments on the production of molecular BECs from fermionic gases was to observe fermionic superfluidity due owing to the famous Cooper-pairing effect, as well as to gain an insight into the mechanisms responsible for superconductivity and superfluidity. This goal was attained in experiments with ultracold ^{6}Li$_2$ molecules under BEC conditions (Chin *et al.* 2004). These researchers studied fermionic pairing in an ultracold two-component mixture (a spin mixture of the two lowermost substates of the electronic 1s^{2}2s ground state), employing the methods of laser cooling and subsequent evaporative cooling and observing energy gap in the radio-frequency excitation spectra. The key results of this experiment are presented in Fig. 8.11. The emergence of a gap with decreasing temperature is clearly visible in the radio-frequency spectra. The marked increase of Δv, the rf offset, with decreasing temperature is in good agreement with theoretical expectations for the pairing-gap energy. The appearance of an energy gap under conditions of moderate evaporative cooling demonstrates that the full evaporation in this experiment brought the strongly resonantly interacting Fermi gas deep into the superfluid state.

It seems that these seminal experiments have marked only the very beginning of the progress of ultracold-matter physics, so that one can expect other remarkable discoveries to be made while this book is being published.

9
Laser photoselective ionization of atoms

In this and subsequent chapters, we consider are other laser methods for controlling atoms and molecules that, in contrast to all the methods discussed above in Chapters 2–8, are *destructive*. They result in the ionization of the atoms or molecules being controlled and the dissociation and fragmentation of molecules. All these methods are united by the capability of the particles to be ionized or dissociated in a photoselective manner, which makes it possible to detect and separate atoms or molecules of certain species. They are based on the fact that all the structural details of atoms and molecules manifest themselves in their spectra, for example, some details manifest in the spectra of atoms are the atomic weight (the number of nucleons), the number of protons (the nuclear charge and the type of element), the number of neutrons (the isotopic state), and even the excitation energy of the nucleus (the isomeric composition). This enables one to excite, and, in some cases, ionize atoms and molecules in a photoselective way and thus accomplish their photoselective excitation and separation. Figure 9.1 lists all types of photoselectivity. All of them have been successfully realized by the method of resonance multistep ionization of atoms that is considered in the present chapter. Our primary attention here is concentrated on the realization of laser control of the photoselectivity of ionization. A detailed description of this method can be found in the monographs by Letokhov (1987) and Hurst and Payne (1988).

9.1 Introduction

The resonance photoionization of atoms came into being in the early 1970s in the course of the development of new methods for exerting selective influence on atoms and molecules by means of laser radiation. Two-step resonance atomic photoionization was proposed to be used for the purposes of separating isotopic atoms and detecting trace elements (Letokhov 1969). The first experiments on the two-step resonance photoionization of Rb atoms were performed at the Institute of Spectroscopy of the USSR Academy of Sciences as far back as 1971 (Ambartzumian et al. 1971; Ambartzumian and Letokhov 1972). The radiation from a tunable dye laser with a wavelength of $\lambda_1 = 7950$Å excited Rb atoms into the $5p^2P_{1/2}$ state. Photoionization of the excited atoms was effected by means of second-harmonic output pulses from the ruby laser used to pump the dye laser. The energy of the second-harmonic quantum was high enough to bring about the photoionization of the excited Rb atoms ($E_i - \hbar\omega_1 = 2.62$ eV) but insufficient for the photoionization of the ground-state

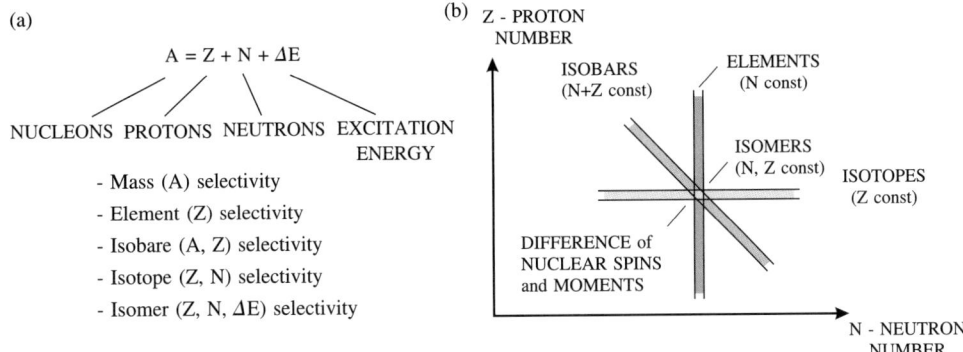

Fig. 9.1 (a) Photoselectivity of laser detection and separation of atoms. (b) Types of selectivity plot of proton number versus neutron number for atoms and nuclei on a.

atoms ($E_i = 4.18\,\text{eV}$). The laser pulse parameters made it possible to attain saturation of the resonance transition and about a 0.1% probability of ionization of the excited atoms. In later years, several two-step photoionization experiments were performed with various atoms. The ionization yield in these experiments was very low because of the impossibility of saturating the transition to the continuum. To ease the task of realization of the absorption saturation condition in the ionization stage, it was suggested that the atoms should first be excited in several steps to move them to high-lying discrete levels in the vicinity of the ionization threshold, and then they should be ionized by means of an electric field (Ivanov and Letokhov 1975). Appropriate experiments on the two-step excitation of Rydberg states in Na atoms, followed by their ionization by an electric field, were carried out in 1975 (Ambartzumian et al. 1975a; Ducas et al. 1975).

In those years investigators' attention was centered primarily, with a few exceptions, on the use of selective ionization precisely for isotope separation purposes. Today we know that the resonance multistep ionization technique is the most sensitive method for studying the isotopic and hyperfine structures of atoms with short-lived nuclei available in very small quantities, for laser on-line separation of nuclear isomers, and for detection of traces of some elements.

9.2 Resonance excitation and ionization of atoms

Numerous schemes can be used to carry into effect the resonance ionization of atoms. A feature common to all such schemes is the preliminary selective laser excitation of one or several intermediate atomic levels, and subsequent ionization of the excited atoms alone. Resonance excitation has been dealt with in the preceding chapters. The present chapter discusses the next step—the ionization of the excited atoms. The molecular case differs substantially from the atomic one and will be considered in the next chapter.

Qualitatively, the basic photoionization schemes may be classified as illustrated in Fig. 9.2. Excited states that are far from the ionization limit can, effectively, be ionized by laser radiation only. The following two possibilities exist in this case: direct

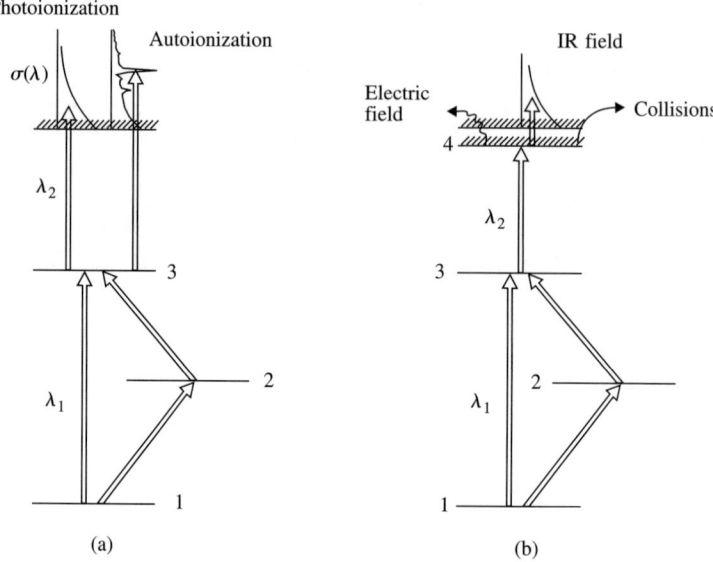

Fig. 9.2 Various ionization schemes for atoms via (a) low-lying excited states and (b) highly excited states.

nonresonance photoionization using a transition to the continuum, and resonance photoionization using a transition to an *autoionization state* (Fig. 9.2(a)). Highly excited (Rydberg) states lying close to the ionization limit are comparatively easy to ionize with a high efficiency by one of the following means: an electric-field pulse, an IR radiation pulse, or collisions with other particles or with walls (Fig. 9.2(b)). The ionization schemes making use of Rydberg and autoionization states apparently involve a greater total number of resonant excitation steps between the ground state and the ionized state. All these schemes have already been tested experimentally, particularly for the isotope-selective and isomer-selective ionization of atoms. Before going into the details of atomic multistep photoionization schemes, let us consider some characteristics of these schemes from the viewpoint of the realization of their ultimate value. These characteristics include the required energy fluence of the laser pulses, the ionization yield, and the ionization selectivity.

In accordance with eqn (2.74), a minimum *energy fluence* at the ionization stage is required in the schemes that make use of those resonance transitions from the excited state upward for which the ionization cross section is a maximum. With the proper choice of the sequence of transitions and intermediate quantum states, it is possible to attain a situation where all of the cross sections of the successive transitions lie in the region $\sigma_1, \sigma_2, \ldots > 10^{-15}$ cm^2 and the critical energy fluences of the pulses are rather low: $\Phi_1, \Phi_2, \ldots < 10^{-5} - 10^{-4}$ J/cm^2 or $< 10^{15} - 10^{16}$ photons/cm^2. The direct photoionization of the excited atoms to the continuum has a very small cross section and, from that standpoint, is far from optimal.

The *photoionization yield* reaches a maximum when all the successive transitions are simultaneously saturated by several laser pulses. If the final excited state of a multistep process decays to the continuum during the laser pulse, it is possible to completely deplete the ground level through excitation and thus achieve an ionization yield of 100%. Such conditions can be realized for all of the schemes illustrated in Fig. 9.2, except for the last one, where the highly excited state is ionized by an electric field. If the electric field is switched on when the laser pulses are over, only the atoms in the last excited state will be 100% ionized. High saturation of all the successive quantum transitions with laser pulses equalizes the populations of all the atomic levels involved (initial, intermediate, and final). Only a fraction of the atoms reach the final state and can be ionized by the electric field. It should be noted that the use of a continuous electric field to deplete the final state is hindered by the strong Stark effect for highly excited states.

9.2.1 Resonance excitation: noncoherence versus coherence

To realize effective multistep excitation of an atom by multiple-frequency radiation to the final bound state, the laser pulse duration must be shorter than the population relaxation times of the intermediate quantum states. In addition the laser pulse fluence $\Phi(\omega_{kn})$ (photons/cm^2) at the transition frequency ω_{kn} must ensure absorption saturation, that is, the following condition similar to eqn (2.74) must be satisfied:

$$\Phi(\omega_{kn}) \gtrsim \Phi_{\text{sat}}^{kn} = \frac{1}{\sigma_{kn}}, \qquad (9.1)$$

where σ_{kn} is the cross section for the stimulated transition $k \to n$ at the frequency ω_{kn}. The multiple-frequency radiation will saturate the entire quantum transition sequence, while the excitation energy remains within the quantum system being excited. Using rate equations, one can easily find the distribution of the populations n_k of the excited levels for specified $\Phi(\omega_{kn})$ and σ_{kn}. This is obvious for the case where the saturation of each resonance quantum transition is so strong that the probability of population of the final state f reaches its maximum:

$$\frac{n_i}{g_i} = \cdots = \frac{n_k}{g_k} = \cdots = \frac{n_f}{g_f}, \qquad (9.2)$$

where g_k is the statistical weight, or degeneracy, of the kth level and the light is naturally polarized. The partial population of the level f will then be defined by the simple relation

$$n_f = \frac{g_f}{\sum_k g_k}. \qquad (9.3)$$

As far as the maximum population of the final state is concerned, it is advantageous to excite a sequence of quantum levels such that its level degeneracies increase progressively. For instance, in the case of a three-step atomic excitation, we can select a sequence of dipole quantum transitions with progressively increasing orbital moments (S → P → D → F), for which the level degeneracies grow as follows: $g_1 = 1$, $g_2 = 3$, $g_3 = 5$, $g_4 = 7$. In this case, the partial population of the final level will be equal to 7/16, or 44%.

It frequently proves possible to effect multistep excitation where the final state suffers decay to the ionization continuum concurrently with excitation. This is, for example, the case with systems in which the final state f either is an autoionization state or lies in the continuum, so that the last quantum transition $n \to f$ is a "level" → "continuum" transition. In this case, if the ionization decay rate is high enough, that is,

$$W_{\text{ion}} \gtrsim \frac{1}{\tau_p}, \qquad (9.4)$$

there occurs depletion of the final state f and, by virtue of the saturation condition (eqn 9.1), of the initial and intermediate levels as well. As a result, a 100% ionization of the multilevel quantum system is ensured.

The physics of the multistep excitation of atoms is, of course, much more profound than may be inferred from the simplest qualitative considerations given above. The whole picture can be found in the two-volume comprehensive monograph by Shore (1990). The approximation of incoherent interaction between a laser field and a real multilevel atom, described by rate equations, is quite acceptable (Ackerhalt and Eberly 1976), especially if account is taken of the degeneracy in the magnetic sublevels m_F. Resonance transitions to various m_F values differ in the projection of the dipole moment d_{12}, and hence in the Rabi oscillation frequency (eqn 2.44). This generally smooths out oscillations and makes the interaction incoherent. It is only in the ideal case of a two-level system free from level degeneracy that one can observe Rabi oscillations, as illustrated in Fig. 9.3.

If it is necessary to transfer 100% of the atoms from their initial state to an excited state and further, one can use the adiabatic-passage technique. This method is based on the use of laser pulses of variable frequency $\omega(t)$ slowly passing back and forth through resonance. If the condition

$$\left(\frac{d_{12}\mathcal{E}}{\hbar}\right)^2 \gg \left|\frac{d\omega(t)}{dt}\right| \qquad (9.5)$$

is satisfied, where d_{12} is the resonance excitation dipole moment and $\mathcal{E}$ is the light field amplitude, a 100% excitation of the atom will be reached, as illustrated in Fig. 9.3. This

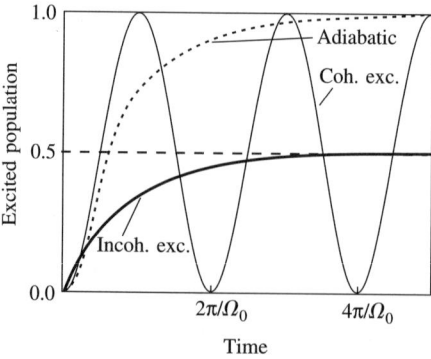

Fig. 9.3 Evolution of the population of the upper level in a two-level system, driven by a coherent radiation field (thin line), by an incoherent radiation field (heavy line), and by an adiabatic-passage process (dashed line). The time scale is in units of the Rabi oscillation period.

method has been extended to a three-level system by using two laser pulses of variable frequency, not passing through exact resonances with both quantum transitions, but somewhat off-resonance with respect to the intermediate level. This method has been called the technique of stimulated Raman adiabatic passage (STIRAP) (Bergmann et al. 1998). What is important is that this method is capable of exciting atoms from all the magnetic sublevels m_F.

9.2.2 Autoionization resonances from excited states

The ionization cross section (typically $\sigma_{2i} \simeq 10^{-17}$–$10^{-18}$ cm^2) is several orders of magnitude smaller than the resonance excitation cross section ($\sigma_{12} = 10^{-11}$–10^{-13} cm^2). Hence it follows that to satisfy the condition in eqn (9.1) for the effective photoionization of the excited state, the energy fluence Φ_{2i} should be around 0.1–1 J/cm^2, that is, 10^5–10^6 times higher than that necessary for resonance excitation. Therefore, efficient resonance multistep photoionization requires higher-efficiency ionization pathways. The autoionization of excited atoms is one such way. Autoionization states are bound atomic states whose energy levels relative to the outer valence electron lie above the ionization limit. There are two ways for the decay of autoionization states to occur: radiative decay and autoionization, that is, nonradiative decay with loss of electrons. The autoionization channel reduces the lifetime of the discrete state and causes broadening of it. Since autoionization can be considered to be a result of superposition of continuous and discrete states, the decay probability increases with increasing coupling between the autoionization state and the continuum. The autoionization resonances that occur when atoms are excited from their ground state into the continuum have been studied both theoretically and experimentally. The theory of such autoionization resonances was developed by Fano (Fano 1961). The presence of a discrete level in the continuum causes interference; the shape of the autoionization resonance becomes asymmetric, and an absorption minimum occurs in the low-frequency wing of the resonance (this is called the Fano profile).

The development of laser spectroscopy methods, particularly the laser multistep photoionization of atoms, has allowed systematic studies to be made of the autoionization states of multielectron atoms whose bound states have a complicated spectrum. Having a choice of various exciting transitions makes it possible to study autoionization states differing in parity. However, atomic absorption spectra can help one study only those autoionization states which combine with the ground state, that is, the states whose parity is opposite to that of the ground state and whose total momentum obeys the selection rules for the corresponding optical transition. Comparatively narrow autoionization states are observed in such noble gases as Ar, Kr, and Xe ($\Gamma_\mathrm{a.i.} \simeq 10$ cm^{-1}), in alkaline-earth elements, such as Ba ($\Gamma_\mathrm{a.i.} \simeq 10$ cm^{-1}), in uranium, and in some lanthanides ($\Gamma_\mathrm{a.i.} \simeq 1$–$2$ cm^{-1}). The widths of these autoionization states can vary by an order of magnitude or two and, at the same time, they tend to decrease for heavier atoms.

Laser spectroscopy methods substantially widen the scope of research into very narrow autoionization states. This can be illustrated by the long-lived autoionization state of the Gd atom (Bekov et al. 1978). Its lifetime is about 0.5 ns, and $\Gamma_\mathrm{a.i.} \simeq 0.05$ cm^{-1}. A simplified energy-level diagram of gadolinium is shown in Fig. 9.4(a). The dye laser with $\lambda_1 = 5618$Å used in the first excitation step raises gadolinium atoms from

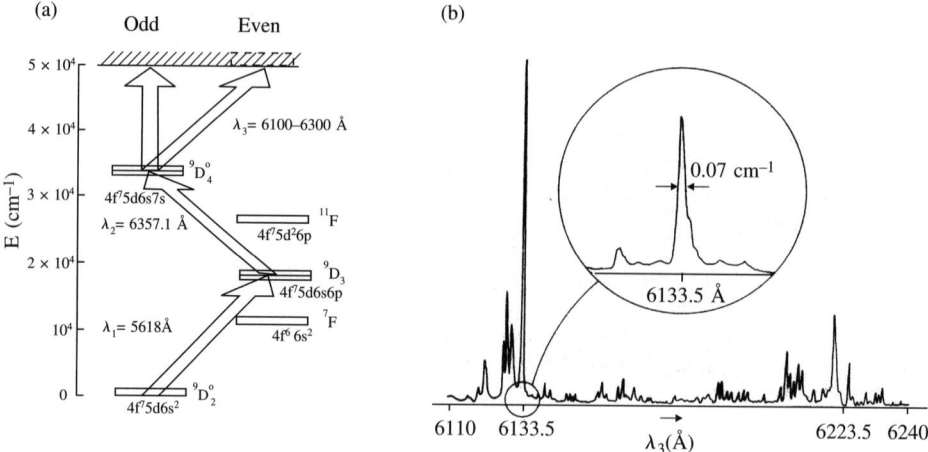

Fig. 9.4. (a) Energy levels of Gd involved in the three-step excitation of the autoionization state, and the relevant quantum transitions. (b) The narrow autoionization resonance for Gd: ion current as a function of the laser wavelength λ_3 in the third excitation step, and the same dependence in the vicinity of the autoionization resonance $\lambda_3 = 6133.5$ Å with higher resolution. (From Bekov et al. 1978.)

their ground state $4f^7 5d6s^2$ ($^9D_2^0$) to the $4f^7 5d6s6p$ ($^9D_3^0$) state. The second-step laser, with $\lambda_2 = 6351.7$ Å, further excites the atoms to move to the $4f^7 5d6s7s$ ($^9D_4^0$) state. And the third-step laser, with λ_3, from 6100 to 6300 Å, excites the Rydberg states and those autoionization states which lie some $300\,\text{cm}^{-1}$ above the ionization limit. Figure 9.4(b) shows the ion signal from gadolinium atoms as a function of λ_3, finely tuned between 6110 and 6240 Å. The sharp autoionization resonance at $\lambda_3 = 6133.5$ Å is of particular interest. A high-resolution record of this region is shown in the top part of Fig. 9.4(b). In this case, the excitation in the third step was effected by a tunable dye laser with a linewidth of $\Delta\nu_3 = 0.03\,\text{cm}^{-1}$. The autoionization resonance half-width is about $0.07\,\text{cm}^{-1}$, which yields an estimated lifetime of 0.5 ns. Since the detected state is only $230\,\text{cm}^{-1}$ above the ionization limit, its decay to the state "electron plus gadolinium ion in the ground state" is most probable. Evidently the selection rules forbid such a decay, and so lead to an increase in the lifetime of the state. This restriction can be removed, for example, by an electric field. A broadening of up to $0.35\,\text{cm}^{-1}$ was observed in a relatively weak electric field ($E = 100\,\text{V/cm}$).

Subsequent high-resolution experiments revealed that the true widths of numerous autoionization resonances in Gd were in the range 0.4–160 MHz, that is, they could be thousands of times as narrow as in Fig. 9.4(b) (Bushaw et al. 2003). The lifetimes of such autoionization states, $\tau_{\text{a.i.}} = (2\pi\Gamma_{\text{a.i.}})^{-1}$, range between 1 and 440 ns, that is, they are comparable to the lifetimes of the bound states lying below the ionization limit. As the widths of the autoionization resonances decrease, the cross sections for the respective transitions to the autoionization states grow larger. The cross section for such a transition to an autoionization state can be as large as 10^{-13}–$10^{-15}\,\text{cm}^2$. This is several orders of magnitude greater than the cross section for direct photoionization into the continuum. This makes a 100% absolute yield possible with very modest requirements on the energy fluence of the laser pulses used, which is especially important for the processes of laser separation of isotopes and of nuclear isomers.

At first glance it may seem that the existence of narrow autoionization states, which provides for optimal ionization with a 100% yield, is quite a rare occurrence. Experiments performed recently with heavy atoms have shown that this is probably not so. For a heavy atom with many valence electrons, we can always choose a final intermediate state for a transition to a narrow autoionization state.

9.2.3 Electric-field ionization of highly excited (Rydberg) states

An effective and universal approach to the resonance ionization of atoms is one where the atoms are first excited in a multistep manner to high-lying states just below the ionization limit and then ionized by an electric field, IR radiation, or collisions (Fig. 9.2(b)). This approach is based on the unique properties of highly excited atoms in states near the ionization limit (called Rydberg states) (Stebbings and Dunning 1983).

For atoms with one optical electron (the hydrogen atom and alkali metal atoms), the position of the energy levels in the vicinity of the ionization limit E_i is described by the following simple expression:

$$E_n = E_i - \frac{\text{Ry}}{(n^*)^2}, \tag{9.6}$$

where $\text{Ry} = 13.6\,\text{eV}$ is the Rydberg constant, and n^* is the effective principal quantum number (for hydrogen, n^* is an integer, while for alkali metal atoms, $n^* = n - \Delta$, where Δ is the quantum defect). As the atomic excitation increases, that is, as n^* grows larger, the atomic characteristics and properties, such as the binding energy, the atomic size, the lifetimes of excited states, and the polarizability, change drastically. These changes are due to the large electron orbits in highly excited atomic states. Table 9.1 lists the properties of Rydberg states as a function of the principal quantum number

Table 9.1 Properties of Rydberg states as a function of the principal quantum number n, with numerical data for the hydrogen atom[a]

Property	$n = 1$	$n = 100$	Arbitrary n (scaling)
Radius of Bohr orbit a_0 (cm)	0.53×10^{-8}	5.3×10^{-5}	n^2
Geometrical cross section πa_0^2 (cm^2)	8.8×10^{-17}	8.8×10^{-9}	n^4
Binding energy E_n (eV)	13.6	1.36×10^{-3}	$1/n^2$
Root-mean-square velocity of electron v_e (cm/s)	2.2×10^8	2.2×10^6	$1/n$
Separation of adjacent levels (eV)		2.7×10^{-5}	$2/n^3$
Optical cross section $(n \to n+1)$			n^4
Quadratic Stark coefficient (shift $\ll$ smallest energy difference)			n^7
Linear Stark coefficient (shift $\simeq$ smallest energy difference)			n^2
Radiative lifetime			n^3

[a]From Stebbings (1976).

(n). The fact that the lifetime of a highly excited atom increases approximately with n^{*^3} (Bethe and Salpeter 1957) is of importance, for this facilitates its ionization. The lifetimes of the d states of the sodium atom, for instance, are given by $\tau_d = 0.95\,(n^*)^3$ ns for $n^* \gtrsim 4$ (Gallagher et al. 1975).

The excitation cross section of a high-energy state depends on n^* and decreases with increasing n^* as n^{*^3}. On the other hand, the higher the value of n^*, the easier the ionization of the atom is. Therefore, an optimal value of n^* can be chosen for each ionization scheme and each particular case. For example, ionization by an electric field (10 kV/cm) can easily be realized with a maximum yield for states with $n^* \simeq 14$. The excitation cross sections of such states are about 10^{-14} cm^2.

The energy structure of highly excited states of an atom, and hence the optical properties of the atom, can be controlled over wide limits by means of an external electric field. This is due to the fact that the quadratic Stark shift for states with large effective quantum numbers changes into a large linear shift similar to the shift for hydrogen atoms. This is due to the presence of a great number of closely spaced states with different orbital angular momenta that easily become mixed in comparatively weak electric fields and exhibit a very strong Stark effect. In addition, under such conditions, the selection rules related to the change in the parity of states break down, and transitions from states with a small orbital angular momentum, for example, s, p, or d states, to states with any orbital angular momentum for a given n become possible. As a result, the number of potentially useful spectral lines in the absorption spectrum of an atom increases drastically. The binding energy of electrons in Rydberg states is proportional to $1/n^{*^2}$, and for $n^* = 20$ the binding energy equals 30 meV, which under normal conditions is comparable to the energy of thermal motion of atoms. This property can be used to ionize highly excited atoms with a continuous or pulsed electric field, or with electromagnetic radiation in the microwave or IR range.

An electric field changes the electronic spectrum of an atom, so that some of the discrete levels nearest to the ionization limit fall into the continuum and some undergo autoionization (Fig. 9.5), the probability of decay through ionization increasing quickly with growing principal quantum number (Ivanov and Letokhov 1975). One can easily estimate the electric-field strength at which a state with a principal quantum number n will fall into the continuum (Bethe and Salpeter 1957). The potential energy of a Rydberg-state electron in an external electric field is $V(r) = -(1/r) - Ez$ (in atomic units), where E is the electric-field strength along the z-axis, and r is the distance from the center of mass. The potential energy has a maximum at a distance $r_{\max} = E^{-1/2}$ along the z-axis and equals $V(r_{\max}) = -E^{1/2}$. The energy of an electron in a state with a principal quantum number n is given by $E_n = -(1/2)n^{-2}$ (neglecting the Stark shift and level splitting). A state with a given n will fall into the continuum if $E_n \geq V(r_{\max})$, or

$$E \geq \frac{1}{(16n^4)} = E_{\mathrm{cr}} \quad (1 \text{ atomic unit} = 5 \times 10^9 \text{ V/cm}). \tag{9.7}$$

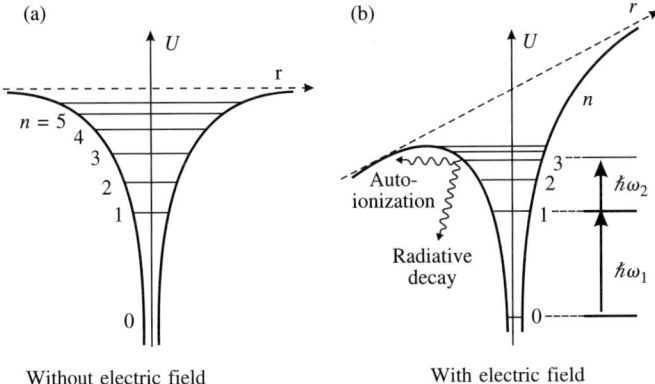

Fig. 9.5 Ionization of a highly excited atom in an external electric field: (a) field-free electron levels and (b) modification of the potential in an external electric field and the formation of autoionization levels.

The use of several properly tuned lasers can excite an atom to a state from which the 100%-probability ionization yield is η_i. Let the ionization cross section for the combined action of a laser and an electric field be given by $\sigma_i = \sigma_R \eta_i$, where σ_R denotes the cross section for the absorption of radiation in the last excitation step. The total atomic ionization yield η_i can rather easily be made equal to unity by choosing an appropriate electric-field strength. As a result, the ionization cross section will depend mainly on the excitation cross section for the radiation in the last step, which ranges from 10^{-12} to 10^{-15} cm^2, depending on the width of the absorption line and the oscillator strength of transitions involving large changes in the principal quantum number n. It is evident that the optimal case occurs when an atom is excited by laser radiation to a state whose ionization rate in the electric field is higher than the rate of its radiative decay.

It is interesting to consider electric-field ionization involving an autoionization state. Figure 9.6 shows the results of experiments with Gd conducted in accordance with the excitation scheme presented in Fig. 9.4. In the top part of the figure, the ion yield as a function of the wavelength λ_3 of the laser in the third step is shown for a case where an electric-field pulse was applied just after the laser pulses. The bottom part of the figure shows the corresponding dependence of the ion yield in a weak electric field unable to ionize the highly excited atoms. For this reason, only the autoionization resonances can be seen in the bottom spectrum. The autoionization peak with the maximum amplitude (its relative amplitude equals 85 and is reduced in the figure) is very narrow. It is shown separately with high resolution in the top part of Fig. 9.4(b). In the top spectrum, the Rydberg states can be seen below the ionization limit (to the right), while the autoionization states can be seen above the ionization limit (to the left). It is clear that they have amplitudes of the same order of magnitude.

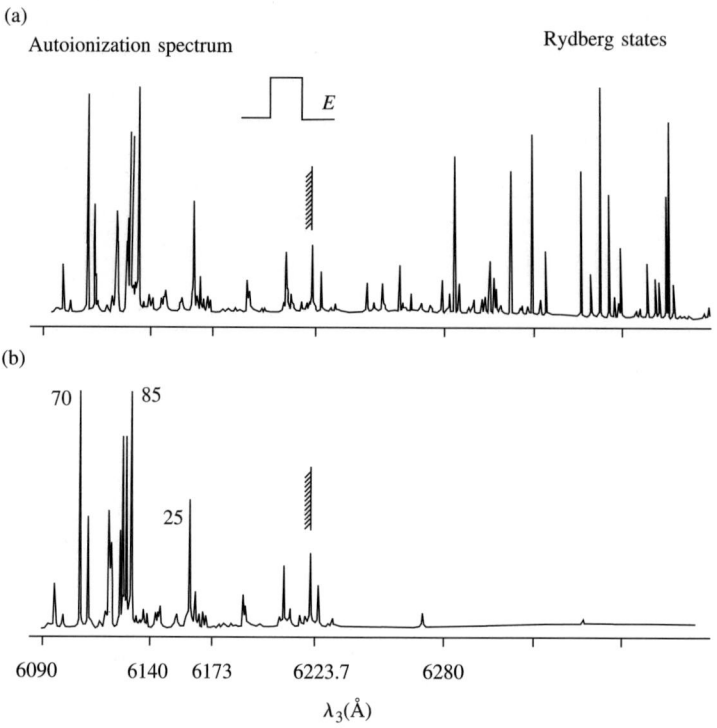

Fig. 9.6 Yield of Gd atomic ions as a function of the laser wavelength in the third stage of excitation of Rydberg and autoionization states from the intermediate quantum state $^9D_4^0$: (a) ionization with a pulsed high electric field ($E = 13\,\text{kV/cm}$); (b) ionization with a weak dc field ($E = 30\,\text{V/cm} = \text{const}$). (From Bekov *et al.* 1978.)

9.3 Photoionization detection of rare atoms and radioactive isotopes

Resonance multistep photoionization has been successfully used to detect rare trace atoms, rare actinides, and, particularly, radioactive isotopes. This method has an important specific feature: the production of photoions that can be isolated and sorted according to their mass. This makes the method very versatile and compatible with nuclear-physics experiments. Figure 9.7 illustrates various ways to realize photoionization detection with (a) a thermal atomic beam obtained by heating and vaporizing a sample containing the rare species of interest, (b) rare atoms in a buffer gas, (c) atoms trapped in a hot cavity, and (d) an accelerated atomic beam under collinear irradiation conditions. Each of these techniques is effective for certain rare species under certain experimental conditions. Some examples are presented below.

9.3.1 Trace elements in a thermal atomic beam

Resonance three-step ionization has been used successfully, for example, for detecting trace amounts of platinum-group elements in natural samples by the scheme shown in

(a) TRANSVERSE ATOMIC BEAM

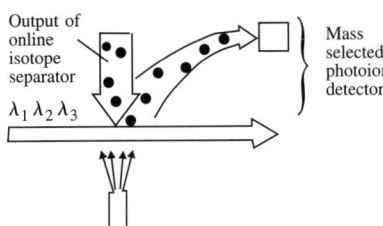

(b) ATOMS IN A BUFFER GAS

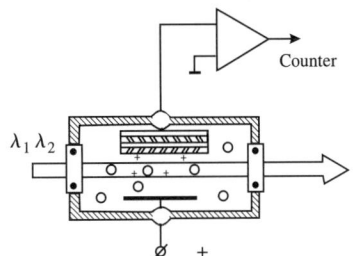

(c) ATOMS TRAPPED IN HOT CAVITY

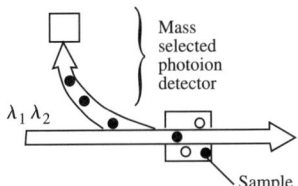

(d) ACCELERATED ATOMIC BEAM

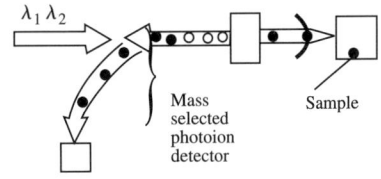

Fig. 9.7 Various resonant photoionization techniques for ultrasensitive spectroscopy of very rare atoms and isotopes: (a) ionization of a transverse thermal atomic beam; (b) ionization of atoms in a buffer gas; (c) ionization of atoms trapped in a hot cavity; (d) ionization of accelerated atoms.

Fig. 9.8(a), that is, by use of thermal atomization of the sample. Ruthenium, one of the rarest metals of the platinum group, had not previously been investigated in ocean waters and sediments because of its low concentration there and the lack of sensitive analytical methods. However, the study of the ruthenium distribution in the ocean is of interest not only in oceanology but also in geology, geochemistry, and cosmochemistry, as it may help in understanding the formation processes of the lithosphere. Bekov et al. (1984) reported the results of ruthenium trace determination in various oceanic environments, carried out using a combination of methods, including fire assay enrichment and laser three-step photoionization spectroscopy in conjunction with vacuum atomization. The ruthenium concentration in marine environments fluctuates over the range covering 4–5 orders of magnitude: $\sim$1 ppt (parts per 10^{12}) in marine water, a few tens of ppt in marine biogenic materials, 10–100 ppt in bottom sediments and marine phosphorites, and >1000 ppt in ferromanganese nodules and the Red Sea metalliferous sediments.

The efficiency of this method was again demonstrated by Bekov et al. (1988) when detecting Rh traces in natural samples. Extraordinarily high concentrations of iridium and other siderophiles were discovered in the Cretaceous/Tertiary (K/T) boundary deposits and interpreted (Alvarez et al. 1980) as a result of a large extraterrestrial body falling upon the Earth, causing mass extinction of the biota dominant in the

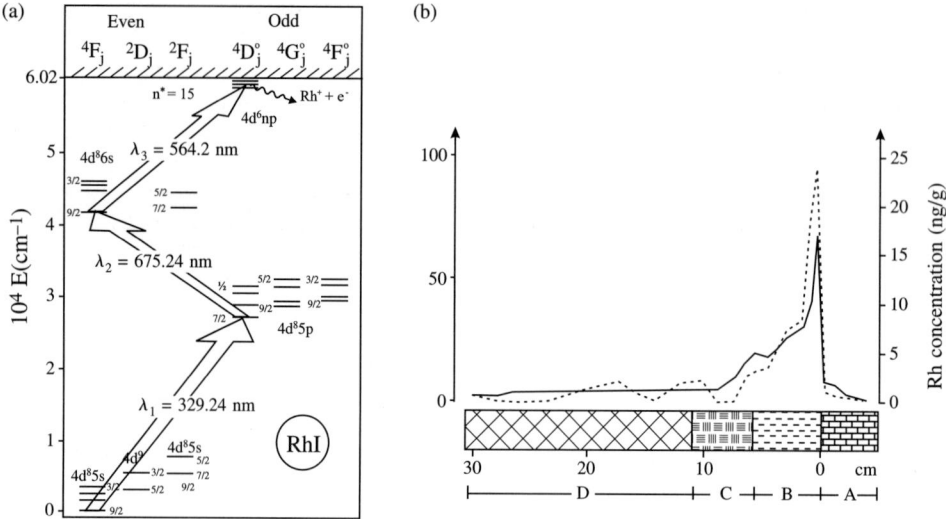

Fig. 9.8 (a) Energy-level diagram of Rh atom, and laser transitions used for selective ionization of this atom. Three dye-laser pulses raise $\sim 10\%$ of Rh atoms to a Rydberg state with an effective principal quantum number of $n^* \approx 15$; 20 ns later, the highly excited atoms are ionized with an efficiency close to unity by means of an electric field pulse. (b) The first data on rhodium concentrations at the K/T boundary (in the Sumbar-SM-4 section, Turkmenia) obtained by the ultrasensitive laser photoionization spectroscopy technique. The maximum Rh concentration in the sample studied was 24.2 ng g^{-1}. The Rh/Ir ratio was 0.34 ± 0.06, which is close to the cosmic ratio of these elements. (From Bekov et al. 1988.)

Cretaceous. The existing alternative models of the Cretaceous terminal event try to explain the anomaly of siderophile elements by their being concentrated through sedimentation or volcanic eruptions. To understand the nature of the K/T event, it was important to establish the proportions of siderophiles and their isotopes in the boundary deposits of that age, especially the proportions of platinum-group elements, their concentrations in extraterrestrial material being high and occurring in ratios different from those typical of terrestrial material. Bekov et al. (1988) presented the first data on rhodium concentrations at the K/T boundary in the Sumbar-SM-4 section (Turkmenia) obtained by the ultrasensitive laser photoionization spectroscopy technique. The atomic–molecular beam formed was irradiated with three dye lasers to effect selective three-step photoionization of rhodium atoms (Fig. 9.8(a)). To improve the selectivity of the analysis, the ions were mass-separated. The detection limit for Rh was found to be equal to 3 pg/g. Figure 9.8(b) shows the results of the Rh determination, together with data on Ir. The maximum Rh concentration (24.2 ng/g) was found in a pocket of a yellow clay material directly at the K/T boundary. Higher up the section, the concentration of Rh gradually fell to 0.5 ng/g at 23 cm from the boundary and remained at this level up to at least 40 cm from the boundary. Downwards from the boundary, the Rh concentration decreased rapidly to 0.3 ng/g at 4 cm below the boundary. These results indicate that Rh, as well as other platinum-group elements, clearly marks the K/T boundary. Its concentration in the K/T boundary deposits may be at least two orders of magnitude higher than the background concentration ($\sim 0.1\,\text{ng/g}$). The Rh concentrations observed in the K/T boundary deposits can readily be explained by the presence of extraterrestrial material, which on average (chondritic material) contains 150–500 ng/g of Rh, that is, at least three orders of magnitude more than the rocks of the Earth's crust. The data obtained point unambiguously to an extraterrestrial origin for the K/T boundary anomaly and bear out the impact theory of the Cretaceous terminal event (Alvarez et al. 1980).

9.3.2 Very rare atoms in a buffer gas

The photoionization detection of atoms in a buffer gas is advantageous over such detection in a thermal atomic beam because the atoms being detected reside for a long time in the region irradiated by the laser pulses. For example, the time it takes for atoms to diffuse out of a spherical region with a diameter of a is in this case given by

$$\tau_{\text{dif}} = \left(\frac{a^2}{16D}\right)\left(\frac{p}{760}\right), \qquad (9.8)$$

where D is the diffusion coefficient at a pressure of 1 atm, and p is the buffer gas pressure (in Torr). In the case of Rb atoms in argon at 1 atm, for example, $D = 0.25\,\text{cm}^2\,\text{s}^{-1}$. When $a = 0.1$–1 cm and $p = 760$ Torr, τ_{dif} is in the range $(0.25$–$2.5) \times 10^{-3}$ s. In order that all the atoms in this region should be irradiated, the repetition rate of the laser pulses must satisfy the modest condition $f \gtrsim (\tau_{\text{dif}})^{-1}$. A successful experiment on the detection of single atoms by means of two-step photoionization in a buffer gas was carried out under such conditions by Hurst et al. (1977a, b). In this experiment, Cs atoms were excited in a buffer to the $7^2 P_{3/2}$ state by the radiation ($\lambda_1 = 4593\,\text{Å}$) of a lamp-pumped pulsed dye laser. The same laser pulse ($\lambda_2 = \lambda_1$) photoionized the excited atoms, as is done in the case of two-color excitation (Ambartzumian et al. 1971; Ambartzumian and Letokhov 1972). The laser radiation fluence ($\Phi_1 = \Phi_2 = 0.2\,\text{J/cm}^2$) during focusing was sufficient to saturate both transitions, which ensured 100% conversion of the Cs atoms into photoions in the irradiated volume ($0.05\,\text{cm}^3$).

The photoionization detection of single atoms in a buffer gas is of great interest in detecting nuclear-reaction products. The first experiment aimed at detecting single daughter atoms formed several seconds after cessation of nuclear fragmentation was carried out at Oak Ridge (Kramer et al. 1978). These authors investigated the spontaneous fission of ^{252}Cf into nonsymmetrical nuclear fragments, one of which was the Cs nucleus. They applied the photoionization technique to detect Cs atoms in a buffer gas. A piece of foil with Cf ions implanted was placed in front of the inlet of a surface-barrier detector responsive to nuclear decay fragments. The fission intensity was about one event per second. A heavy fragment of the nuclear fission of Cf produced 10^6 electrons in its ionization track, which were collected on a plate for $2\,\mu\text{s}$. The collecting plate was charged negatively $40\,\mu\text{s}$ after the nuclear decay came to an end and $10\,\mu\text{s}$ before the laser pulse was fired. The photoelectrons formed by selective two-step photoionization drifted through a hole in the plate into a proportional counter that detected them. Resonance excitation was effected by a lamp-pumped pulsed dye laser whose spectrum covered the wavelengths of two transitions: $6^2S_{1/2} \to 7^2P_{3/2}$ (4555Å) and $6^3S_{1/2} \to 7^2P_{1/2}$ (4593Å). The same pulse concurrently caused nonresonant photoionization of the excited Cs atoms. The experiment demonstrated that a considerable proportion of the Cs nuclei were thermalized to neutral atoms. Thus, given an optimal buffer gas composition, one can realize conditions wherein almost all high-speed ions are neutralized, and hence apply laser methods of detecting single atoms to particles produced initially in the form of high-speed ions.

Photoionization detection in a buffer gas has also been used to study the properties of superheavy (transuranium) elements with charge numbers $Z > 92$. Isotopes of such elements can only be produced by fission reactions in heavy-ion collisions or by transfer reactions using radioactive targets. The elements produced can be placed in an optical buffer-gas cell for the purpose of laser resonance photoionization spectroscopy. This was successfully demonstrated with atoms of such radioactive elements as americium ($Z = 95$) (Backe et al. 2000), einsteinium ($Z = 99$) (Köhler et al. 1997), and fermium ($Z = 100$) (Sewtz et al. 2003).

The photoionization detection of atoms in a buffer gas has definite pros and cons, as compared with detection in a vacuum. Some of the pros can be summarized as follows.

1. With the buffer gas composition and pressure properly chosen and a potential applied across the gas cell, one can detect photoelectrons with a proportional counter. This makes it possible to detect even a single atom in the irradiated volume (Hurst et al. 1977a, b).

2. The buffer gas limits the free escape of atoms from the irradiated volume for the diffusion time τ_{dif}. This property may come in handy if one has to detect atoms with a laser of insufficiently high pulse repetition rate.

The cons of photoionization detection in a buffer gas are as follows.

1. It is necessary to use a very pure buffer gas in order to prevent the atoms being detected from participating in chemical reactions with impurities during the course of detection. For example, in the case of metal atoms, even a negligible admixture of oxygen is dangerous. Understanding this disadvantage is important for analytical applications where the background impurities in the buffer gas restrict the detection sensitivity for trace atoms.

2. The substantial buffer gas pressure causes collisional broadening of the spectral lines, thus concealing such important details of the spectrum as its hyperfine and isotopic structures. So, the detection of atoms in a buffer gas may be Z-selective (selective with respect to the type of the atom) but not A-selective (that is, it may be nonselective with respect to the number of neutrons, that is, the mass number A).

9.3.3 Rare isotopes in a hot cavity

The high sensitivity of the photoionization method is very useful for the spectroscopy of very rare elements whose spectral properties are not well known. One such element is francium (Fr), the natural abundance of which is extremely low (1 Fr atom is found in 3×10^{18} atoms of natural uranium), and which is formed as a result of the decay of ^{235}U. In our experiments (Andreyev et al. 1987, 1988), the resonance photoionization technique was used to detect Fr atoms and study their Rydberg states. The atoms were formed at a rate of about 10^3 atoms/s in a sample of 10^9 ^{225}Ra atoms implanted into a tantalum foil. Such an ultrahigh sensitivity was achieved thanks to two-step ionization of the Fr atoms inside a hot cavity (Fig. 9.7(c)).

The experiment is illustrated schematically in Fig. 9.9(a). A sample in which Fr atoms were formed as a result of the radioactive decay ^{225}Ra $\rightarrow$ ^{225}Ac $\rightarrow$ ^{221}Fr (4.8 min) was placed in the hot cavity. The quasi-closed cavity had two small holes in its walls to introduce a two-color laser beam and extract the Fr photoions produced. While moving inside the cavity, the Fr atoms released from the sample could cross the irradiation region many times before leaving through the exit hole. If the laser pulse repetition frequency was sufficiently high (10^4 Hz), practically every released atom would be ionized. The Fr atoms were ionized through two-step excitation of their Rydberg states $ns^2S_{1/2}$ and nd^2D via the intermediate state $7p^2P_{3/2}$, which was

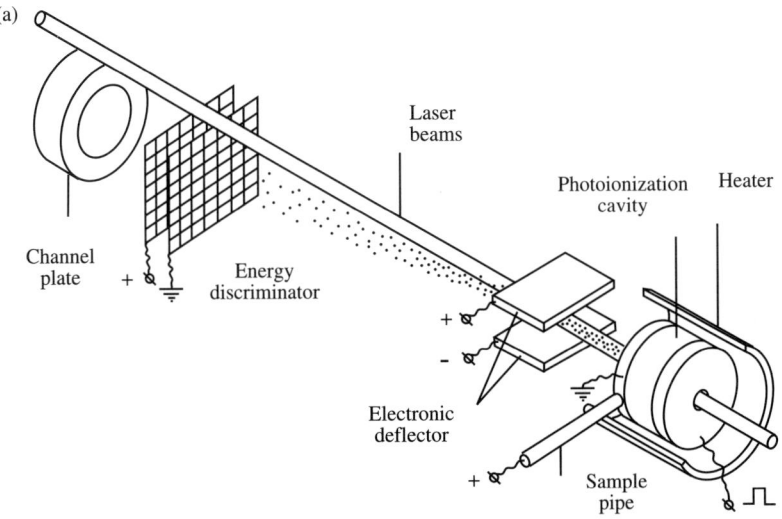

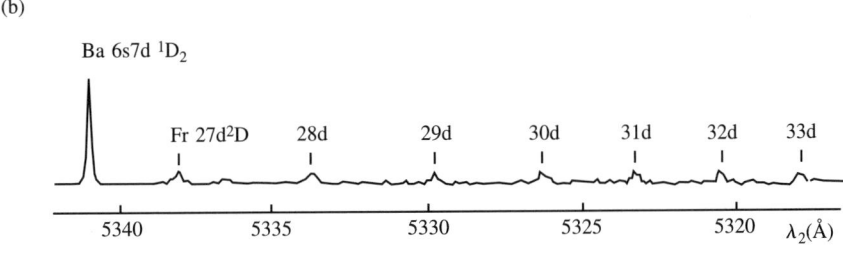

Fig. 9.9 (a) Experimental scheme for laser photoionization detection and spectroscopy of very rare Fr atoms; (b) resonant ionization spectrum of Fr Rydberg states for $7p^2P_{3/2} \rightarrow nd^2D$ transitions. (From Andreyev et al. 1987, 1988.)

excited by laser radiation at $\lambda_1 = 718$ nm. The dye lasers employed were pumped by the output pulses from a Cu vapor laser (at 8.7 kHz). The photoions generated were extracted from the cavity through the exit hole in its wall by the electric-field pulses used to ionize the Rydberg atoms and were detected by a channel multiplier. To reduce the background noise signal due to thermal ions, all ions were mass- and energy-analyzed in a time-of-flight mass spectrometer and an electrostatic analyzer. As an example, Fig. 9.9(b) shows the ion signal obtained while the laser wavelength λ_2 in the second excitation stage was varied. The spectral lines correspond to transitions from the state $7p^2P_{3/2}$ to the states $nd^2D_{5/2}$ and $nd^3D_{3/2}$; the laser bandwidth was $\Delta\nu_2 = 1$ cm^{-1}. The measured quantum defect was $\Delta = 3.4 \pm 0.1$. The data obtained give an ionization limit $I = 32848.25 \pm 0.25$ cm^{-1}, which is close to the theoretical value $I = 32841$ cm^{-1}. The ultrahigh sensitivity of the photoionization spectroscopy of atoms trapped in a hot cavity paves the way for systematic investigations into the optical spectra of very rare atoms with the use of low-radioactivity sources.

9.3.4 Ultratrace isotopes in a collinear atomic beam

The scheme of photoionization detection of atoms in a collinear accelerated atomic beam (Fig. 9.7(d)) is of interest for the ultrasensitive and ultraselective detection of ultratrace, very rare radioactive long-lived isotopes with a relative abundance of 10^{-10}–10^{-20}. There are a fairly large number of rare isotopes of cosmic origin, particularly those formed in the upper atmosphere as a result of nuclear reactions under the effect of cosmic rays. These isotopes form in the upper atmosphere, precipitate, and accumulate on the earth's surface and the ocean bottom. The rate of their precipitation in the ocean can be considered to remain constant for a long period of time, exceeding their half-life $T_{1/2}$.

The selectivity of resonance excitation using the isotope shift in a single (usually the first) step is insufficient to selectively ionize only the rare isotope to be detected. Therefore, detection methods have been suggested on the basis of multistep resonance excitation of the atoms in a multiple-frequency laser field, wherein use is made of the isotope shifts in several consecutive resonance transitions. As a result of such a multistep resonance excitation, the selectivities S_i attained in each excitation and ionization stage are multiplied (Letokhov and Mishin 1979). However, the idea of multiplication of the isotopic selectivities attained in each excitation stage is difficult to realize for the most interesting cosmogenic isotopes, because it is hard to find a series of consecutive upward transitions in these isotopes with noticeable isotope shifts, such shifts being characteristic of the ground state only.

A universal way to overcome this difficulty and make the multistep resonance photoionization method actually applicable to the detection of rare isotopes was suggested by Kudriavtzev and Letokhov (1982). The idea of the method is based on the collinear stepwise photoionization of a beam of accelerated atoms. The acceleration of the atoms in the form of ions under a given potential difference U with subsequent neutralization of the ions into atoms leads to bunching of the longitudinal ionic velocities, and hence to narrowing of the Doppler width of all the spectral lines of the given atomic species (if viewed in a collinear fashion), as compared with the initial Doppler width $\Delta\nu_D^{(0)}$ at an ion source temperature of T (Anton et al. 1978):

$$\frac{\delta\nu_D(U)}{\delta\nu_D^{(0)}} = \frac{1}{2}\left(\frac{kT}{eU}\right)^{1/2}. \tag{9.9}$$

This relation can be derived in a simple way from the following considerations. Since the energy width of the ions remains constant throughout the course of electrostatic acceleration, that is,

$$\delta E = \delta\left(\frac{1}{2}Mv_z^2\right) = Mv_z\,\delta v_z, \tag{9.10}$$

an increase in the longitudinal velocity component v_z must cause a decrease in the velocity spread δv_z. In other words, faster atoms span a given acceleration interval more rapidly, and so the constant force exerted by the electric field imparts a lower velocity to them than to slower atoms. At $U = 10^4$ V, the narrowing factor reaches 10^3, which has been observed experimentally (Anton et al. 1978). What is important is that the atoms in this case gather into a smaller volume of phase space, and their Doppler-free laser excitation is therefore effected without any loss in sensitivity. Along with the narrowing of the Doppler width, there also occurs a Doppler shift of all the spectral lines of the accelerated atoms, which depends on the mass. As a result, there occurs an artificial "mass" shift in any spectral transition of the atom:

$$\frac{\Delta\nu_{sh}}{\nu_0} = \left(\frac{1}{c}\right)\left(\frac{\sqrt{2eU}}{M}\right), \tag{9.11}$$

which is different for different isotopes. This method was successfully demonstrated in detecting the rare isotope ^{3}He at relative concentrations as low as 10^{-9} (Aseyev et al. 1991). Figure 9.10(a) schematically illustrates the two-step excitation of the n^3D Rydberg state from the 2^3S triplet metastable state, and Fig. 9.10(b) shows the isotopic selectivity of the process of laser photoionization of accelerated He atoms.

The parallel progress in the laser cooling and trapping of atoms has opened up a new possibility for ultrasensitive isotope trace analysis (see Section 6.5.4).

9.4 Laser photoionization separation of isotopes, isobars, and nuclear isomers

9.4.1 Stable isotopes

When the first tunable dye lasers made their appearance late in the 1960s (see Stuke 1992), suggestions were put forward as to the use of resonance stepwise ionization for separating isotopes on the basis of isotope shifts in atomic spectra (Letokhov 1969). Following the first successful experiments on the selective ionization of Rb atoms and their isotopes (Ambartzumian et al. 1971), programs were initiated in a number of countries on laser separation of uranium isotopes (^{235}U/^{238}U) by a method that came to be known as the atomic-vapor-laser-isotope-separation (AVLIS) technique (Paisner

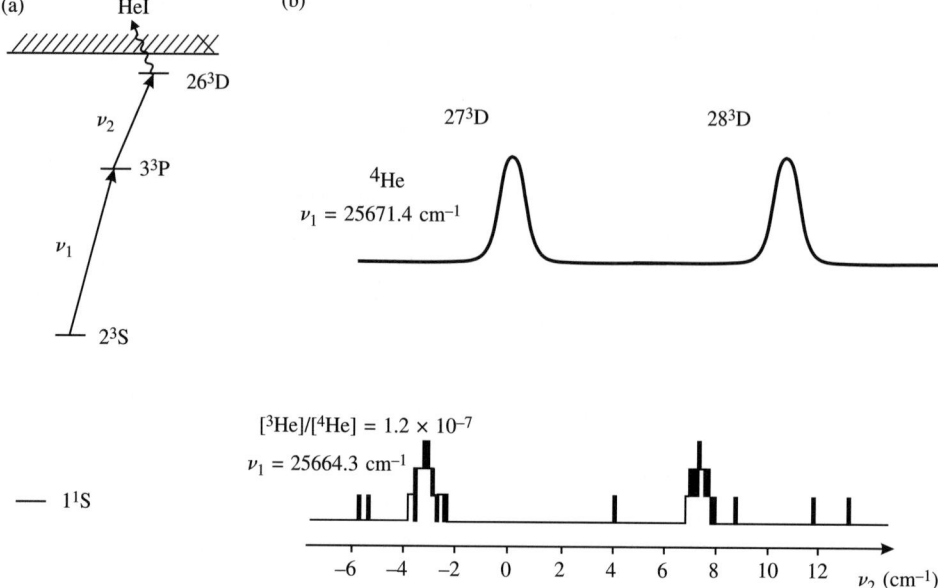

Fig. 9.10 Laser multistep ionization detection of the rare isotope ^{3}He. (a) Scheme of two-step isotopically selective excitation of ^{3}He; (b) ion signal as a function of the second-step laser frequency ν_2 with the first-step laser frequency in resonance with the ^{4}He isotope; (c) first-step laser frequency in resonance with the ^{3}He isotope with a concentration [^{3}He] of 1.2×10^{-7}. (From Aseyev et al. 1991).

1988; Robieux 2000). The start of these programs was rather fast but premature. It is only now, 30 years later, now that tunable solid-state lasers pumped by semiconductor laser array are available, that the realization of industrial-scale AVLIS facilities looks much more realistic.

The general scheme of laser photoionization separation of isotopes looks simple enough. It includes several successive processes:

1. *Production of vapor* from the solid-phase working element. This vapor contains a natural mixture of isotopes. In the case of high-melting elements, for example, uranium and gadolinium, use is made of electron-beam vaporization. Low-melting metals, such as ytterbium, are heated in an oven.

2. *Isotope-selective multistep ionization* of the vapor by means of several tunable lasers. To provide predominant ionization of the desired isotope, these lasers should have a very narrow linewidth and should be tuned very accurately in resonance with the necessary transition frequencies of this isotope. Also, the laser frequencies should be sufficiently stable.

3. *Extraction of ions* with enriched isotopic composition from the atomic vapor, and their transportation to the collector. The substance thus obtained is processed further by well-known chemical methods.

This sequence works ideally in the case of optically thin absorption conditions. In practice, much more rigorous requirements are imposed upon this scheme. First, it is necessary to use the laser radiation with maximum efficiency. After all, it carries light energy of high quality and hence of high cost. For example, in the first transition the entire amount of radiation is absorbed, while in the last ionizing transition only a hundredth of the energy introduced is absorbed. Second, the process of extraction of ions from a low-temperature laser-produced plasma is rather difficult to control. Third, there is the problem of isotopic scrambling in the ion collector.

All this makes the resonance photoionization technique much more difficult to use for separating isotopes in large quantities and at low cost. This is true, for example, for the ^{235}U/^{238}U isotopes, which can be separated by other methods on an industrial scale. Therefore, in the first stage of its application, this method is being used for separating isotopes of elements which are currently obtained in small quantities of a few grams by the expensive electromagnetic separation technique, for example, the separation of isotopes of rare earth elements (Tkachev and Yakovlenko 2003). In that work, laser enrichment of the ^{168}Yb isotope was performed using three-step isotope-selective (at the second step) ionization. The production of the highly enriched ^{168}Yb isotope in weighable quantities was reported. The rate of production of enriched ytterbium was 5–10 mg h^{-1}. The results of experimental studies into the selective photoionization of palladium were also reported. A substantial enrichment of various palladium isotopes was achieved.

These are only the first steps that have been taken on the road to the production of monoisotopic materials for many applications by means of laser selective ionization, which will probably become a very important laser technology in the twenty-first century.

9.4.2 Radioactive isotopes and isobars

The ultrahigh sensitivity and high selectivity of laser resonance ionization are of great interest as regards the application of this technique in nuclear-physics experiments, where it is very important to detect and study short-lived radioactive isotopes obtained in very small quantities. Of great importance in such experiments is the selectivity of resonance ionization with respect to various elements (Z-selectivity), isotopes (N-selectivity), isobars (($N+Z$)-selectivity), and even nuclear isomers (($\Delta E, N, Z$)-selectivity), as shown in Fig. 9.1. All these types of selectivity can be attained, owing to the fact that the hyperfine and isotope shifts of spectral lines depend on the nuclear spin I, the magnetic moment μ_I, the spectroscopic nuclear quadrupole moment Q_s, and changes in the mean-square nuclear charge radius $\delta \langle r^2 \rangle^{A,A'}$ between isotopes with mass numbers A and A'. For this reason, highly selective ionization of atoms enables one, first, to obtain information about these characteristics of short-lived atoms; second, to separate isotopes, isobars, and nuclear isomers; and third, to study the nuclear characteristics of exotic nuclei isolated from an enormous number of other, abundant nuclei.

Successful experiments on the photoionization detection and separation of radioactive isotopes were conducted by Zherikhin et al. (1984). Later on it became clear that atoms could be ionized most effectively not in a beam by the scheme presented in Fig. 9.7(a), but in a hot cavity in accordance with the scheme shown in Fig. 9.7(c),

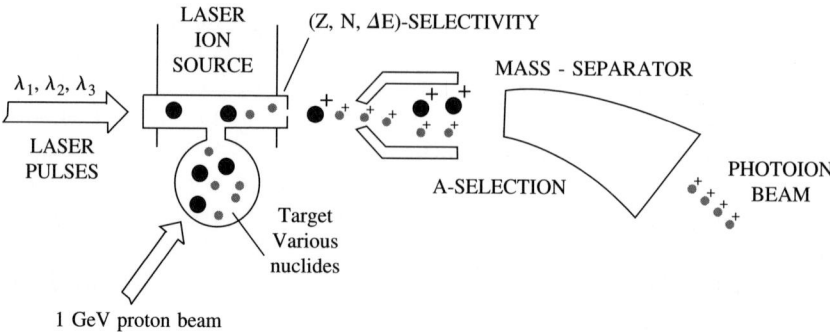

Fig. 9.11 Schematic illustration of resonant multistep ionization of radioactive atoms in a hot cavity (laser ion source) and following mass separation with selectivity by element (Z), isotope (N), and isomer (ΔE).

where the laser beams repeatedly interacted with the atoms in the cavity, thus providing for a much higher ionization yield. This is of principal importance in the ionization of rare atoms. The ionization technique combines naturally with the mass separation of the photoions produced, which makes the entire process even more selective and, what is most important, allows the detected rare atoms to be physically isolated.

A Z-selective photoion source of radioactive atoms, demonstrated by Alkhazov et al. (1989) in a hot metal cavity with three-step ionization by tunable high-repetition-rate (11 kHz) lasers, was used in an online regime with the proton accelerator at the ISOLDE facility at CERN (Mishin et al. 1993). A schematic diagram of this setup, a most successful application of the resonance ionization method in the practice of fundamental physical research, is presented in Fig. 9.11. The key element in this setup is the resonance ionization of atoms inside a hot cavity. The role of the cavity is to keep atoms for a certain time in a volume where they can be irradiated by laser light and to confine the ions during their drift towards the extraction region. The ion confinement is a natural consequence of the emission of electrons by high-temperature metal surfaces, which produces a negative potential well of a few electron volts. The potential prevents the photoions from hitting the cavity walls and recombining into neutral atoms. The ions are pushed towards the exit by the internal field set up by the DC heating of the tubular cavity. The temperature of the cavity must be kept high enough to provide ion confinement. Consequently, thermal ionization of atoms can take place on the hot surfaces, particularly if the ionization potential of the atoms is low. The ratio between the isotope yields attained with the laser tuned on and off resonance can be as high as 104 or more. This method was realized successfully in the resonance ionization laser isotope source setup ISOLDE RILIS at CERN (Fedoseyev et al. 2003). The RILIS setup can also be used for atomic-spectroscopy studies of exotic radioactive isotopes produced at rates of only a few atoms per second.

9.4.3 Nuclear isomers

As far back as the early 1970s, it became clear that resonance stepwise ionization could help solve the less practical but more fundamental problem of the separation of nuclear isomers (Letokhov 1973b). Nuclear isomers cannot be separated by any of the existing non laser methods used to separate isotopes, because of the small difference in mass between excited (isomeric) and ground-state nuclei. The high value of the nuclear spin of the excited isomeric nuclei provides for a rich hyperfine structure of the optical spectral lines of the atoms. As soon as a setup was developed for the photoionization detection of short-lived isotopes online with a proton accelerator (Zherikhin et al. 1984), the first successful experiments were performed on the laser separation of the ^{141}Sm and ^{164}Tm nuclear isomers (Mishin et al. 1987). The photoionization of thulium atoms was realized by three successive resonance transitions into an autoionization state. The transition sequence was as follows: $4f^{13}6s^2F_{7/2} + \hbar\omega_1$ (5896Å) → $4f^{12}5d6s^2(6,7/2)_{7/2} + \hbar\omega_2$ (5712Å) → $4f^{12}5d6s6p + \hbar\omega_3$ (5755Å) → autoionization state. Figure 9.12 presents experimental photoionization spectra and the results of their interpretation. The spectrum in Fig. 9.12(a) was obtained for a mixture of ^{164}Tm isotopes in the ground and excited states. The fluxes of ^{164}Tm emerging from the target in the ground and the excited state were practically the

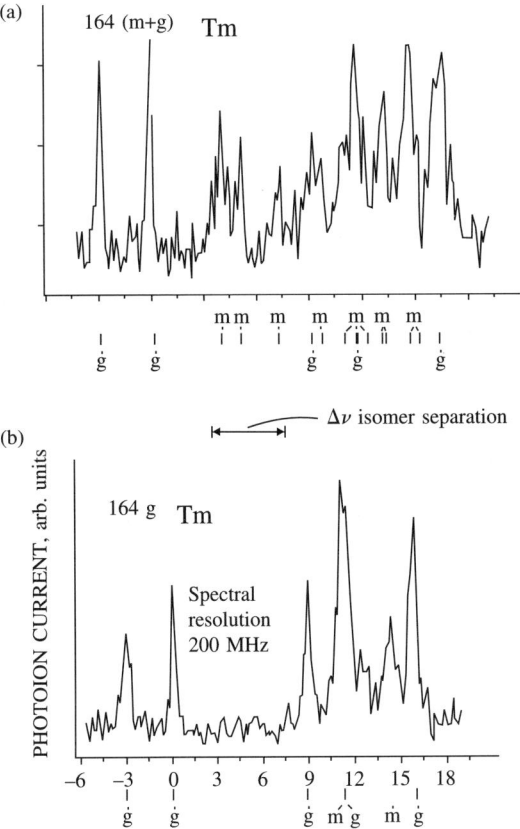

Fig. 9.12 Photoionization yield as a function of the frequency ν_1 of the first resonant transition of the ^{164}Tm atom in (a) a mixture of atoms in the ground and isomeric nuclear states and (b) the ground nuclear state. (From Mishin et al. 1987.)

same. To identify atomic lines in the hyperfine structure, one needs a spectrum belonging to the isotope either in the ground or in the excited nuclear state. For this purpose, the above authors obtained ^{164}Tm in the ground state as follows. The ion beam at the exit from the mass separator contained a number of isobars, ^{164}Yb in particular. ^{164}Yb ($T_{1/2} = 76$ min) decays to yield ^{164}Tm ($T_{1/2} = 2$ min) in the ground state, the isomeric thulium state remaining practically unpopulated. Isobars with a mass number $A = 164$ were accumulated in a cold tantalum cylinder for 1 hour. Then a 15 min time interval was used to allow the ^{164}Tm isomer to decay ($T_{1/2} = 5$ min). After that the cylinder was heated. Figure 9.12(b) shows the photoionization spectrum of ^{164}Tm in the nuclear ground state.

An analysis of the photoionization spectra in Fig. 9.12 shows that for ^{164}Tm there exist frequency regions containing excitation lines belonging only to atoms with isomeric nuclei. Inasmuch as the laser photoionization scheme is such that an isomeric photoion can be detected only after it has been extracted from the atomic beam and deposited onto the multiplier cathode, one can speak of isomer separation in this experiment. The experimental setup, consisting of a mass separator, a laser spectrometer, and a photoionization chamber arranged in line with a proton accelerator, allows one to study and separate isomers with a lifetime of the order of the time of their extraction from the target. Typical targets allow extraction times of around 1 s. A similar setup can be built around other sources of isomeric nuclei, such as a high-energy ion beam or a nuclear reactor.

The ISOLDE RILIS setup at CERN was also used to conduct more experiments on the separation of isomers and the investigation of the characteristics of the separated isomers of a series of elements (Cu, Pb, Ag, and others). Figure 9.13 presents experimental results for the laser separation of ground-state ^{68g}Cu and isomeric ^{68m}Cu (Köster et al. 2000). In this experiment with ^{68m}Cu, the intensity ratio between the

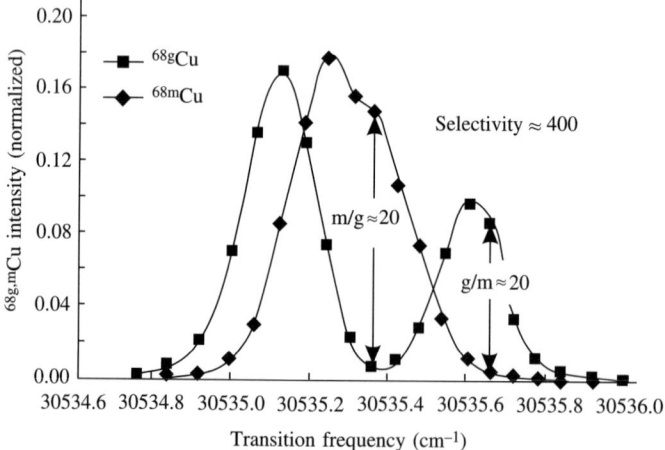

Fig. 9.13 Laser resonant-photoionization separation of ^{68g}Cu and ^{68m}Cu. The intensities were normalized individually. In reality, the ^{68m}Cu intensity is four to five times the ^{68g}Cu intensity. (Reprinted with courtesy and permission of Kluwer Publ. Co. from Köster et al. 2000.)

low- and high-spin states can be as high as 400, given the proper choice of the hyperfine components for resonance excitation.

The development of the rather universal and effective laser photoionization method for separating nuclear isomers opens up a new field of research in low-energy nuclear physics. One of the possibilities, namely the production of samples with excited nuclear isomers, that is, with an inverted population of the nuclear states, for use in a γ-laser, was discussed in the early years (Letokhov 1973d). However, it will probably take many years to realize this possibility. Another challenging line of inquiry has been demonstrated using as an example the high-resolution (10^{-8}) mass spectrometry of trapped nuclear isomers (Blaum et al. 2004). According to the energy–mass equivalence $E = mc^2$, excited long-lived nuclear states (isomers) are distinguished from the ground state by additional mass. Ultrahigh-accuracy Penning traps for ions allow one to "weigh" nuclear excitation energies.

10
Multiphoton ionization of molecules

The multiphoton ionization techniques for molecules were developed in parallel with the studies into the resonance multistep ionization of atoms, but their aim was to achieve photoselective laser detection of molecular traces in combination with mass spectrometry of the photoions produced, that is, to create what is called a "laser two-dimensional optical mass spectrometer" (Letokhov 1976). However, the more complex structure of molecular spectra, the high photoionization potentials of molecules (8–12 eV), and the absence of tunable laser sources operable in the UV region of the spectrum made it necessary to use the available UV excimer lasers capable of photoionizing molecules by "brute force." In that case, the low selectivity of the optical channel was compensated by the extra selectivity of the mass spectrometer. Figure 10.1 schematically illustrates the main methods of multiphoton ionization of molecules: (a) resonance stepwise ionization, (b) resonance-enhanced two-photon (or multiphoton) ionization (RETPI or REMPI), and (c) nonresonance multiphoton ionization. The stepwise photoionization technique provides good spectral selectivity; it requires a tunable laser radiation source of moderate power output (kW–MW/cm^2). The REMPI technique is of limited selectivity and requires higher radiation powers (MW–GW/cm^2); it is especially convenient where use is made of ultrashort laser pulses. And, finally, the nonresonance multiphoton ionization (MPI) technique features no spectral selectivity; it is a "brute-force" method, where the necessary chemical selectivity is attained solely in the mass-spectrometer channel. Therefore, the last two techniques are unsuitable for the detection of trace molecules in a mixture of organic molecules, because many of the molecules simultaneously suffer photoionization and photofragmentation. However, when combined with the cooling of molecules in a pulsed supersonic jet, which prevents their distribution among many sublevels and thus makes their spectra sharper, the REMPI technique proves very effective (Reisler and Wittig 1985; Lubman 1990).

When they exposed a surface sample of organic molecules to intense UV radiation pulses, Antonov *et al.* (1980*a*) observed the formation of molecular ions without any noticeable fragmentation, which had the character of a nonthermal process. A detailed investigation led to an understanding of the two-stage character of the process in the form of thermal desorption of the molecules and their subsequent photoionization. A study of the laser thermal-desorption (ablation) process revealed the possibility of soft chemoionization in the dense cloud of laser-desorbed molecules (Karas and Hillenkamp 1987). This process was called matrix-assisted laser desorption/ionization

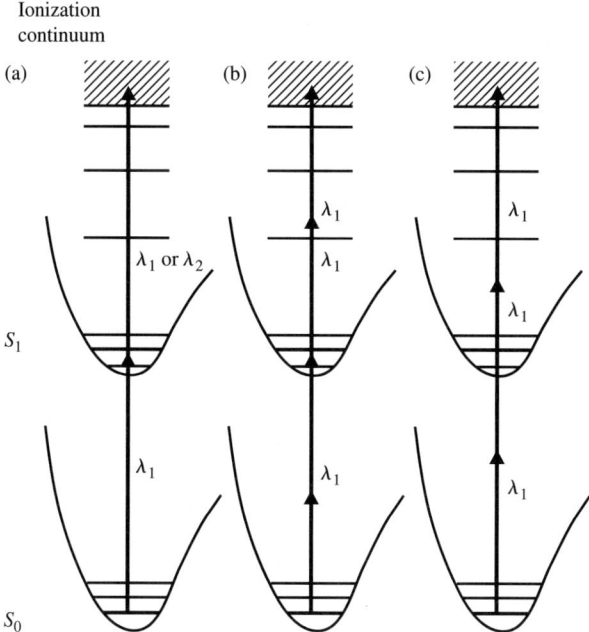

Fig. 10.1 Energy-level diagram of multiphoton ionization of molecules: (a) resonance two-photon ionization; (b) resonance-enhanced multiphoton ionization (REMPI); and (c) nonresonance multiphoton ionization (MPI).

(MALDI) and made it possible, when combined with mass spectrometry, to develop a powerful technique to effect the mass spectrometry of large biomolecules without the perceptible fragmentation that substantially complicated the identification of biomolecules by the standard mass-spectrometric means.

As in the case of resonance photoionization of atoms (Chapter 9), the photoionization of molecules is carried out under various conditions (in a thermal molecular beam, a pulsed jet-cooled molecular beam, or a laser-desorbed molecular cloud). But in all cases, use is made of mass spectrometry of the photoions produced, for the combination of the optical and mass spectra makes the method much more informative. All these methods for laser ionization and laser desorption/ionization of molecules are briefly discussed below.

10.1 Photoselective resonance ionization of molecules

The earliest experiments on the resonance stepwise photoionization of molecules (H_2CO) were conducted by Andreyev et al. (1975) in an ionization chamber without using any mass separation of the photoions produced. The next natural step was the two-stage resonance photoionization of molecules by the scheme of Fig. 10.1, involving mass analysis of the photoions produced (Antonov et al. 1977, 1978). The experimental setup for studying the stepwise photoionization of polyatomic molecules in a mass spectrometer consisted of a static magnetic mass spectrometer and time-synchronized

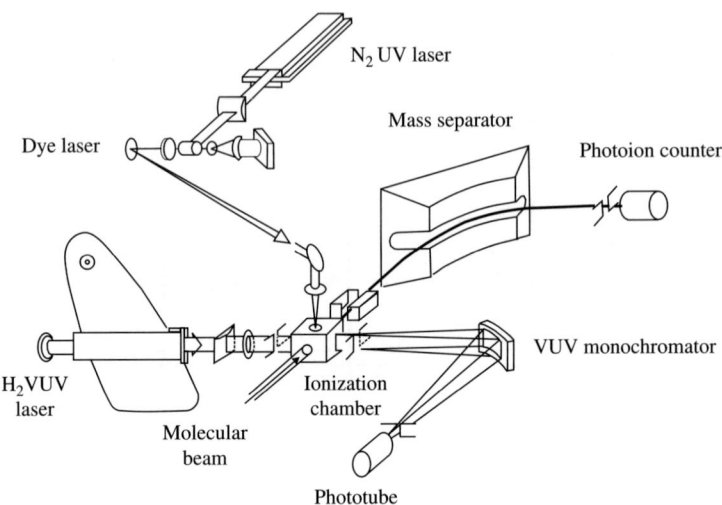

Fig. 10.2 Experimental setup for the study of the two-step resonance photoionization of molecules in a molecular beam with a mass spectrometer. (From Antonov et al. 1977, 1978.)

exciting and ionizing lasers (Fig. 10.2). The intermediate electronic states of the polyatomic molecules (benzaldehyde, benzophenone, etc.) were excited using a pulsed nitrogen laser ($\lambda_1 = 337$ nm) or a dye laser. In all cases, the excited molecules were photoionized by a pulsed H_2 VUV laser (Fig. 10.3(a)). Both laser pulses were directed into the ion source of the mass spectrometer. The photoions produced passed through a cylindrical ion optics system and a magnetic separator and were detected by a secondary-electron multiplier. The molecular vapor was injected into the ionization chamber through a glass multichannel inlet.

The mass spectrum obtained from the two-step photoionization of benzaldehyde (Fig. 10.3(b)) exhibits a peak for the molecular ion and a peak for the fragment $C_6H_5CO^+$ resulting from the detachment of the hydrogen atom from the aldehyde group. Figure 10.3(c) illustrates the dependence of the molecular and fragment photoion yields on the delay time of the ionizing pulse. The increase in the signal in the zero-delay range is due to the finite duration of the laser pulses. A slow decrease in the submicrosecond region is caused by the escape of the excited molecules from the interaction region. The absence of any rapid decrease in the ion current at delay times shorter than 1 μs agrees with the fast ($\tau_{ST} \simeq 10^{-12}$s) conversion of the excited molecule to the triplet state (Fig. 10.3(a)). Another scheme for the mass analysis of photoions, by means of a quadrupole filter, was utilized in experiments on two-step molecular photoionization by Boesl et al. (1978).

The stepwise resonance ionization of molecules differs materially from the resonance photoionization of atoms by, first, the small resonance excitation cross sections of the electronic states in polyatomic molecules, owing to their distribution among many rotational–vibrational levels, and, second, the fact that the absorption bands of many molecules occur in the UV region of the spectrum. For this reason, the soft resonance stepwise photoionization technique has proved unsuitable for the majority of

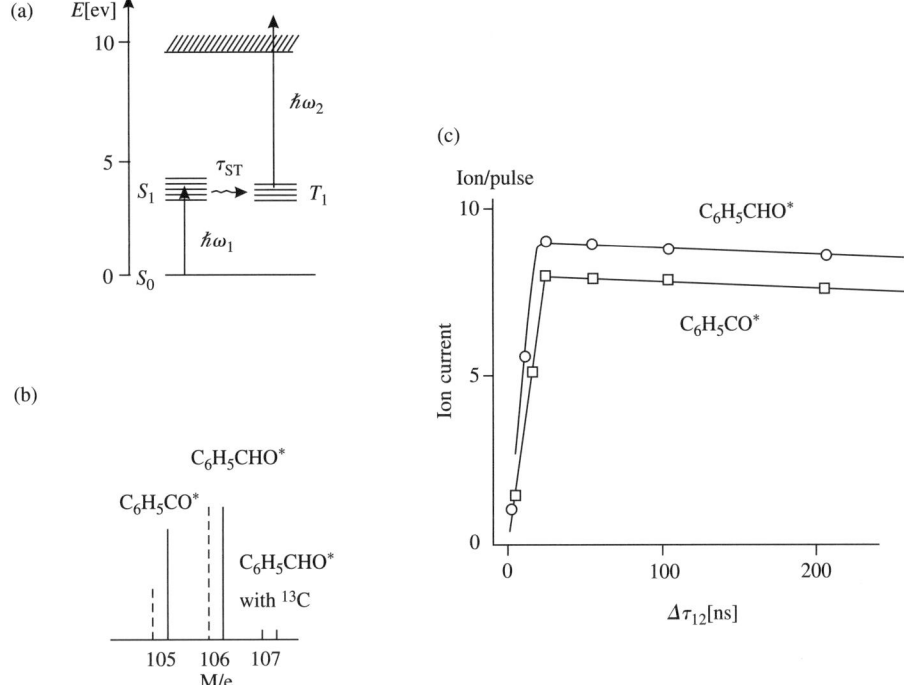

Fig. 10.3 Two-step photoionization of benzaldehyde: (a) quantum transitions for two-step photoionization; (b) mass spectrum of two-step photoionization (solid lines) and single-photon ionization (dashed lines); (c) dependence of molecular and fragment photoion yield on the delay time between the two laser pulses. (From Antonov *et al.* 1978.)

molecules. Further progress was achieved by two methods: (1) the resonance-enhanced multiphoton ionization technique, notwithstanding its limited spectral selectivity and (2) the jet-cooled pulsed molecular-beam technique, which provides a dramatic simplification of the molecular spectra (Levy 1980).

10.2 Resonance-enhanced multiphoton ionization (REMPI) of molecules

It is quite natural to carry out the resonance photoionization of molecules via their electronically excited states. But this is not the only possibility. Polyatomic molecules can also be excited via their highly excited vibrational states (Bagratashvili *et al.* 1983). The former approach has found the widest application (Lubman 1990), thanks to the availability of high-power UV lasers, whereas the latter is not so widespread, as powerful tunable IR lasers are scarce. However, both these methods deserve a brief consideration.

10.2.1 REMPI via excited electronic states

Successful experiments on the resonance-enhanced multiphoton ionization of polyatomic molecules by intense UV laser pulses with a tunable frequency were performed by Zandee and Bernstein (1979). Such experiments commonly use a time-of-flight mass spectrometer, which makes it possible to record with moderate resolution the entire mass spectrum during a single laser pulse. The mass spectrum is very rich in this case because of the strong fragmentation of the molecules. The resolution of pulsed-photoionization mass spectra can be dramatically improved by means of the time-of-flight reflectron. This instrument is capable of increasing the resolution of time-of-flight photoionization spectrometry up to around 10 000 Da (Schlag and Neusser 1983), and it is ideally fitted for use with the pulsed-photoionization technique.

A high ionization yield with moderate molecular fragmentation can be attained in two-step photoionization if the intermediate electronically excited state remains stable throughout the duration of the laser pulse and the laser quantum energy in the second step exceeds the vertical ionization potential of the molecule in the excited state. The detection of single polyatomic molecules was demonstrated under these optimal conditions (Antonov et al. 1981a). The experiment was performed with naphthalene molecules, using an excimer KrF laser (λ = 248 nm) to excite and ionize the molecules via intermediate electronic states. The photoions were recorded by means of a time-of-flight mass spectrometer. In this experiment, the average value of the naphthalene molecular-ion signal was about 1 ion/pulse. Considering the fact that the yield of molecular ions at 100% ionization efficiency is about 50% and that the photoion recording efficiency is equal to 50%, we may conclude that the signal observed corresponds to four molecules in the photoionization volume. This value corresponds to a partial naphthalene pressure of 10^{-14} Torr and a relative concentration in the air of 10^{-9}. The results obtained are almost the ultimate as regards the efficiency of detecting molecules in the photoionization volume, and correspond to the detection of a few molecules over the course of a single pulse at low pressure.

Polyatomic molecules placed in an intense UV laser field undergo not only multiphoton photoionization, but also strong photofragmentation owing to multiphoton photoionization of the molecular photoions produced. As a result, as the laser pulse intensity is increased, the mass spectrum of photoions becomes richer and richer. To illustrate, let us present the results obtained with the benzene molecule, which is a very convenient object for investigations into the photoionization and photofragmentation processes (Rockwood et al. 1979; Antonov et al. 1980b; Reilly and Kompa 1980). The energy of the first excited electronic state S_1 of the benzene molecule is higher than half its ionization potential ($I = 9.25$ eV), and this permits the photoionizing of excited molecules by a second photon in a single-frequency laser field ($2\hbar\omega > I$). Figure 10.4 shows mass spectra of benzene photoions formed under the action of KrF excimer laser pulses with $\hbar\omega = 5$ eV. At low laser energy fluences $\Phi < 10^{-4}$ J/cm^2, only the molecular ions formed by way of two-step photoionization (Fig. 10.4(a)) are observed in the mass spectrum. As the laser pulse energy is increased, progressively lighter fragment ions appear in the mass spectrum (Fig. 10.4(b)). At laser energy fluences $\Phi > 5$ J / cm^2, the fragmentation becomes very intense (Fig. 10.4(c)) and exceeds the fragmentation under bombardment by electrons with energies of 70 eV (Fig. 10.4(d)). Under these conditions, one of the strongest peaks in the mass spectrum is the ionic component C^+ (see Fig. 10.4(b)). The formation potential of the atomic carbon ion is about 27 eV, and therefore, in order for this ion to form, the molecule (or its fragments) must absorb no less than six photons. The total yield of ions is very high and approaches 10% at a laser energy fluence of about 10 J/cm^2 (Antonov et al.

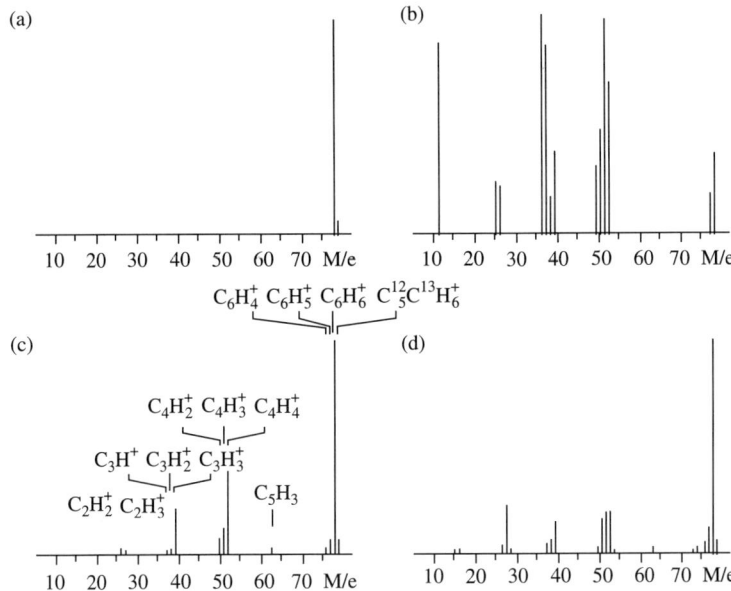

Fig. 10.4 Mass spectra of the benzene molecule: (a) photoionization mass spectrum for a radiation intensity $I = 10^5$ W/cm^2 ($\lambda = 248$ nm, $\tau_p = 20$ ns); (b) for $I = 2 \times 10^6$ W/cm^2 (from Antonov et al. 1980a); (c) for $I = 3 \times 10^8$ W/cm^2 (adapted from Reilly and Kompa 1980); and (d) electron-impact mass spectrum ($E_{el} = 70$ eV) for comparison.

1980b). Aside from being highly efficient, the photoionization of benzene by excimer laser radiation is selective. For example, even at the maximum radiation intensities, such atmospheric components as H_2 and H_2O are practically completely absent in the mass spectra of the photoions.

The mechanism of the multiphoton photoionization and fragmentation of polyatomic molecules by intense (>10^9 W/cm^2) UV laser pulses was studied in detail by Boesl et al. (1980). The spectroscopic features of resonance-enhanced multiphoton excitation and ionization were reviewed by Ashfold and Howe (1994).

10.2.2 REMPI via highly excited vibrational states

To achieve highly selective and sensitive detection of molecules, it was suggested that one should use laser resonance excitation of molecular vibrations, followed by photoionization of the vibrationally excited molecules, that is, vibrationally mediated photoionization (Ambartzumian and Letokhov 1972). As a matter of fact, it proved very difficult to realize this idea because the shift of the VUV absorption bands of the molecules caused by their vibrational excitation was too small. However, there exists another possibility for polyatomic molecules, namely multiphoton (MP) excitation of high-lying vibrational states by high-power IR laser pulses tuned to resonance with the pertinent vibrational transitions (Chapter 11). The highly excited vibrational states can be photoionized by another UV or VUV laser pulse. This molecular-photoionization experiment was performed in accordance with the scheme

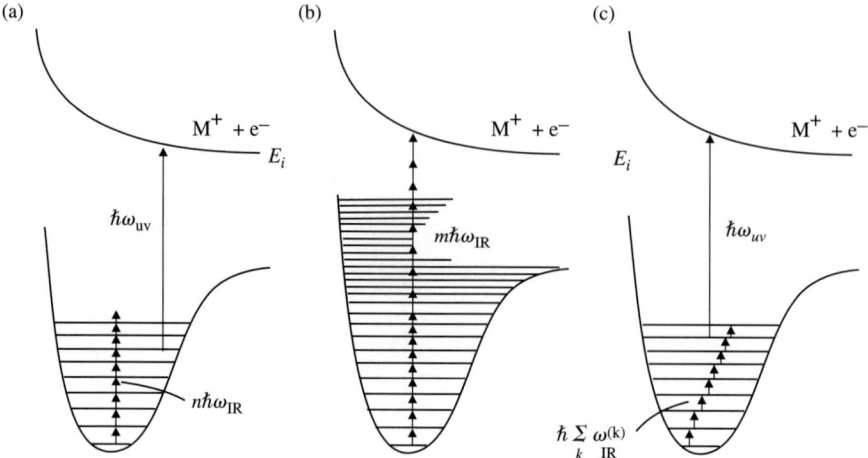

Fig. 10.5 Schemes for resonance molecular photoionization via high-lying vibrational states by way of multiphoton resonance vibrational excitation with IR laser radiation: (a) multiphoton IR + VUV excitation, (b) IR multiphoton excitation, and (c) IR multistep + VUV excitation.

shown in Fig. 10.5(a). A powerful CO_2-laser pulse excited high-lying vibrational levels of the SF_6 molecule in a molecular beam. Owing to the presence of highly excited molecules, the molecular photoionization spectrum suffered a slight deformation in the long-wavelength wing of the VUV region (Sudbo et al. 1979). That the magnitude of the deformation of the spectrum was small was explained by the fact that the proportion of highly excited SF_6 molecules was small.

Large polyatomic molecules can even be ionized by powerful IR laser radiation alone (Bagratashvili et al. 1983). It is customary to assume that the maximum vibrational energy $E_{\text{vib}}^{\max}$ of a molecule is limited by its dissociation energy D_0. This is true only for small molecules or for molecules subject to collisions. However, when the IR field is strong enough to provide a sufficiently high multiphoton excitation rate (see Chapter 11), a large polyatomic molecule with a limited monomolecular decay rate can be vibrationally overexcited well above the dissociation threshold, so that $E_{\text{vib}}^{\max} - D_0 > D_0$. Such a competition between the two processes is quantitatively described by the statistical (RRKM) theory of monomolecular decay (Robinson and Holbrook 1972). In the semiclassical approximation of this theory ($E_{\text{vib}}^{\max} \gg E_0$, where E_0 is the energy of the zero-order molecular vibrations), the following estimates can be obtained for $E_{\text{vib}}^{\max}$ at a given multiphoton excitation rate W:

$$\frac{E_{\text{vib}}^{\max} - D_0}{D_0} = \left[\left(1 - \frac{W}{K}\right)^{1/(s-1)}\right]^{-1} - \frac{E_0}{D_0} - 1, \qquad (10.1)$$

where s is the number of vibrational degrees of freedom, and K is the frequency factor appearing in the expression for the monomolecular decay rate in the semiclassical approximation, whose magnitude is of the order of 10^{15} s^{-1}. With IR multiphoton excitation, an excitation rate of $W = \sigma_{\text{vib}} I \simeq 10^9$ s^{-1} can easily be ensured. Then, according to eqn (10.1), strong vibrational overexcitation is possible for a molecule whose number of vibrational degrees of freedom s is much greater than $\ln(K/W)$. Because of the nonadiabatic coupling between vibrational and electronic motions, the molecule can be expected to move from very high vibrational levels within the electronic ground state (but at an energy of $E_{\text{vib}} \gg D_0$) to excited electronic terms leading to the ionization decomposition of the molecule.

An experiment of this kind was performed (Bagratashvili et al. 1983) with anthracene ($C_{14}H_{10}$) molecules ($s = 66$, $D_0 = 4.8$ eV), for which the estimate in eqn (10.1) gives $E_{\text{vib}}^{\max} - D_0 \simeq 3D_0$. With so strong an overexcitation, ionization of the molecule ($I = 7.4$ eV) is quite possible. In fact, the formation of anthracene molecular ions was observed when the molecules were irradiated by sufficiently powerful (about 10^9 W/cm^2) short (70 ns) CO$_2$-laser pulses under collisionless conditions (pressure 4×10^{-6} Torr) in a time-of-flight mass spectrometer. These ions can be believed to appear as a result of the IR multiphoton ionization of molecules in accordance with the scheme of Fig. 10.5(b). Of course, such a multiphoton ionization technique is applicable only to large polyatomic molecules, since for polyatomic molecules with a small number of atoms the maximum possible degree of overexcitation above the dissociation threshold is comparatively low.

In subsequent experiments conducted with the C$_{60}$ fullerene molecule using a CO$_2$ laser (Hippler et al. 1997) and a tunable free-electron laser (von Helden 1998), it was found that this molecule could absorb a few hundred IR photons and then undergo autoionization. The formation of ions was unexpectedly slow, occurring some 0.1 ms after the exciting laser pulse. The time-delayed photoions and electrons in the IR-REMPI process are of thermionic origin. Resonant ionization using IR light (IR-REMPI) is becoming a new tool to study the spectroscopy and dynamics of gas-phase large polyatomic molecules and clusters (von Helden et al. 2003).

It is expected that methods will be developed for the multistep soft vibrational excitation of molecules by means of multiple-frequency IR laser pulses (as shown in Fig. 10.5(c)). This will allow one to deposit substantial amounts of energy (1–3 eV) into the molecules. In that case, it will probably prove possible to realize the idea of highly selective ionization of molecules by way of their strong vibrational excitation by short chirped-frequency IR laser pulses and subsequent electronic excitation.

10.3 Laser desorption/ionization of biomolecules

Photoselective ionization combined with mass spectrometry seemed from the outset to hold the greatest promise for the study and identification of complex organic molecules, especially biomolecules. This is due to the fact that two-photon ionization at moderate laser pulse powers causes no strong fragmentation of the molecules, whereas ionization by electron impact causes substantial fragmentation. This is evident from a comparison between the mass spectra presented in Figs. 10.4(a) and (d). Several ways have been found to achieve this goal by means of soft laser desorption of molecules from a surface, followed by their photoionization, specifically in a jet-cooled stream, and chemiionization in a dense cloud of desorbed (ablated) biomolecules.

The processes leading to the formation of free ions accompanying the action of laser radiation on the surface of a solid have for a long time attracted a great deal of attention. In investigations conducted in the past few years in various laboratories, use was made of laser radiation with parameters varying over a very wide range: pulse durations varied from continuous-wave radiation to femtoseconds, wavelengths varied from the IR to the UV region, and intensities varied from a few W/cm^2 to 10^{11} W/cm^2. The action of radiation with various parameters on a surface must give rise to very diverse processes that lead to the formation of ions. The investigation of these processes is of very practical interest. For example, the laser-stimulated formation of ions is already being used today in the mass-spectrometric analysis of a wide class of materials, ranging from refractory materials to biological molecules.

10.3.1 Formation of molecular ions by laser irradiation of a surface

The mechanism of ionization accompanying the action of laser radiation with an intensity of $I = 10^9$–10^{11} W/cm^2 on the surface of a solid, leading to strong heating ($T \gtrsim 1000°C$) of the surface, intense vaporization of the material, and the formation of a plasma, has been studied in detail (Ready 1971). In this case, the ionization process has a thermal character and is described by the Langmuir–Saha equation. This method is of little use in analyzing specimens containing complex organic molecules, because it leads to complete decomposition of these molecules due to the strong heating of the material. Starting in 1976, a number of laboratories began to intensively study the mechanisms of laser-stimulated formation of ions of organic molecules on the surface of solid specimens (see Letokhov 1987).

The formation of molecular ions upon irradiation of crystalline nucleic-acid base and anthracene specimens with UV laser pulses of moderate intensity ($I = 10^4$–10^6 W/cm^2) was reported (Antonov et al. 1980a, 1981b). An important feature of the observed process is the absence of fragment ions. Only the molecular ions M^+ were observed to be present in the mass spectra. The ion yield was observed to increase with the photon energy. In the case of adenine, estimates showed that when the recorded ionic signal was at its minimum, the heating of the surface did not exceed 100°C, which was much lower than the melting point of adenine ($T_m = 360°C$) and could not lead to any intense vaporization of the specimen. The characteristic mass spectra of the ions formed upon irradiation of crystalline adenine and anthracene powders are shown in Fig. 10.6. Comparison between the molecular-ion yields obtained upon irradiation of adenine by nanosecond ($\tau = 20$ ns, $\lambda = 249$ nm) and picosecond ($\tau = 30$ ps, $\lambda = 266$ nm) pulses (Antonov et al. 1982) shows that the formation of ions is determined by the density of absorbed energy, $\Phi\chi$, where χ is the coefficient of absorption of the crystal per unit length, and not by the radiation intensity. If one assumes that the energy absorbed in the crystal is completely thermalized, then, neglecting the thermal conductivity, the maximum possible heating of the surface at a characteristic energy fluence of $\Phi = 4$ mJ/cm^2 is $\Delta T = \Phi\chi/c\rho = 200°C$, where c and ρ are the specific heat capacity and the specific weight, respectively, of adenine. Under such heating conditions, the rate of desorption of molecules from the surface can be expected to increase considerably.

The formation of molecular ions upon irradiation of adenine was observed to be accompanied by the escape of neutral molecules from the surface of the specimen

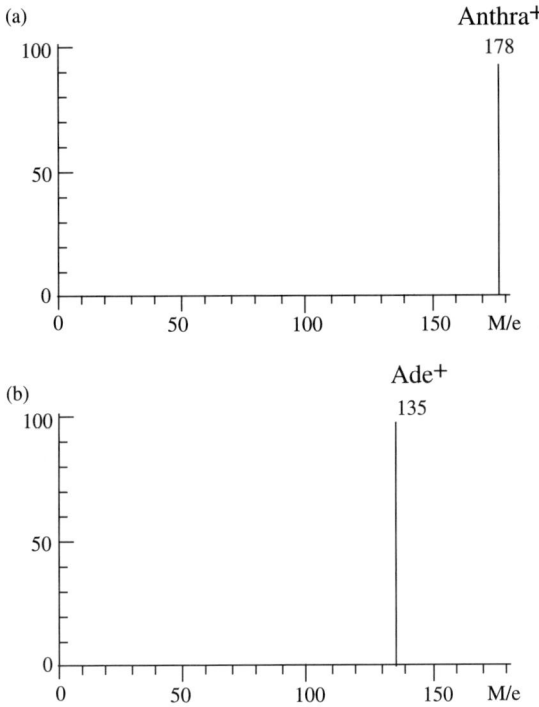

Fig. 10.6 Characteristic mass spectra of ions formed upon irradiation of crystalline powders of (a) anthracene and (b) adenine by KrF laser radiation. (From Antonov et al. 1980a, 1981b.)

(Antonov et al. 1982). The surface of the specimen was in that case irradiated with picosecond pulses, and the neutral molecules were detected with the help of a photoionizing pulsed KrF laser irradiating a small region above the surface of the specimen. This observation suggested the following elementary ion formation mechanism: pulsed laser heating of the surface of the specimen → desorption of neutral molecules → photoionization of free molecules above the surface. With the thermal mechanism of desorption of molecules from the surface and their subsequent photoionization in the gas phase, it is quite possible to observe ions of any molecule adsorbed on a surface, provided that the molecules are susceptible to multiphoton ionization. However, since the thermal nature of desorption is entirely different from the character of photoionization, in order to optimize these processes it is most advantageous to effect them by means of different laser pulses. The molecules are desorbed under the effect of a laser pulse irradiating the surface and are then ionized at some distance from the surface under the action of another laser pulse (or pulses) of a certain wavelength. The use of pulsed desorption allows the thermal decomposition of the molecules to be materially reduced, and laser stepwise photoionization is practically the only technique that enables desorbed molecules to be ionized with a probability close to unity (Egorov et al. 1984a).

To observe the formation of photoions directly on the specimen surface, with the desorption of neutral molecules being excluded, use should be made of subpicosecond laser pulses, for in that case one can guarantee that no neutral molecule will have enough time to move away from the surface by more than 0.3 nm during the laser pulse. Experiments by Chekalin et al. (1988) revealed some specific features that are especially evident from comparison of the mass spectra resulting from irradiation with femtosecond and nanosecond laser pulses in the UV (308) nm and visible (620 nm) regions of the spectrum (Fig. 10.7). The degree of fragmentation depends on both the laser pulse duration and the wavelength. Figure 10.7 shows the mass spectra obtained upon irradiation of a pure tryptophan (Trp) powder specimen with laser pulses differing in parameters. The UV pulse effects multiple-photon ionization of Trp through the resonant S_1 state, while the visible pulse effects the multiple-photon ionization and two-photon excitation of the S_1 state. As can be seen from comparison between the mass spectra of Figs. 10.7(a) and 10.7(c), the ionization of the Trp chromophore is predominant in the latter case.

Of particular interest is the pulse-duration dependence of the threshold energy density E_{thr} for the appearance of mass spectra (Fig. 10.8) in the case of multiphoton UV ionization of the TrpGlyAla peptide. It is clearly seen that the threshold energy fluence is reduced by a factor of 10 in the case of irradiation with femtosecond pulses. The same effect is observed with pure tryptophan powder. The threshold energy density for the appearance of the 300 fs pulse mass spectrum, $E_{th}^{fs} = 10^{-3}\,\text{J/cm}^2$ ($I \simeq 3 \times 10^9$ W/cm^2), is an order of magnitude lower than that for the 15 ns pulse spectrum, $E_{th}^{ns} = 10^{-2}\,\text{J/cm}^2$ ($I \simeq 6 \times 10^5$ W/cm^2). It should be noted

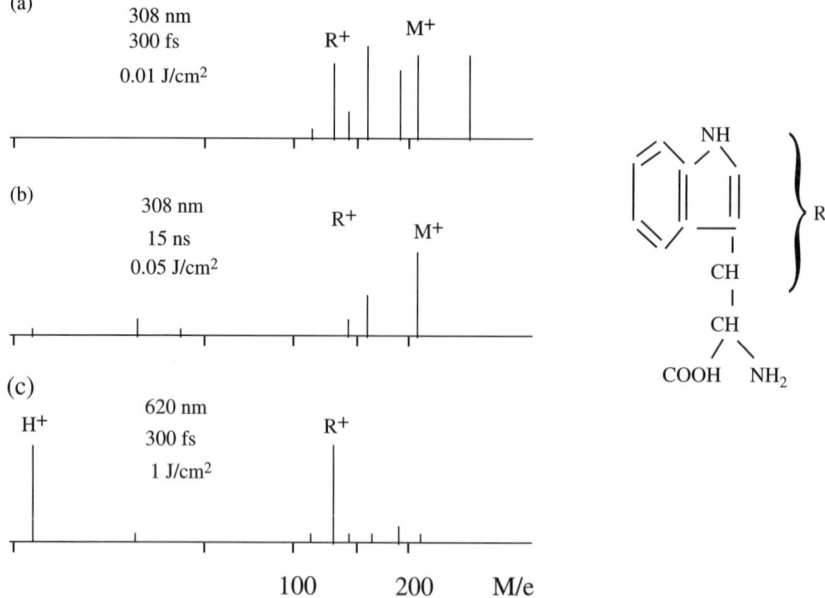

Fig. 10.7 Positive-ion mass spectra obtained upon irradiation of tryptophan with single laser pulses at wavelengths of 308 nm and 620 nm. The pulse duration and radiation energy fluence are indicated on the left. (From Chekalin et al. 1988.)

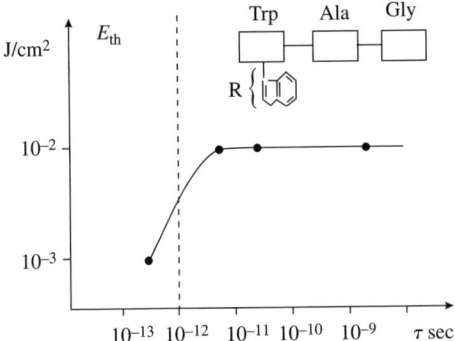

Fig. 10.8 Threshold energy fluence for the appearance of the TrpAlaGly peptide mass spectrum as a function of the UV laser pulse duration. (From Chekalin et al. 1988.)

that the mass spectra observed are of photoionization origin and have no relation to the formation of a plasma due to the strong heating of the irradiated surface. The formation of a plasma at high energy densities (observed only upon irradiation with nanosecond pulses) is clearly manifest in the character of the mass spectrum: the spectrum loses its sharply defined structure. Besides, the surface heating was estimated as $10-100°C$. This temperature is insufficient for plasma formation. The kinetic energies of the ions observed upon irradiation with femtosecond pulses are much higher than in the case of irradiation with nanosecond pulses; the measured width of the distribution in the latter case does not exceed 1 eV (which corresponds to the resolution of the instrument), while for 300 fs pulses, this width amounts to 25 eV. What is more, the maximum of the distribution in the case of excitation with femtosecond pulses is shifted a few tens of electronvolts toward the high-energy side, the shift reaching its maximum for the lightest mass (R^+) and diminishing with increasing mass. The experimental results presented above suggest that the mechanisms whereby ions are produced upon irradiation with nanosecond and femtosecond pulses differ.

One of the main factors governing the sensitivity of detecting surface-adsorbed polyatomic molecules by this method is the maximum proportion of the molecules that can be desorbed from the surface without being decomposed. This matter was investigated in experiments on detecting naphthalene and anthracene molecules adsorbed on the surface of graphite (Antonov et al. 1983). The experimental setup (Fig. 10.9(a)) consisted of a time-of-flight mass spectrometer with a graphite substrate at $T = 200$ K held on the repulsion electrode of the ion source, and time-synchronized pulsed CO_2 ($\lambda_1 = 10.6$ μm) and KrF ($\lambda_2 = 248$ nm) lasers. The CO_2-laser emission was made to impinge upon the substrate at a small angle ($7°$), and the ionizing radiation of the excimer laser was directed parallel to the substrate, close to its surface. The delay time τ_d between the desorbing and ionizing laser pulses was 1 μs. The naphthalene and anthracene molecules were deposited on the substrate from the gaseous phase and were in equilibrium with their vapors at partial pressures (as estimated from photoionization measurements) of $10^{-12}-10^{-13}$ Torr.

Photoionization of the molecules desorbed under the effect of the IR radiation was carried out through intermediate electronically excited states by means of UV

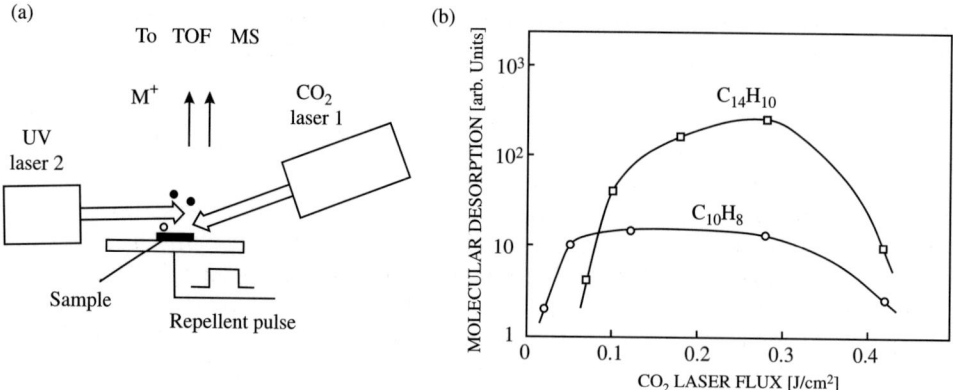

Fig. 10.9 (a) Schematic diagram of an experiment on the laser detection of adsorbed molecules by means of their IR-laser-induced pulsed desorption (laser 1) and subsequent multistep UV-laser resonance ionization (laser 2). (b) Yield of naphthalene and anthracene molecules desorbed from a graphite surface as a function of the CO_2-laser energy fluence. The desorbed molecules were detected by way of REMPI effected with the aid of a pulsed KrF laser operating at $\lambda = 248$ nm with an energy fluence of 0.015 J/cm^2. (From Antonov et al. 1983.)

radiation pulses from the KrF laser, the energy of two photons ($2\hbar\omega = 10$ eV) being sufficient to ionize both the naphthalene and the anthracene molecules. When the KrF laser pulse fluence was as high as 0.2–0.3 J/cm^2, the ionization yield for the naphthalene molecules reached 100% (Antonov et al. 1981a). In the experiment, to reduce the amplitude of the photoionization signal, the pulse fluence of the ionizing laser was kept at 0.015 J/cm^2, which provided a 2% ionization yield. Figure 10.9(b) shows the dependence of the naphthalene and anthracene photoionization signals on the desorbing CO_2-laser pulse fluence for a fixed fluence of the ionizing KrF laser pulse in the range 0.05–0.3 J/cm^2; the dependence is less distinct in the case of anthracene. The maximum ion signal in the experiment was around 10^3 ions/pulse for naphthalene and 10^4 ions/pulse for anthracene. A method of increasing in the detection sensitivity for low molecular concentrations has been demonstrated (Egorov et al. 1984b). The idea is that molecules are accumulated on a cooled surface owing to adsorption, and then pulsed-laser desorption and stepwise photoionization of these molecules above the surface are performed. The detection sensitivity for naphthalene and anthracene molecules was increased by more than 10^3 times.

10.3.2 Laser REMPI of biomolecules in a supersonic jet

Desorbed polyatomic molecules feature broad electronic absorption bands, and hence low spectral selectivity in multiphoton excitation and ionization. However, the use of the supersonic-molecular-beam technique radically changes the situation (Levy 1980). With this technique, the molecules are deeply cooled by injecting them into a carrier gas that is then allowed to expand into a vacuum through a small hole (a nozzle). The size of the hole should be large compared with the mean free path of the expanded carrier gas. The expansion process converts the kinetic energy of the random motion of the gas into kinetic energy of directed mass motion, which results in a narrowing of the velocity distribution of the molecules. With monatomic gases, temperatures as low as 1 K are quite attainable. The jet cooling of polyatomic molecules is much less effective. For this reason, a small amount of polyatomic molecules is injected into an atomic gas. At sufficiently low concentrations, the velocity of the polyatomic molecules (biomolecules) becomes as high as that of the carrier gas molecules, and their temperature becomes the same as the carrier gas temperature. The internal degrees of freedom of the polyatomic molecules begin coming into equilibrium with the translational degrees of freedom as a result of collisions with the carrier gas particles. This method can help one to attain rotational and vibrational temperatures as low as 0.5 K and 20–30 K, respectively. The supersonic-jet cooling of polyatomic molecules simplifies their spectra, because almost all of them relax to their lowermost vibrational state and a few lower rotational states.

The REMPI of biomolecules in a supersonic molecular beam is illustrated in Fig. 10.10 (Lubman and Li 1990). A CO_2-laser pulse with an energy of 10–40 mJ desorbed the molecules into a pulsed supersonic molecular beam, about 50 μs in pulse duration. REMPI of the desired molecular species was effected directly at the entrance of a time-of-flight mass spectrometer by means of a tunable dye laser of about 10^6 W/cm^2 in power output. Under these conditions, injected biomolecules suffer practically no fragmentation, provided that the wavelength of the ionizing laser pulse falls within the long-wavelength wing of the absorption band of the molecules and the radiation power is moderate. Figure 10.10(b) illustrates the REMPI of tryptophan molecules by UV laser pulses differing in wavelength, their power being the same.

The laser REMPI technique possesses many valuable properties that make it very effective in the discrimination and analysis of biological isomers in complex mixtures of organic molecules. Its enhanced spectral selectivity and low degree of fragmentation, combined with the capabilities of mass spectrometry, are especially valuable for the study of biomolecules (Lubman 1990). The need to use high-power tunable UV laser pulses with an intensity of the order of 10^6 W/cm^2 can be considered a disadvantage of this technique.

10.3.3 Matrix-assisted laser desorption/ionization of biomolecules

The laser control of molecules is a very versatile method. This fact is exemplified by the development of the a fairly efficient analytical technique of matrix-assisted laser desorption/ionization (MALDI) time-of-flight (TOF) mass spectrometry. The roots of this technique lie in the first experiments on the observation of photoionization of organic molecules on a surface under the effect of UV excimer-laser pulses (Section 10.3.1). In these experiments, photoions of large molecules that suffered no strong fragmentation were detected. The photoionization process was found to proceed in two stages. A small proportion of the molecules with thermal velocity v_{th} are first desorbed from the surface, and then they undergo photoionization at a distance of $\Delta x = v_{th} \times \tau_{pulse} \simeq 1\,\mu$m from the surface under the effect of laser radiation, other desorbed molecules also taking part in the process.

The MALDI technique uses laser-induced pulsed threshold evaporation of a sample matrix, containing the biomolecules to be analyzed as an impurity. The role of the laser

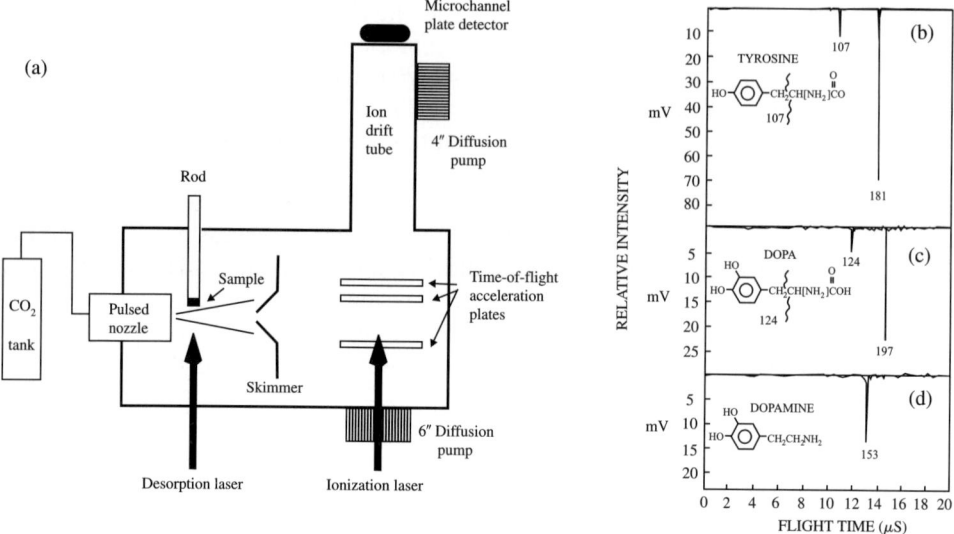

Fig. 10.10 (a) Schematic diagram of an experimental setup for the REMPI of biological molecules volatilized by way of pulsed laser desorption in a supersonic jet; (b)–(d) laser desorption time-of-flight mass spectra of various molecules ionized by UV pulses at (b) 280 nm, (c) 266 nm, and (d) 280 nm. (Adapted from Lubman and Li 1990.)

radiation is reduced solely to the pulsed evaporation of the sample that absorbs it. The role of laser control here is minimal. The material of the matrix is selected so as to make the desorbed biomolecules in the dense cloud of vaporized matrix material undergo chemiionization without the participation of the laser radiation. This method can help one to obtain ions from very large molecules without perceptible fragmentation.

The MALDI technique was suggested by Karas et al. (1987) and Karas and Hillenkamp (1988), who used an organic matrix for ionization purposes, whereas Tanaka et al. (1988) used an inorganic matrix. These authors reported UV-laser desorption of bioorganic compounds above 10 kDa for the first time. A schematic setup of a linear MALDI-TOF instrument is presented in Fig. 10.11. Typically, the sample molecules are mixed with an organic compound that acts as a matrix (protonated at an acidic pH) to facilitate their desorption and ionization. The ions under analysis are accelerated by means of a high voltage (15–25 kV), separated in a field-free flight tube, and detected as an electric signal at the end of the tube. The matrix is normally a weak aromatic acid that strongly absorbs energy at the wavelength of the laser used. Most commonly, this is an N_2 laser emitting radiation at 337 nm. However, Nd-YAG (266 nm) and IR (2.94 μm) lasers have also been successfully applied. The matrix should not modify or react with the substance under analysis prior to being exposed to laser radiation. Further, it must be possible to dissolve the matrix in the same solvent as the substance to be analyzed in order to obtain proper mixing of them. The substance is mixed with the matrix material in solution in a molar ratio ranging from 1:100 to 1:50 000, which is optimal for ion production. The mixture is then allowed to dry as a crystalline coating on a metal probe. The matrix molecules

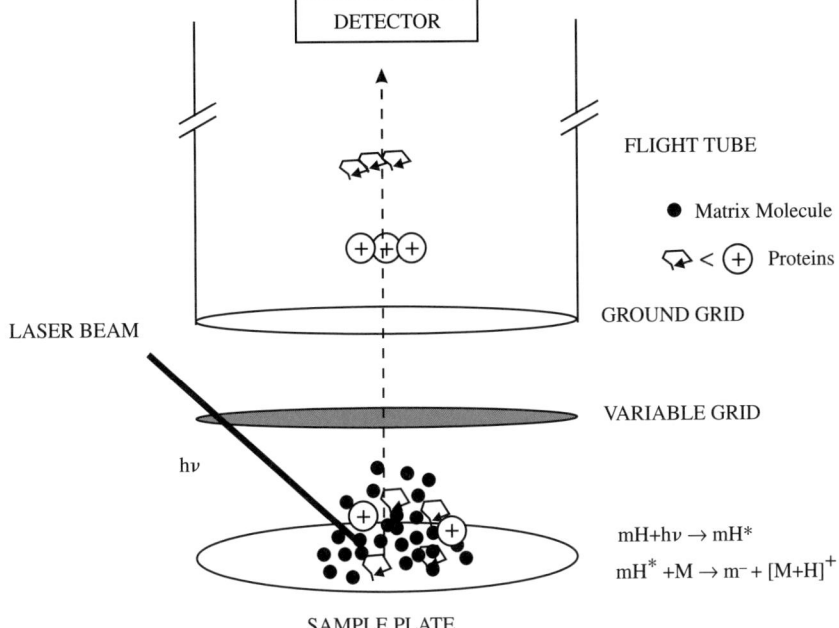

Fig. 10.11 Principles of the MALDI-TOF MS technique (mH = matrix, M = molecule under analysis, mH* = excited matrix molecule). Ions are separated in a flight tube, where small ions travel faster than their larger counterparts and thereby reach the detector sooner.

absorb laser energy, causing translational motion and ionization of the molecules of the substance under analysis, which are then accelerated toward the detector. Several advantages have been demonstrated with the MALDI technique, including spectral simplicity due to singly charged ions, a wide mass range (up to 900 kDa), a low noise level, high sensitivity, little sample consumption, a short measurement time, moderate salt tolerance, and minimal fragmentation.

The matrix in the MALDI technique plays an important role in several ways: (1) it absorbs energy and protects the substance being analyzed from excessive energy exposure, that is, against decomposition; (2) it enhances the production of the ions of the substance under analysis by way of proton transfer following the photoexcitation or photoionization of the matrix molecules; and (3) it prevents the association of the embedded molecules of the substance to be analyzed (Hillenkamp et al. 1991). Today a wide variety of compounds can be analyzed by the MALDI-TOF mass spectrometry technique, including peptides, proteins, synthetic polymers, oligonucleotides, oligosaccharides, drugs, and metabolite systems.

11
Photoselective laser control of molecules via molecular vibrations

The vibrational–rotational spectra of a polyatomic molecule contains information about its main structural properties. Information about the bonds and structures formed by the atoms in the molecule can be extracted from the IR and Raman spectra. Vibration frequencies that characterize a definite bond are nevertheless very sensitive to the rest of the molecular components. This circumstance provides potentially high selectivity in the effect that resonance laser radiation can have on molecular vibrations, as well as the possibility of implementing laser control of molecules via their vibrations. The following tasks have been achieved:

(1) excitation of a specific component of a gas mixture, including isotope- or isomer-selective excitation, that is, the realization of intermolecular selectivity;

(2) excitation of molecules up the ladder of their vibrational states to the stage of their dissociation into fragments;

(3) development of novel spectroscopic methods for studying the fundamental properties of highly excited molecular states;

(4) the finding of conditions that allow high intermolecular selectivity in multiple-photon excitation and dissociation (MPE/D) of molecules by resonance infrared radiation;

(5) development of laser isotope separation methods using intermolecular (isotopic) selectivity.

The photoselective laser-induced excitation and dissociation of molecules have been described in detail in a number of monographs and reviews (Jortner et al. 1981; Letokhov 1983; Bagratashvili et al. 1985b; Letokhov 1989; Quack 1998). Therefore, we shall present below only brief information about the progress made in this field. The main schemes for isotope-selective dissociation of molecules via vibrational states are shown in Fig. 11.1. Accordingly, the problem of photoselective laser control of molecules will be considered below in consecutive order, starting with the simplest scheme of Fig. 11.1(a) and ending with that of Fig. 11.1(c).

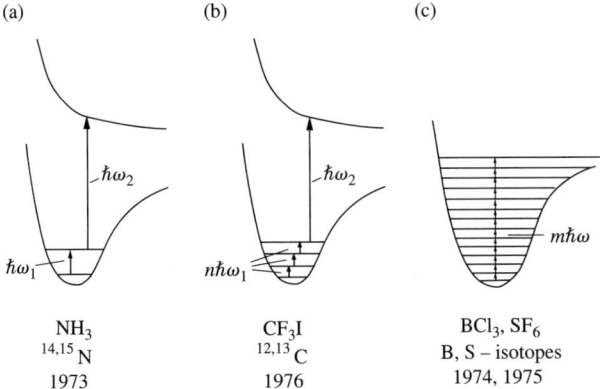

Fig. 11.1 Various schemes for isotopically selective photodissociation of molecules via vibrational states, and the first experiments (molecules, isotopes, and years).

11.1 Vibrationally mediated photodissociation of molecules via excited electronic states

Probably the first suggestion for utilizing the properties of laser light (the high intensity and short duration of radiation pulses) was (Letokhov 1969) to use the vibrationally mediated photodissociation of molecules via an excited repulsive electronic state with noncoherent isotope-selective saturation of the vibrational transition (Fig. 11.2). The isotope-selective two-step photodissociation of molecules consists of pulsed isotope-selective excitation of a vibrational state in the molecules by IR laser radiation and subsequent pulsed photodissociation of the vibrationally excited molecules via an excited electronic state by a UV pulse (Fig. 11.2(a)) before the isotope selectivity of the excitation is lost in collisions. Selective two-step photodissociation of molecules is possible if their excitation is accompanied by a shift of their continuous-wave electronic photoabsorption band. In that case, the molecules of the desired isotopic composition, selectively excited by a laser pulse of frequency ω_1, can be photodissociated by a second laser pulse of frequency ω_2 selected to fall within the region of the shift where the ratio between the absorption coefficients of the excited and unexcited molecules is a maximum (Fig. 11.2(b)).

The first experiment on isotope-selective dissociation was carried out with the $^{14}NH_3$ and $^{15}NH_3$ molecules, which could be selectively excited by CO_2-laser radiation (Ambartzumian et al. 1973). The IR and UV absorption spectra and the photochemical decomposition of these molecules were well known. The lowest frequency, ν_2, is in the 10 μm region. The absorption bands of $^{14}NH_3$ and $^{15}NH_3$ overlap but the spectrum exhibits a rich structure consisting of hundreds of vibration–rotation lines. This structure has been analyzed for both $^{14}NH_3$ and $^{15}NH_3$. In the spectrum of the ν_2 bands, there are some nonoverlapping rotational–vibrational lines, which coincide closely with CO_2-laser lines. In the UV region, the absorption spectrum of $^{14}NH_3$ consists of a vibronic progression of ν_2, which starts in the region of 2168 Å and ends at shorter wavelengths. The electronically excited Ã state is unstable because of predissociation. Molecular transitions from the ground state to an excited electronic

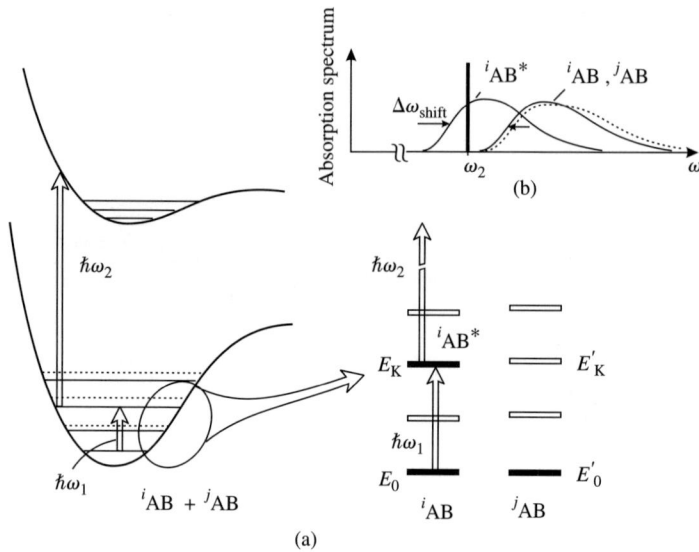

Fig. 11.2 (a) Schematic diagram of the IR-UV isotope-selective photodissociation of molecules ^{i}AB in a mixture with molecules ^{j}AB of different isotopic composition via an intermediate vibrational state (for example, $\nu = 2$); (b) change of absorption spectrum. (Modified from Letokhov 1973a.)

state and the corresponding spectral lines are shown schematically in Fig. 11.3. The vibrational spectrum is somewhat different for NH$_3$ molecules vibrationally excited to the $v''=1$ level of the ν_2 band. The most important feature is a new absorption line due to the $(\tilde{X}, v''=1) \to (\tilde{A}, v'=0)$ transition. It exhibits a "red" shift equal to $\hbar\omega_1$. When the $v''=1 \leftarrow v''=0$ vibrational transition is fully saturated by IR laser radiation, the molecules are equally distributed between the levels. Figure 11.3(b) shows the theoretical intensity distribution of the vibrational spectrum of ammonia for two cases: (i) 99% of the molecules are in the ground state $(\tilde{X}, v''=0)$ and (ii) the molecules are equally distributed between the levels $(\tilde{X}, v''=0)$ and $(\tilde{X}, v''=1)$. The new line at 45 250 cm^{-1} ($v'-v''=-1$ in Fig. 11.3) was used to photodissociate excited molecules. The separation factor in this experiment ranged between 2.5 and 6. The experiment demonstrated that the photodissociation method could in principle be used to separate isotopes, and revealed its potential advantages and restrictions.

The isotope-selective excitation of several vibrational levels in a polyatomic molecule through multiple-photon absorption of intense IR radiation can provide a much greater shift of the UV absorption band of the molecule toward the long-wavelength region (Fig. 11.1(b)) than in the case of one-photon absorption (Fig. 11.1(a)). This extends the possibilities of using the IR–UV photodissociation method for isotope separation purposes. The method was studied in detail in experiments with the ^{12}CF$_3$I and ^{13}CF$_3$I molecules (Knyazev et al. 1978). Vibrationally mediated photodissociation has some potential for bond-selective chemistry of simple molecules (Crim 1993) and, more importantly, of polyatomic molecules.

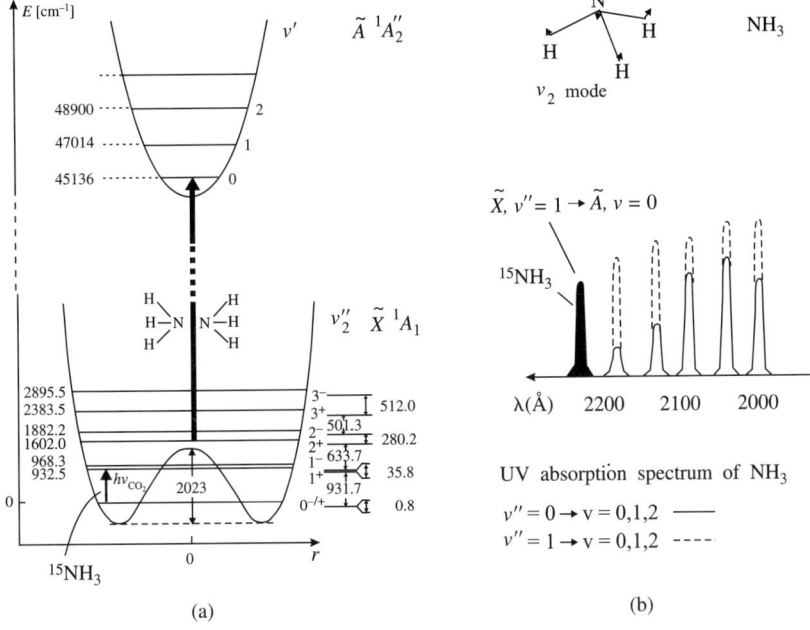

Fig. 11.3 (a) Energy-level diagram and vibrational transitions for the two-step dissociation of NH$_3$. (b) Spectral lines for normal (dashed lines) and laser-excited (solid lines) NH$_3$.

11.2 Basics of IR multiple-photon excitation/dissociation of polyatomic molecules in the ground state

In addition to the photodissociation of molecules via excited vibrational and electronic states under the effect of IR and UV radiation (Section 11.1), isotope-selective photodissociation can also take place in the electronic ground state as a result of absorption of a large number (20–100) of IR photons. This phenomenon was observed for the first time when the ^{10}B and ^{11}B isotopes in the BCl$_3$ molecule were separated (Ambartzumian et al. 1974). Thereupon successful experiments were performed on separating the ^{32}S and ^{34}S isotopes in the SF$_6$ molecule (Ambartzumian et al. 1975b). Figure 11.4 illustrates the results of these experiments, where it proved possible to directly observe isotope enrichment from the change in the IR absorption spectrum of the natural isotopic mixture under IR irradiation. By tuning the pulsed CO$_2$ laser used to resonance with the absorption band of the ^{32}SF$_6$ molecules, Ambartzumian et al. managed to decrease the concentration of these molecules to that of the ^{34}SF$_6$ isotopic molecules. The results of these experiments were soon confirmed by Los Alamos Laboratory researchers (Lyman et al. 1975). Experiments with BCl$_3$ and SF$_6$ molecules demonstrated the resonant feature of the IR MPE/D process, that is, its intermolecular selectivity, which could not be attained under conditions of thermal heating, gas discharge, flames, etc. Later on, this effect was demonstrated for a few tens of isotopes in around a hundred different molecules.

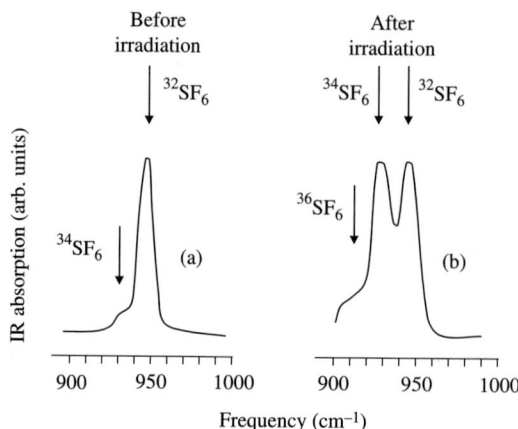

Fig. 11.4 Macroscopic 20-fold enrichment of SF$_6$ gas with the ^{34}S isotope by means of the IR multiphoton dissociation of ^{32}SF$_6$, observed in the IR spectrum of SF$_6$ before (a) and after (b) irradiation. (From Ambartzumian *et al.* 1975*b*.)

11.2.1 Selectivity of resonance multiple-photon excitation

The resonance MP excitation of molecular vibrations in an intense IR field forms the basis for several essentially different approaches to laser control. They can be classified according to the relations between the various relaxation times for the excited vibrational level interacting with the IR field:

$$\tau_{\text{stoch}} \ll \tau_{\text{transf}} \ll \tau_{\text{relax}}. \tag{11.1}$$

Here τ_{stoch} is the time for the intramolecular vibrational redistribution (IVR) of the absorbed energy among various vibrational modes of the excited molecule, τ_{transf} is the time for the intermolecular vibrational energy transfer between molecules of different kinds in the given molecular mixture (for example, a mixture of molecules differing in isotopic composition), and τ_{relax} is the relaxation time, that is, the time it takes for the molecular vibrational energy to be transferred to the translational degrees of freedom and for complete thermal equilibrium to be reached in the molecular mixture.

The rate of vibrational excitation of a molecule through multiple-photon absorption, W_{exc}, depends on the radiation intensity I and the vibrational-transition cross sections:

$$W_{\text{exc}} \simeq \langle \sigma_{\text{abs}} \rangle I, \tag{11.2}$$

where $\langle \sigma_{\text{abs}} \rangle$ is the average cross section for the successive vibrational upward stimulated transitions in the molecule. Considering the inequality (11.1), we can distinguish between four different conditions for the relations between W_{exc} and the vibrational-energy relaxation rates in various circumstances. Accordingly, there are four different approaches to IR MP laser chemistry, depending on how far from equilibrium the vibrational excitation in the molecule and the molecular mixture lies (Fig. 11.5): (1) mode-selective laser chemistry, (2) intermolecularly selective laser chemistry,

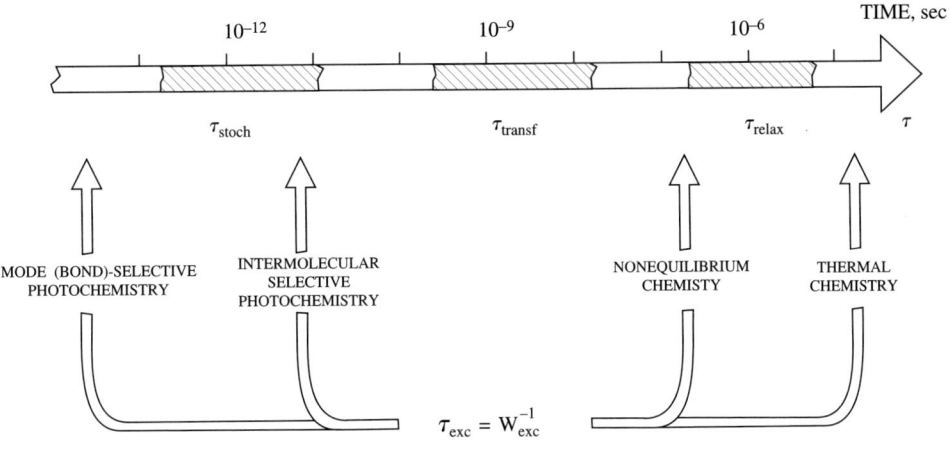

Fig. 11.5 Various rates of laser photoexcitation or deposition of light energy into a molecule can provide various types of laser-induced chemistry: from mode-selective photochemistry for very fast, subpicosecond rates of excitation to ordinary thermal chemistry for low-rate, millisecond excitation. The real values of relaxation times depend on the density of the irradiated substance.

(3) nonequilibrium laser chemistry, and (4) thermal laser chemistry. All the above-mentioned laser chemistry methods can be applied, though the degrees of development of various applications differ widely. IR MP selective photochemistry for isotope separation is the most advanced application.

The intense IR radiation of a pulsed laser at a frequency tuned to resonance with the vibrational absorption band of any polyatomic molecule causes a very fast vibrational excitation of the molecule. In other words, the vibrational degrees of freedom of a molecule in a resonant IR field are subjected to strong heating, which depends on the number of IR photons absorbed by the molecule. Using IR pulses with an energy fluence from 1 to 10 J/cm^2, one can relatively easily deposit an energy of 1–10 eV into a polyatomic molecule, for example by making it absorb 10–100 photons from a CO_2 laser operating in the 10 μm region. For such a strong vibrational excitation to take place, the photoexcitation rate W_{exc} and the laser pulse duration τ_p must satisfy the following condition:

$$\tau_p^{-1}, W_{exc} = \langle \sigma_{abs} \rangle I \gg W_{relax}, \qquad (11.3)$$

where W_{relax} is the relaxation rate of the excited vibrational states. This efficient excitation condition is satisfied when the laser energy fluence $\Phi = I\tau_p$ is greater than $\langle \sigma_{abs} \rangle^{-1}$ (photons/cm^2). For a single molecule to effectively absorb n photons, the laser pulse energy fluence Φ evidently must satisfy the more stringent condition $\Phi \gtrsim n/\langle \sigma_{abs} \rangle$ (photons/cm^2). The IR MP absorption cross sections $\langle \sigma_{abs} \rangle$ of various polyatomic molecules lie in the range 10^{-18}–10^{-20} cm^2.

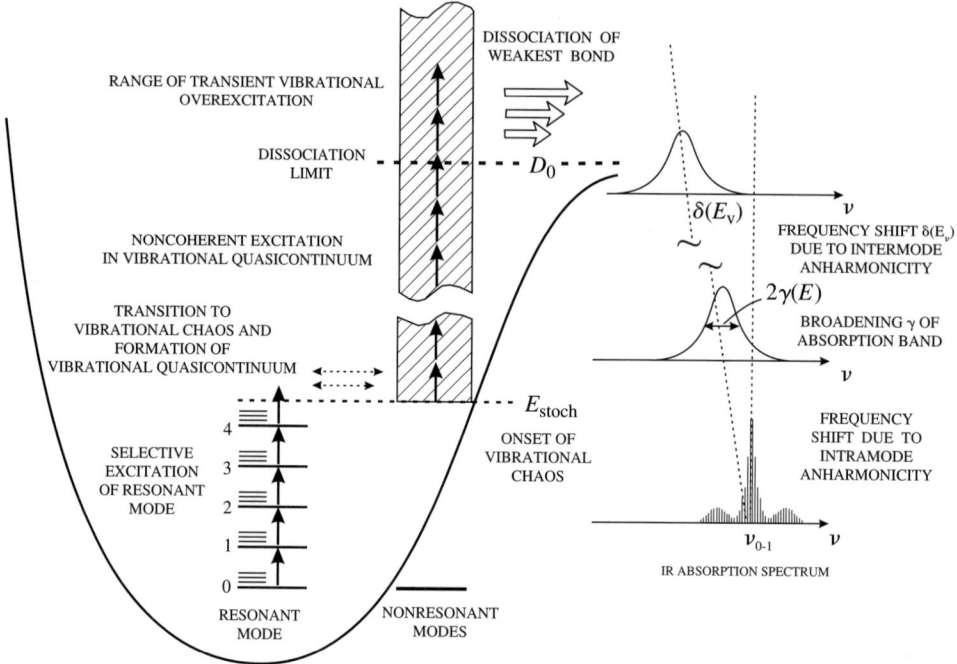

Fig. 11.6 Model of photodissociation of a polyatomic molecule by an intense IR field: *left*, scheme of vibrational-energy acquisition by the molecule in the regions of "mode-selective" and "mode-nonselective" excitation; *right* evolution of the fundamental IR absorption band spectrum with increasing vibrational energy of the molecule. Even at the dissociation limit the molecule is capable of absorbing, in a quasi-resonant fashion, IR radiation at a laser frequency ν_L tuned to the long-wavelength wing of the fundamental absorption band.

11.2.2 IR MP excitation/dissociation model

A qualitative model of the MP vibrational excitation and dissociation of an isolated molecule under the action of an intense IR radiation pulse was developed in the 1970s (Ambartzumian and Letokhov 1977; Bloembergen and Yablonovich 1978; Quack 1978; Schultz *et al.* 1979). The molecule, in the course of vibrational excitation, passes through the following three qualitatively different vibrational-energy regions: (1) the region of low-lying discrete vibrational–rotational levels, where single- or two-photon transitions are possible from a given quantum state; (2) the vibrational quasi-continuum (QC) region, where many close vibrational levels interact and many transitions are possible from a given state; and (3) the real continuum region above the dissociation limit, where unimolecular decay of the vibrationally overexcited isolated molecule becomes possible. At the same time, a deeper insight was gained into the MP excitation process, and a more detailed model, illustrated in Fig. 11.6, was formulated as a result of the development of the simple model. The most essential addition here is the understanding of how the IR absorption spectrum of the vibrational mode in resonance with the exciting IR radiation evolves with the increasing vibrational energy of the molecule as a whole.

The molecule absorbs the first few IR photons as usual via quantum transitions between the lower discrete vibrational–rotational levels. The intramodal anharmonicity of the resonant vibrational mode in this region gives rise to a red shift of the center of the IR absorption band. The interaction of the exciting IR field with molecular vibrations has a resonant or mode-selective, but not necessarily coherent, character. The interaction of the modes themselves for the present does not cause their mixing, unless there is an accidental coincidence of the frequencies of several modes, which gives rise to *intermode resonances*. Such mixing, however, becomes quite inevitable when the vibrational energy of the molecule becomes high enough. Of essence here is the fact that a polyatomic molecule has many vibrational degrees of freedom. Even a weak intermode interaction is therefore sufficient for many intermode resonances, otherwise known as Fermi resonances, to overlap. The overlapping of the intermode resonances allows the *stochastization* of the molecular vibrational energy. As the Fermi resonances are approximate, that is, they take place at frequency detunings of tens of wavenumbers, to make them overlap at a given intermodal anharmonicity, it is necessary that their density should be sufficiently high, and this is possible only with a certain vibrational-energy store. For this reason, the stochastization of vibrational energy starts at some threshold energy, referred to as the stochastization onset energy E_{stoch}, that can be considered the lower boundary of the QC. When the vibrational-energy store in the resonant mode exceeds E_{stoch}, the energy spreads over the other modes in accordance with the character of the intermodal resonances. This process may be repeated several times until the total vibrational energy of the molecule exceeds the dissociation energy D_0, to open up the unimolecular decay channel. In the case of an equilibrium energy distribution among all modes, the vibrational-energy store in each nonresonant mode of such a highly excited molecule amounts to a mere D_0/S, where $S = 3N - 6$ is the number of vibrational degrees of freedom of an N-atom molecule.

In the stochastic region, the shape of the absorption band of the molecule changes radically in the vicinity of the frequency of the resonant mode. Owing to the intermode interaction, a broadened absorption band is formed, whose width depends on the intramolecular vibrational-energy redistribution rate, and whose central frequency is shifted toward the long-wavelength region by an amount depending on the average *intermode anharmonicity*. Both the width and the shift of the vibrational band increase as the vibrational energy of the molecule grows higher, but as a rule, the shift does not perceptibly exceed the width. This is due to the fact that, far from resonance, all vibrational modes of the molecule contribute to the intermodal anharmonicity (in the model of the active and passive modes of a vibrational reservoir of Bagratashvili *et al.* 1985a). It is exactly this relationship between the width and the shift of the fundamental IR absorption band of a polyatomic molecule that ensures the possibility of its absorbing a larger number of IR photons.

The IR MP excitation model presented in Fig. 11.6 elaborates the initial, simple model in two respects. First, it introduces the notion of the stochastization energy E_{stoch}, which is determined by the density of the intermodal Fermi resonances and the intermode interaction. And second, it introduces the notion of a homogeneously broadened fundamental absorption band at a molecular vibrational energy $E_v > E_{\text{stoch}}$, with the bandwidth 2γ and the shift δ increasing with increasing vibrational energy of the molecule. Theoretical work has made it possible to develop models for calculating

E_{stoch}, $\gamma(E_v)$, and $\delta(E_v)$ and to relate them to the anharmonicity parameters, which is an essential step in the quantitative description of the IR MP excitation of polyatomic molecules (Bagratashvili et al. 1985a; Stuchebrukhov et al. 1986; Lokhman et al. 1999).

A highly vibrationally excited molecule can first undergo various monomolecular reactions (such as dissociation, fragmentation, and isomerization) and then its decomposition products, such as active radicals, can themselves take part in subsequent reactions.

11.2.3 Properties of IR MP excitation

Let us summarize some features of the processes of IR MP excitation and dissociation of polyatomic molecules that are most important as far as industrial applications are concerned.

1. *Versatility.* IR MP processes can be observed in polyatomic molecules of any complexity and symmetry if they have more than four or five atoms and have IR absorption bands in the lasing spectral region of high-power IR lasers. Most experiments on MP excitation and MP dissociation have been carried out with CO_2 lasers ($\lambda = 9$–10 μm). The development of free-electron lasers tunable over a wide IR range made it possible to observe IR MPE/D of molecules when the 3 μm C–H stretch vibrational absorption band and the 6 μm C=O band were acted on (Petrov et al. 1997).

Resonance IR MPE/D is a universal phenomenon for polyatomic molecules, although the average MP absorption cross section depends on the degree of complexity of the molecule. Figure 11.7 presents experimental data on the absorption cross section σ_{abs} and, accordingly, the MP dissociation threshold energy fluence Φ_0 for a number of simple and complex molecules. On the left of Fig. 11.7, approximate values of the vibrational level density are presented, whose increase substantially lowers the threshold IR MP dissociation energy fluence.

2. *Intermolecular selectivity.* Intense IR radiation can deposit the energy necessary for dissociation into the desired molecules in a gas mixture without depositing so high an energy into the other molecules in the mixture. This is due, first, to the great difference in MP absorption spectra between different molecules and, second, to the possibility of realizing collisionless excitation conditions during the course of an IR pulse, where there is no significant collisional energy transfer from the excited molecules to the other molecules in the mixture. This most important feature of IR MP processes is being used successfully in laser isotope separation.

It should be noted that the MP vibrational heating of polyatomic molecules is sometimes misunderstood as simple heating of them. A typical statement is "An early attempt at laser-controlled chemistry amounted to nothing more than an expensive way to heat up the test tube. The energy went in, but it just broke the weakest link in the chain—exactly like a Bunsen burner." In actual fact, this is not the case at all. A Bunsen burner cannot provide the intermolecular selectivity that IR MPE/D processes are capable of under appropriate conditions. In the course of collisionless IR MP excitation, laser radiation increases the vibrational energy of only the chosen molecules, their translational energy remaining almost constant. This situation differs from the case of purely thermal initiation, in which all of the molecular degrees of

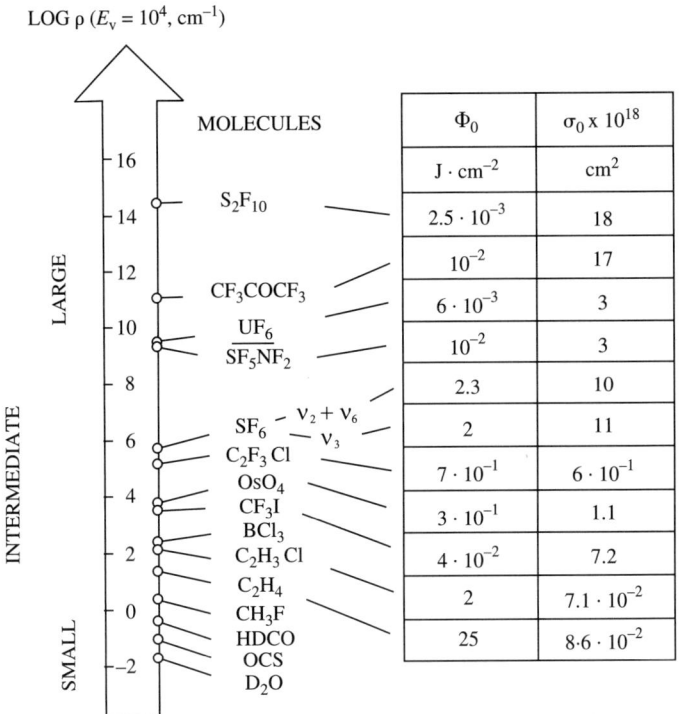

Fig. 11.7 The growth of the density of vibrational levels of a molecule in line with its increasing complexity leads to an increase in its average IR MP excitation cross section and a decrease in its IR MP dissociation threshold energy fluence.

freedom (vibrational, translational, and rotational) have the same high temperature and all the molecules in the mixture are excited.

3. *Vibrational-energy stochastization.* A high-power IR field involves all or at least many modes of a polyatomic molecule placed in it in the process of vibrational excitation. This conclusion follows from numerous direct and indirect experiments concerned with intramolecular vibrational-energy redistribution. As the MP excitation of polyatomic molecules is effected with IR radiation pulses 10^{-6}–10^{-8} s long at a rate of $W_{exc} \sim 10^7$–10^9 s^{-1}, the vibrational motion of the molecules excited to the dissociation limit becomes stochastic and can be described by statistical methods. Indeed, in all reliable experiments on the IR dissociation dynamics of molecules, no essential difference from the predictions of the statistical unimolecular-decay (RRKM) theory could be observed. This, among other things, means that the dissociation of MP-excited molecules generally occurs as a result of a break in the weakest bond.

The stochastization limit E_{stoch} and hence the necessary order of the multiple-photon process depend strongly on the number of atoms in the given molecule and the relationship between the frequency ν_{res} of the resonant mode and the frequencies of the other modes. For simple molecules, the stochastization limit lies high, and such molecules are therefore difficult to excite so that they rise to the quasi-continuum up the ladder of the lower levels. In contrast, for large-sized molecules, even the first vibrational level is strongly coupled to the QC, and so the required order of the multiple-photon process is in this case equal to unity, that is, it is a single-photon process. For simple molecules with a high QC boundary, a high-order multiple-photon single-frequency process is ineffective. In this case, it is most advisable to use multiple-frequency IR radiation whose frequencies are suitably selected to raise the molecule up the ladder of the lower vibrational levels by way of resonance multiple-step excitation.

11.3 Characteristics of the IR MPE/D of polyatomic molecules

The characteristics of the multiple-photon vibrational excitation and dissociation of polyatomic molecules can be conveniently considered in an order corresponding to the above qualitative picture of the three stages of this process.

11.3.1 Multiple-photon excitation of lower vibrational levels

Systematic studies into resonant transitions between the low-lying ($v = 0, 1, 2, 3$) vibrational–rotational levels of polyatomic molecules were undertaken precisely because of the need to explain the spectra of resonant multiple-photon excitation (MPE) and multiple-photon dissociation (MPD) of polyatomic molecules by IR laser radiation. Despite the large amount of experimental and theoretical work, up to now no comprehensive qualitative description of the MPE process for the lower levels of polyatomic molecules has been achieved, which is probably explained by the lack of data on the complex structure of the vibrational levels $v = 2, 3, 4$ of such molecules. Nevertheless, the qualitative picture seems quite clear.

Data on the spectra of multiple-photon excitation of the lower vibrational transitions are obtained by measuring the dependence on the laser frequency of the absorbed energy or of the energy of the excited molecules at moderate laser energy fluences or intensities, when the molecules absorb only a few photons and are still far from the limit of photodissociation. The main features of MPE of the lower levels may be summarized as follows:

1. MPE spectra are *wider* and *shifted toward the red region*, as compared with the linear IR absorption spectrum. The first data on MPE spectra were obtained in experiments on the photodissociation of the SF_6 molecule under the effect of two IR pulses differing in frequency (Ambartzumian et al. 1976). The IR pulse frequencies ω_1 and ω_2 were selected so that the first pulse, of moderate intensity at the frequency ω_1, effected MPE of the lower levels, whereas the second, more intense pulse at the frequency ω_2 brought about the MPD of the molecules vibrationally excited by the first pulse. By varying the frequency ω_1 while keeping fixed the frequency ω_2 lying outside the absorption band of the lower vibrational transitions, the researchers could measure the MPE spectrum of these transitions.

2. MP absorption and excitation spectra may have a *sharply resonant* structure. This becomes manifest when the molecules are cooled (Alimpiev et al. 1979) and when a continuously tunable high-pressure CO_2 laser is used for MP excitation (Alimpiev et al. 1979). The intensity of some peaks showed a clear-cut power-law dependence on the laser pulse fluence Φ or intensity I ($\sim I^n$, where $1 < n \leq 2$), which points to

a contribution of two-photon resonances to the MPE process. As the laser intensity grows higher, the positions of the peaks in the MPE spectrum remain the same, but their contrast diminishes because of the appearance of other MP resonances, and so the width of the MPE spectrum increases. More detailed information on MPE spectra is presented in the book by Letokhov (1989).

3. MPE of the lower vibrational transitions is accompanied by the *depletion of most rotational levels* of the vibrational ground state. At low IR pulse intensities, when the contribution of MPE is negligibly small, the monochromatic IR field, in the absence of rotational relaxation during the pulse, interacts only a with small fraction of the molecules in the rotational levels J_{res}, for which the $(v=0; J_{res}) \to (v=1; J_{res} \pm 1)$ resonance transition at the field frequency ω is possible. The thermal relative population of these rotational levels determines the maximum fraction f of molecules that can be excited by a short pulse so as to rise to the level $v=1$ when the vibrational–rotational transition is saturated. This effect, referred to as the *rotational bottleneck*, was considered by Letokhov and Makarov (1973). The value of f, estimated for various molecules, ranges between 0.1 and 0.001 and depends strongly on the temperature of the molecular gas.

When the bleaching of the IR absorption band of many vibrational–rotational transitions was observed simultaneously, it was inferred that a great number of initial rotational states became depleted under the action of an intense IR pulse (Alimpiev et al. 1977). This experiment demonstrated that the fraction of molecules involved in the MPE process, $q(\Phi)$, at $\Phi = 0.01$–1 J/cm² is much greater than the fraction f of molecules interacting linearly with the IR field, that is, $q(\Phi) \gg f = q(0)$. Thus, under typical conditions, the MPE process produces two molecular ensembles. The fraction q of molecules involved in MPE forms an ensemble of vibrationally "hot," that is, highly excited, molecules. The rest of the molecules, $1 - q$, remain in the lower vibrational levels and form an ensemble of vibrationally "cold" molecules (Fig. 11.8(a)). Figure 11.8(b) shows the dependence on the laser fluence of the fraction q of molecules excited into the vibrational QC for three different molecules. The conclusion as to the depletion of many rotational levels was later confirmed by probing the populations of individual rotational levels of the SF_6 molecule with a tunable diode laser (Apatin et al. 1983).

To give a theoretical interpretation of the spectra of MPE of the lower vibrational levels, it is necessary to know the actual structure of the vibrational levels $v = 0, 1, 2, 3$ of the vibrational mode being excited over the entire vibrational-energy range up to the stochastization limit. The most detailed data are available for the ν_3 vibration of the SF_6 molecule, mainly on account of research on the spectrum of its second overtone (0–$3\nu_3$). It has been demonstrated that, apart from the anharmonic shift and the rotational-level structure, the octahedral splitting of the levels of the triply degenerate ν_3 vibration is also of importance. The anharmonic splitting increases as the principal quantum number increases, which compensates for the small anharmonic shift and opens up many possible pathways for multiple-photon resonances in the ladder of lower vibrational levels (Cantrell et al. 1980). Allowing even for the simplest two-photon resonances makes it possible to explain a considerable proportion of the peaks in the MPE spectrum observed. Figure 11.9 shows the results of experimental measurements

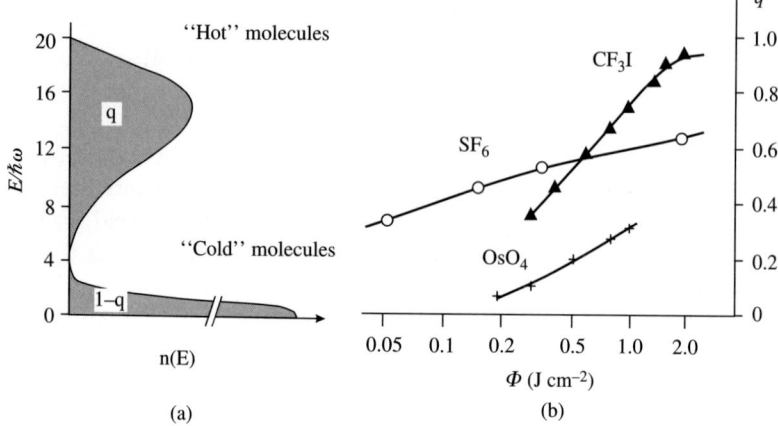

Fig. 11.8 Formation of two ensembles of vibrationally "hot" and "cold" molecules on account of the rotational "bottleneck" effect, upon MP excitation of molecules with an intense IR pulse: (a) vibrational-energy distribution for MP excitation; (b) relative fraction of highly excited molecules as a function of the CO_2-laser pulse fluence. (From Letokhov 1983.)

(Alimpiev et al. 1979) of the MPE spectrum of the SF_6 molecule, its linear absorption spectrum (bottom), and the theoretical two-photon absorption spectrum (top). As can be seen, the experimentally observed spectrum fits quite satisfactorily the results of a calculation considering only the single- and two-photon transitions to the split sublevels of the $2v_3$ state.

The multiple-photon molecular excitation mechanism should not include multiple-photon processes of too high an order, because the molecule needs to be excited only to an energy level $n_{res}\hbar\omega_{res} < E_{stoch}$. The absorption of the next photon will now lead to the excitation of a vibrational level interacting with the other nonresonant vibrational modes, stochastization of the vibrational energy of the molecule, and a corresponding broadening of the vibrational level $(n_{res}+1)\hbar\omega_{res}$. The number of IR photons absorbed by the resonant mode, n_{res}, depends on the magnitude of the stochastization energy. The stochastization limit E_{stoch} and hence the necessary order of the multiple-photon process depend strongly on the number of atoms in the given molecule and the interaction between the resonant-mode frequency and the frequencies of the other modes.

11.3.2 Vibrational-energy stochastization

In the classical vibrational spectroscopy, the subject of investigation is the vibrational–rotational motion of polyatomic molecules near the very bottom of their potential-energy surface in the electronic ground state. In this case, the normal-mode approximation proves quite applicable. Indeed, one can expand as a Taylor series the potential energy of the molecule near the equilibrium position (the potential-energy minimum) and write down the molecular Hamiltonian in the form

$$H = \sum_{i=1}^{s} \omega_i \left[\left(\frac{p_i^2}{2}\right) + \left(\frac{q_i^2}{2}\right) \right] + \left(\frac{1}{3!}\right) \sum_{i,j,k=1}^{s} \Phi_{ijk} q_i q_j q_k + \cdots, \qquad (11.4)$$

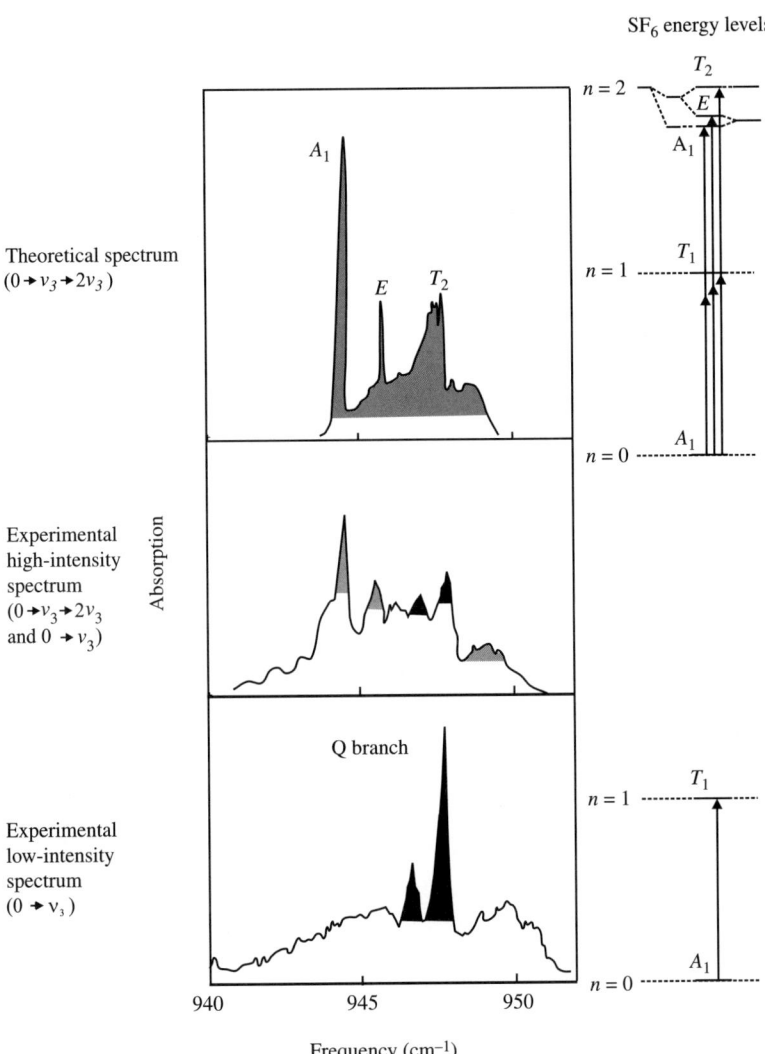

Fig. 11.9 Contribution of two-photon transitions to absorption by the lower vibrational levels of the SF$_6$ molecule. The linear absorption spectrum of the ν_3 vibration at low IR radiation intensities is shown at the *bottom*. When the intensity is raised to 3×10^6 W/cm^{-2}, there appear in the spectrum (shown in the *center*), in addition to one-photon peaks, also two-photon absorption peaks (shaded). The theoretical two-photon absorption spectrum is shown at the *top*. The low-frequency peaks of the A_1 and E transitions appear at high radiation intensities, whereas the T_2 transition coincides with the linear absorption peak. (Adapted from Alimpiev *et al.* and Letokhov 1989.)

where q_i and p_i are the normal coordinates and momenta. Assuming that all the terms except the first one in eqn (11.4) are small, we at once get s independent harmonic oscillators with frequencies ω_i. In this normal-mode approximation, the state $|v_1, v_2, \ldots, v_s\rangle$ is described by a set of numbers v_i, with v_i denoting the number of vibrational quanta in the ith mode, and the selection rules in the case of dipole interaction with the external field allow transitions involving a change in only one of these numbers: $v_i \to v_i \pm 1$. The small terms in eqn (11.4) can be taken into consideration by perturbation theory, which will lead, first, to an anharmonic shift and splitting of the normal modes as a result of their interaction and, second, to the occurrence of vibrational transitions at combination frequencies and overtones with intensities lower than those of the transitions at the fundamental frequencies.

This simple picture holds true for polyatomic molecules only at low vibrational energies, no more than, say, a few thousands of wavenumbers, and becomes invalid as the molecular energy is increased. This is due to the fact that as the vibrational energy of the molecule grows higher, the higher-order terms in the expansion in eqn (11.4) increase faster than the quadratic ones and cannot be neglected any longer. The normal modes start to mix, giving rise to a qualitative change in the intramolecular dynamics, that is, a transition from quasi-periodic (or regular) to stochastic (or irregular) motion. This transition was observed for the first time by Henon and Heiles (1964) in numerical experiments carried out within the framework of classical mechanics on a system of two oscillators with a cubic interaction. Numerous subsequent investigations, both numerical investigations and rigorous mathematical ones, have demonstrated that such a chaotic motion is typical of nonlinear systems at energies exceeding some threshold value.

Starting at a certain critical molecular energy E_{stoch}, in any polyatomic molecule there takes place the intramolecular vibrational-energy redistribution (IVR) effect. This effect is due to what is known as dynamical chaos, which is inherent in vibrational systems with many degrees of freedom. The magnitude of E_{stoch} differs between different molecules, ranging, for example, from ≤ 3300 cm^{-1} for H≡C–CH$_2$CH$_3$ (McIlroy and Nesbitt 1990) to 7500 cm^{-1} for CF$_3$Br (Doljikov et al., 1986), and higher for lighter and/or simpler molecules. However, the IVR process itself takes some time to proceed. Typical IVR times range from a fraction of a picosecond to a hundred picoseconds, depending on the following three factors: the molecular species being excited, the vibrational mode therein subject to excitation, and the total vibrational energy of the molecule (Nesbitt and Field 1996).

Considering the IVR effect, the excitation of a molecule by a laser pulse can be represented schematically as a competition between two processes—the pumping of some mode and the transfer of the pump energy to the other modes (Fig. 11.10). Under ordinary conditions, where the IVR rate exceeds the pumping rate, the pump energy is distributed among the modes in a statistical equilibrium manner. Accordingly, molecular reactions, both unimolecular (dissociation and isomerization) and bimolecular (collisional) ones, proceed by the statistical mechanism described within the framework of the RRKM (Rice–Ramsperger–Kassel–Marcus) theory. It is only in the opposite limiting case, where the IVR process has not enough time to redistribute the pump energy all over the molecule, that one can actually reckon on the possibility of a

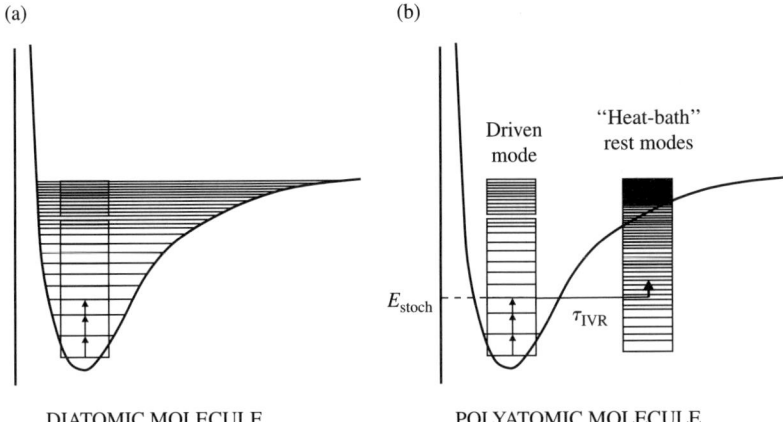

Fig. 11.10 Illustration of the crucial role of the intramolecular vibrational redistribution (IVR) above the threshold of stochastization for IR MPE/D of a polyatomic molecule (b), which does not occur for a simple diatomic molecule (a).

nonstatistical (mode-selective) process. The absence of the IVR process in a simple diatomic molecule (Fig. 11.10(a)) explains the impossibility of realizing IR MPD of such molecules in IR fields of moderate intensity (10^6–10^{10} W/cm^2).

That the vibrational-energy stochastization limit actually exists is also shown by experiments. Here we shall present for illustration the convincing results of experiments conducted by Doljikov et al. (1986) on the CF$_3$Br molecule excited with a powerful CO$_2$-laser pulse. By using a Raman-scattering technique to probe the vibrational energy in the various vibrational molecular modes (the mode in resonance with the IR pulse and the other, nonresonant modes), these authors managed to measure independently the following two parameters immediately after the IR pulse: (1) the proportion q of the molecules that found their way into the stochastization region where the vibrational energy was distributed among all the modes, and (2) the average vibrational-energy store $\langle E_q \rangle$ of these molecules. Figure 11.11 presents $\langle E_q \rangle$ and q measured in this work as a function of the energy fluence of the exciting IR pulse. It can be seen from the curves that as the laser pulse energy fluence is increased, the proportion q of the molecules in the QC grows rapidly at first and then more slowly and, later on, the vibrational-energy store of the molecules excited into the QC rises. What is usually measured is the average vibrational energy level per molecule, that is, the quantity $\langle E \rangle = q \langle E_q \rangle$, which rises monotonically with increasing laser pulse energy fluence. In the case illustrated in Fig. 11.11, one can clearly see the stochastization limit, that is, the minimum energy of the molecules in the vibrational QC.

The vibrational-energy stochastization time in a molecule with many degrees of freedom is determined by the anharmonic interaction therein, and lies in the picosecond region. There are also a number of possibilities here for direct measurements with high time resolution. Specifically, one situation of great interest is that where not all but only some of the vibrational degrees of freedom are interacting. Such a situation may, for example, occur in complex molecules that have different spatially separated functional groups. In this case, the anharmonic interaction between the vibrations

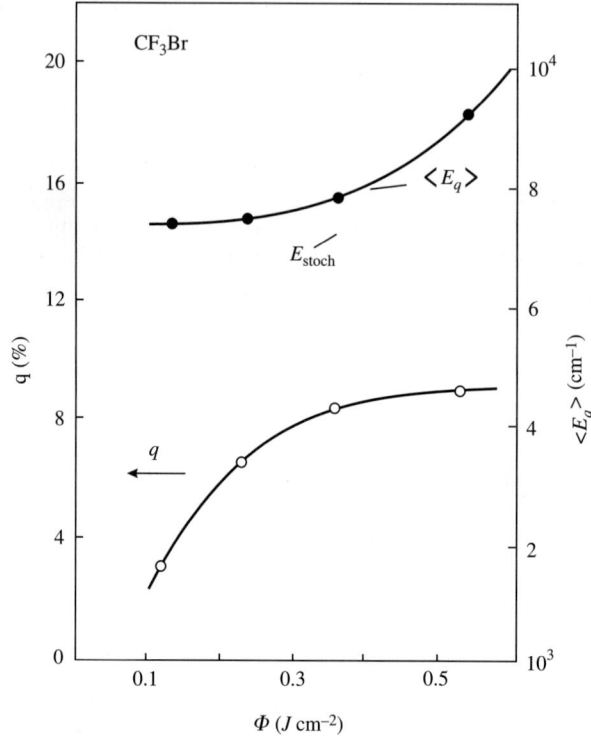

Fig. 11.11 IR MP vibrational excitation of the CF_3Br molecule: the fraction q of molecules that have found their way into the region of stochastization of vibrations (*bottom*) and their average energy $\langle E_q \rangle$ (*top*) as a function of the CO_2-laser pulse fluence. (From Doljikov et al. 1986.)

of these functional groups may prove sufficiently weak. It is also possible that some molecular mode may lack any suitable Fermi resonance that would provide coupling to the other modes. In such situations, one can observe a stochastization of the vibrational motion in a molecule for some 10^{-10} s (Malinovsky et al. 2004).

11.3.3 Excitation of molecules in the vibrational quasi-continuum

Referring to the realistic model of MP excitation of vibrations in a polyatomic molecule (Fig. 11.6), it is obvious that the excitation of the line shape of the molecule above the stochastization limit is governed by the evolution of the line shape of the fundamental absorption band $\sigma(\omega_i, E_v)$ with the rising vibrational energy E_v of the molecule. The evolution of the absorption spectrum, including homogeneous and inhomogeneous broadening of the contour of the absorption band, and the vibrational-energy distribution in highly excited molecules were considered in Letokhov (1989). Here we shall only emphasize that because of the high intramolecular vibrational-energy relaxation rate, the process of molecular excitation in the QC can be considered to be incoherent. This is always possible if the stimulated transition rate is less than the phase relaxation rate $1/T_2$ in the initial and/or final quantum transition state:

$$\Omega_0 = \frac{dE}{\hbar} \ll 1/T_2, \tag{11.5}$$

where Ω_0 is the Rabi frequency, equal to the stimulated transition rate, d is the transition dipole moment, and E is the field strength of the light wave. The incoherent interaction of molecules in the QC with an IR field may be described by simple rate equations. For example, the irreversible transition of a molecule from the last, kth discrete level in the lower vibrational-level system into the QC states, confirmed by a fast IVR process, may be described by the equation

$$\frac{dP_k}{dt} = -W_{k,k+1} P_k, \tag{11.6}$$

where P_k is the population of the last, kth, discrete level excited upon absorption of k photons in the lower transitions, and $W_{k,k+1}$ is the probability of a stimulated transition into the QC states upon absorption of the $(k+1)$th photon. There is no need to take account of the reverse transitions from the QC to the lower discrete levels, because of the large difference between the numbers of states of the discrete levels and the spectrum interval being excited into the QC.

The kinetics of the population P_n of the QC levels is described by the equation

$$\frac{dP_n}{dt} = \sigma_{n-1,n} I \left[P_{n-1} - \left(\frac{\rho_{n-1}}{\rho_n}\right) P_n \right] - \sigma_{n,n+1} I \left[P_n - \left(\frac{\rho_n}{\rho_{n+1}}\right) P_{n+1} \right], \tag{11.7}$$

where ρ_n is the density of vibrational states of a molecule with vibrational energy $E_{\text{vib}} = n\hbar\omega$, I is the radiation intensity (in photons/cm^2 s), and σ is the cross section for successive transitions, which is related to the dipole moment $d(E)$ and the line shape of the respective transition. When the energy of a polyatomic molecule exceeds 5–10 vibrational quanta, the density of vibrational states becomes very high. However, it should be emphasized that the IVR rate $\gamma_{\text{IVR}} = 1/\tau_{\text{IVR}}$ depends not on the density of vibrational states, but on the density of the intermode Fermi resonances (Kuz'min and Stuchebrukhov 1989).

The properties of the solution of eqn (11.7) depend on the ratio between the rates of direct and reverse transitions in the QC:

$$\frac{W_{n,n+1}}{W_{n+1,n}} = \frac{\rho_n}{\rho_{n+1}}. \tag{11.8}$$

For example, in the extreme case $\rho_n \gg \rho_{n-1} (n > k)$, for equal cross sections $\sigma = \sigma_{n,n+1}$ the solution of eqn (11.8) has the following simple form, a Poisson distribution with a relatively narrow width (Fig. 11.12):

$$P_n = \frac{(\sigma I \tau_p)^n}{n!} \exp(-\sigma I \tau_p) = \left(\frac{\langle n \rangle^n}{n!}\right) e^{-n}, \tag{11.9}$$

where $\langle n \rangle = \sigma I \tau_p = \sigma \Phi$ is the average vibrational excitation level produced under the effect of a pulse τ_p in duration, and Φ is the photon flux. Obviously, in this extreme case the average number $\langle n \rangle$ of IR photons absorbed by the molecule is a linear function of Φ, which not contradict the experimental data.

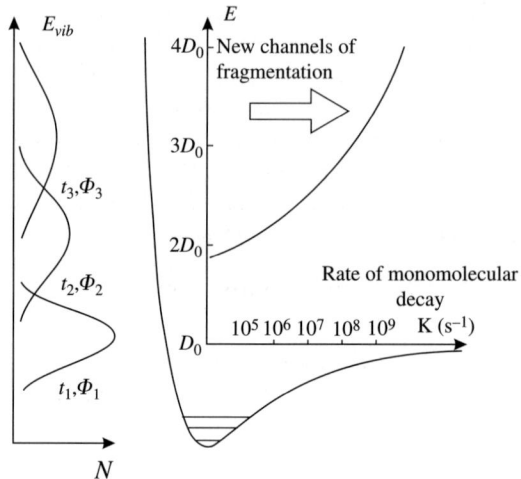

Fig. 11.12 Illustration of vibrational IR MP overexcitation of a polyatomic molecule (14 atoms, $S=36$) above the dissociation limit D_0, which opens new channels of fragmentation.

11.3.4 Unimolecular decay of vibrationally overexcited molecules

The action of an intense IR pulse gives rise to a distribution of molecules among the vibrational levels (Fig. 11.12), the tail of which reaches the boundary of a real continuum of states wherein the molecules are unstable. If the vibrational energy of an isolated molecule exceeds the dissociation energy D_0, or, in other words, the breaking energy of the weakest bond X_0, the molecule can decay spontaneously to produce fragments. The decay rate should apparently depend on the state of the molecule. For example, if the entire vibrational excitation energy $E > D_0$ is localized in the bond X_0, the decay should occur very quickly, approximately in a single vibration period. But if the molecular energy is concentrated mainly in those vibrational degrees of freedom which have no connection with the vibrations of the bond X_0 for the molecule to decay requires that a fluctuation should occur, as a result of which an energy $E \geq D_0$ would be concentrated in the bond X_0. Of course, the realization of such a fluctuation requires a much longer time than a vibration period. For large molecules, this time can be commensurate with the laser pulse duration $\tau_p \simeq 10^{-7}$–10^{-8} s and allow the vibrational-energy E to exceed appreciably the dissociation limit D_0 (this phenomenon may be termed *vibrational overexcitation*), and make it possible to study the properties of vibrationally overexcited molecules.

The MPE rate of a molecule competes with its unimolecular decay rate when its vibrational-energy store E_{vib} exceeds the dissociation energy D_0 of its weakest bond. The unimolecular decay rate $k(E)$ described by the RRKM theory (Robinson and Holbrook, 1972) increases rapidly as the excess energy E_{vib} over D_0 grows higher. Figure 11.12 illustrates the IR MP vibrational overexcitation of a 14-atom molecule ($S=36$ vibrational degrees of freedom). It can be seen that for more complex molecules and/or shorter laser pulses, the overexcitation energy can be much higher than the dissociation energy D_0 (see Section 10.2). The first evidence of a considerable vibrational overexcitation was obtained for several molecules, namely $(CF_3)_3CI$

($S=36$), $C_6H_5CH_2NH_2$ ($S=45$), and anthracene, $C_6H_4(CH)_2C_6H_4$ ($S=66$), by Bagratashvili et al. (1983). Experiments with $(CF_3)_3CBr$ molecules in a molecular beam have shown that their decay rate under not very high overexcitation conditions is $k \simeq 3 \times 10^3$ s^{-1}. The direct time-resolved photoionization detection of a dissociation product of the $(CF_3)_3CI$ molecule, the iodine atom, has made it possible to measure not only the slow decay rate of this molecule and the dependence of its dissociation rate on the vibrational overexcitation level, but also its IR absorption spectrum far above the dissociation limit (Bagratashvili et al. 1989).

11.3.5 Coupling of highly vibrationally excited and electronically excited states

The MP IR excitation of vibrational states in a molecule usually takes place within the limits of the electronic ground state. However, a polyatomic molecule can absorb so great a number of IR photons that its vibrational energy E_v becomes comparable to the energy of an electronically excited state. This is possible, for example, if the minimum energy E_1 of the electronically excited state lies below the dissociation energy D_0 (Fig. 11.13(a)), or even it $E_1 > D_0$ in the case of vibrational overexcitation (Fig. 11.13(b)). In such cases, the approximate description of the molecular dynamics by totally separated vibrational and electronic motions, that is, a description in the Born–Oppenheimer adiabatic approximation, becomes incomplete, and it then becomes necessary to take account of the nonadiabatic coupling or interaction between the highly excited vibrational states of the ground electronic term S_0^*, which has an energy $E = E_v$, and the low-lying vibrational states of the excited electronic state S_1, which has a vibrational energy $E_v = E - E_1$ and an electronic energy $E_{el} = E_1$, that is, the same total internal energy E. Owing to this coupling, a large number of levels of the term $S_0^*(0, E)$, with a density of states $\rho_0(E)$, interact with a comparatively small number of levels of the term $S_1(E_1, E - E_1)$, with having a density of states $\rho_1(E - E_1)$. This manifests itself in the form of prolonged visible or UV fluorescence of the state S_1, observable during a time $\tau_1 \gg \tau_0$. Of course, this holds true only so long as all the other decay channels for the vibrationally excited molecules S_0^* are slower than the radiationless process of "inverse" electronic relaxation, described above, otherwise the duration of fluorescence will be determined by the time for which the "pumping" of the $S_0^* \to S_1$ transition continues. Inverse electronic relaxation, the converse of the well-known direct electronic relaxation $S_1 \to S_0^*$, has been experimentally observed in the MP IR excitation of OsO_4 and other molecules. A detailed study of inverse electronic relaxation in molecules under IR MPE conditions was conducted by Puretzky and Tyakht (1979). A polyatomic molecule in an electronically excited state may continue participating in an MP absorption process. This is possible if the frequency of the fundamental IR absorption band is maintained in the excited electronic state S_1. As a result, the molecule can be excited above the dissociation limit D_1 of the excited electronic term, so that there can appear dissociation products other than those resulting from the breaking of the weakest bond in the electronic ground state, and also products in an excited electronic state (Fig. 11.13(b)).

A polyatomic molecule with a large number of atoms can be overexcited high above the dissociation limit (Section 11.3.4). In that case, there can occur an intermixing of the vibrational states of the molecule above the dissociation limit with an excited

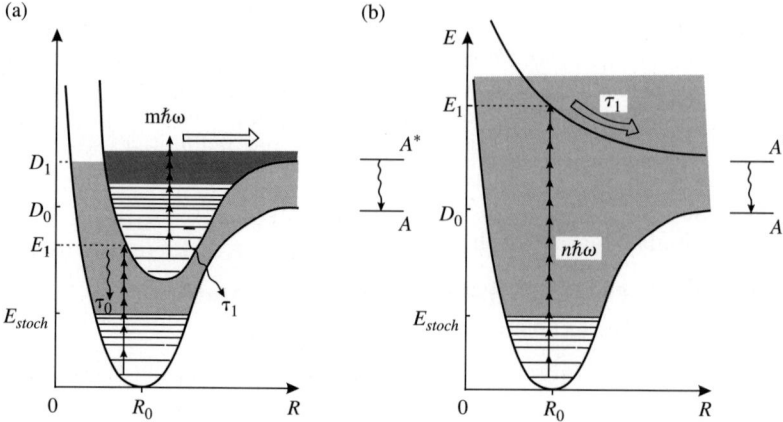

Fig. 11.13 Various schemes for excitation of electronic states in IR MP excitation of vibrations in a polyatomic molecule. The minimum energy of an excited electronic state can lie (a) below or (b) above the dissociation energy D_0.

electronic term. As a result, the molecule can find itself in a repulsive electronic term lying above D_0 and dissociate via a channel leading, for example, to the formation of an electronically excited fragment (Fig. 11.13(b)). Because of the short lifetime of the molecule in the repulsive electronic state, the return process $S_1 \rightarrow S_0^*$ for the molecule is practically impossible. In essence, there occurs electronic predissociation of the molecule on account of mixing of electronic terms that do not intersect unless there is a high vibrational overexcitation. Such a process was observed experimentally in the IR MP excitation of the $(CF_3)_3CI$ molecule and led to the production of iodine atoms in an excited electronic state (Bagratashvili *et al.* 1982).

11.4 Intermolecular selectivity of IR MPE/D for laser isotope separation

The isotopic selectivity of IR MP dissociation was first shown in experiments with the $^{10}BCl_3$ and $^{11}BCl_3$ molecules (Ambartzumian *et al.* 1974). In this case, of isotope shift in the frequency of the 10 μm vibration being excited was 39 cm^{-1}. A great number of subsequent experiments with different molecules showed that the case of BCI$_3$ molecule was not unique and that the selectivity of MP dissociation is a characteristic property of it. In experiments (Horsley *et al.* 1980) on the separation of the uranium isotopes ^{235}U and ^{238}U, isotopic selectivity of dissociation was observed when the isotope shift in the excited mode was just 0.7 cm^{-1}. Laser isotope separation is one of the most important industrial applications of selective MP dissociation. Therefore, the great majority of experiments dealing with the study of the enrichment kinetics and the dependence of the selectivity on the excitation conditions have been carried out with different isotopic molecules, but the results obtained are valid for mixtures of any other types of molecular species.

Many experiments on the isotopic selectivity of MP molecular dissociation have thus far been performed. These experiments have covered many isotopes, from light ones (hydrogen, deuterium, and tritium) to heavy ones (osmium and uranium), contained in very different molecules. Studies into the laws governing the evolution of the vibrational spectra of polyatomic molecules subject to increasing vibrational excitation have helped to develop various IR dissociation schemes capable of maximum isotope selectivity (or separation factor) and dissociation yield. For example, the use of a two-frequency IR field, where the isotope-selective excitation and dissociation functions are separated (Ambartzumian et al. 1976), and the dissociation of comparatively simple polyatomic molecules with a high-lying quasi-continuum boundary in a multiple-frequency IR field can provide a separation factor as high as 10^3–10^4 (Evseev et al. 1985). One more method of multiple-frequency IR dissociation of molecules is based on the excitation of overtones of high-frequency vibrations (C–H, Si–H, etc.) in them as a result of one-photon absorption and subsequent IR MPD of these overtone-preexcited molecules. By virtue of the high selectivity of the first step, this method can provide high dissociation selectivity. To illustrate this, in the IR MPD of $^{13}CF_3H$ molecules preexcited to the second C–H stretch overtone, the separation factor $\alpha(^{13}C/^{12}C)$ reached around 9000 (Boyarkin et al. 2003).

Laser isotope separation (LIS) methods based on the photodissociation of molecules (the MLIS approach) are capable of very high elementary separation act parameters (ESAPs), namely, the yield of the desired isotope, β_i, and the separation factor (selectivity), $\alpha = \beta_i/\beta_j$, when two isotopes i and j are being separated. They have thus opened up possibilities for developing an industrial LIS technology compatible with the traditional isotope separation technologies. However, going over from laboratory experiments to industrial technology has required a great deal of work on the scaling of the process. To develop an efficient industrial LIS technology, it is necessary that a laser radiation source with an average power from 1 to 10 kW, depending on the required output capacity, be available, so that sufficiently high ESAPs for the given working substance and the desired isotope can be achieved. The working molecule must provide the opportunity to operate MLIS at gas pressures as high as possible. Besides, when scaling an LIS process, it is necessary to keep the ESAPs unchanged when the working substance is irradiated with high-repetition-rate pulses of high average power, provide a high radiation utilization factor, and implement an optimal process scheme, including the extraction and subsequent processing of the enriched products. All these problems have been solved, to some degree, by developing MLIS technologies.

In the MLIS approach, the isotope of interest is incorporated into a polyatomic molecule. As compared with the AVLIS technology based on selective photoionization of atoms, this opens up additional possibilities for selecting a starting working compound suitable for existing IR lasers. The most advanced work on the MLIS technology has been done in separating isotopes of uranium and carbon. The two stable carbon isotopes, ^{13}C and ^{12}C, exist in nature in concentrations of around 1.1% and 98.9%, respectively. The ^{13}C isotope is widely used in medicine for the noninvasive diagnosis of a number of diseases and also in biological research. The vibration frequency of the C–F bond falls within the tuning range of the CO_2 laser, and this fact predetermined the search for a suitable working molecule among the fluorocarbons. The CF_2HCl (Freon-22) molecule proved most suitable for large-scale laser separation of the carbon isotopes (Gauthier et al. 1982). The isotope shift in the ν_3 (1107.6 cm^{-1})

and ν_8 (1127.5 cm^{-1}) vibrations between the carbon isotopes is $\Delta\nu_{is} \cong 20$ cm^{-1}. This enables one to easily tune the CO$_2$ laser used to the absorption band of the ^{13}CF$_2$HCl molecule and dissociate it. The IR MPD of CF$_2$HCl gives rise to CF$_2$ fragments, which thereupon combine to form C$_2$F$_4$:

$$^{13}\text{CF}_2\text{HCl} + nh\nu \rightarrow \text{CF}_2 + \text{HCl} \\ \phantom{^{13}\text{CF}_2\text{HCl} + nh\nu \rightarrow{}} \hookrightarrow \text{C}_2\text{F}_4 \tag{11.10}$$

The CF$_2$HCl molecule also provides high ESAP values at moderate laser radiation fluences ($\Phi = 5$–10 J/cm^2). The IR MPD yield of the desired isotope ^{13}C amounts to 1–10%. The selectivity $\alpha(^{13}\text{C}/^{12}\text{C})$ of the process can be as high as 10^2–10^3. What is important is that the working pressure of the CF$_2$HCl gas can in this case be as high as a few kilopascals. Another important circumstance is that Freon-22 is a sufficiently low-priced product, manufactured on a mass scale. Once the ^{13}C isotope has been extracted, the depleted Freon-22 is returned to the manufacturer, which solves the problem of utilization of the used working compound and substantially improves the economic characteristics of the LIS process.

The development of an industrial carbon isotope separation technology based on the IR MPD of CF$_2$HCl has been reported (Letokhov and Ryabov 2004). In this technology, the separation reactor is placed inside the cavity of a high-pulse-repetition-rate (HPRR) TEA CO$_2$ laser. Such an arrangement is equivalent to a multiple-pass cell and provides a high radiation utilization factor. A matching lens ensures the necessary radiation fluence in the reactor. A fast gas circulation rate prevents the heating of the working substance and hence the worsening of the ESAPs. The working substance is a mixture of CF$_2$HCl and N$_2$. Nitrogen is used as a buffer gas to protect the optical elements and improve selectivity. On the basis of this approach, an experimental plant for the commercial production of the ^{13}C isotope has been built in Russia. A block diagram of one of the modules of this plant is presented in Fig. 11.14. The laser separation unit includes an HPRR TEA CO$_2$ laser and a separation reactor, wherein the working mixture is irradiated. The parameters of this CO$_2$ laser are as follows: pulse repetition rate up to 600 pps, pulse energy up to 3 J, and average power up to 1.8 kW. The irradiated mixture is fed to the rectification unit, where the enriched tetrafluoroethylene is extracted. In the conversion unit, this C$_2$H$_4$ is converted into CO$_2$, which is delivered to customers. The plant consists of three such modules with an output (per module) of up to 1 g of ^{13}C per hour, its concentration in the enriched product ranging from 50 to 30%.

A 30–50% concentration of ^{13}C is sufficient for some medical applications. However, for many other applications, the enriched product needs to have a higher ^{13}C concentration, up to 95–99% or even more. At present, the necessary additional enrichment is carried out by conventional methods. Therefore, the development of a purely laser-based separation process is of interest. The natural way to enhance the laser ^{13}C enrichment process is to use one more laser-enrichment stage, in which Freon-22 preenriched in ^{13}C serves as a working substance. In this approach, the ^{13}C-enriched product, C$_2$F$_4$ obtained from the first laser-enrichment stage must be efficiently converted back into CF$_2$HCl in an intermediate chemical cycle. This has been successfully

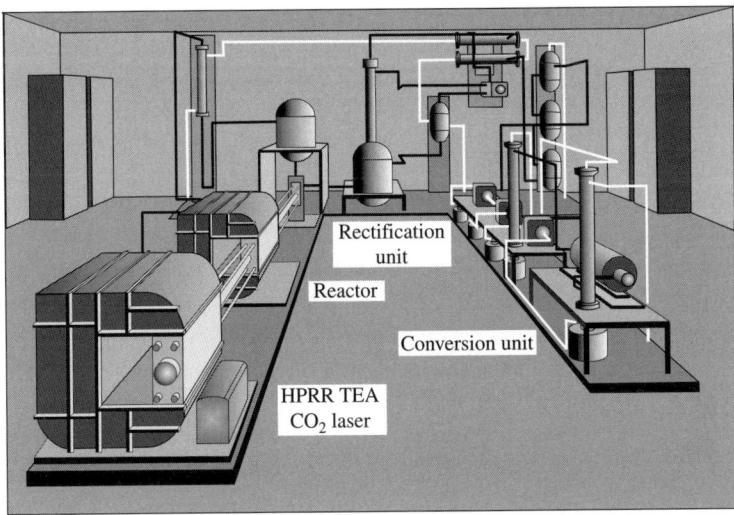

Fig. 11.14 Block diagram of a module of laser-based carbon-isotope separation plant. (From Letokhov and Ryabov 2004.)

demonstrated in experiments (Letokhov and Ryabov 2004), which confirmed the possibility of designing a purely laser-based two-stage process for producing highly enriched carbon-13 with a concentration of $\geq 99\%$. This example demonstrates the potentialities of using the intermolecular (isotopic) selectivity of IR MPE/D for the production of isotopes of various elements to satisfy future demands, in line with the further development of laser technology in the twenty-first century.

11.5 Prospects for mode-selective MPE/D by IR femtosecond pulses

The development of femtosecond lasers tunable in the mid-IR region has opened up the possibility of exciting vibrations in polyatomic molecules on a timescale comparable to or even shorter than the intramolecular vibrational redistribution times, which that lie in the picosecond or subpicosecond range (Schultz et al. 1979; Bloembergen and Zewail 1984). Femtosecond pulses feature a wide radiation spectrum capable of covering the frequency region of many vibrational transitions, vibrational anharmonicity notwithstanding. Indeed, if one bears in mind not the very-high-frequency modes involving a hydrogen atom, but medium-frequency vibrations, the typical magnitude of the anharmonicity is 1–10 cm^{-1}, while the spectral width of a pulse with a duration of 100 fs equals 50 cm^{-1}. The first experiments along these lines were conducted with $Cr(CO)_6$, $Fe(CO)_5$, and similar molecules by Windhorn et al. (2002). Laser pulses in the 5 μm region were used to excite the vibrational levels $v = 7, 8$ of the CO stretch vibration. Naturally, to obtain effective excitation, the pulses were focused into a cell containing the gas-phase metal carbonyl of interest so as to reach the dissociation limit with a fluence of 10^{-2} J/cm^2 at an effective vibrational absorption cross section $\langle \sigma \rangle \simeq 10^{-18}$ cm^2. From comparison with an analytical model, these authors concluded that only one to three out of 27 degrees of freedom were involved in the excitation of $Fe(CO)_5$.

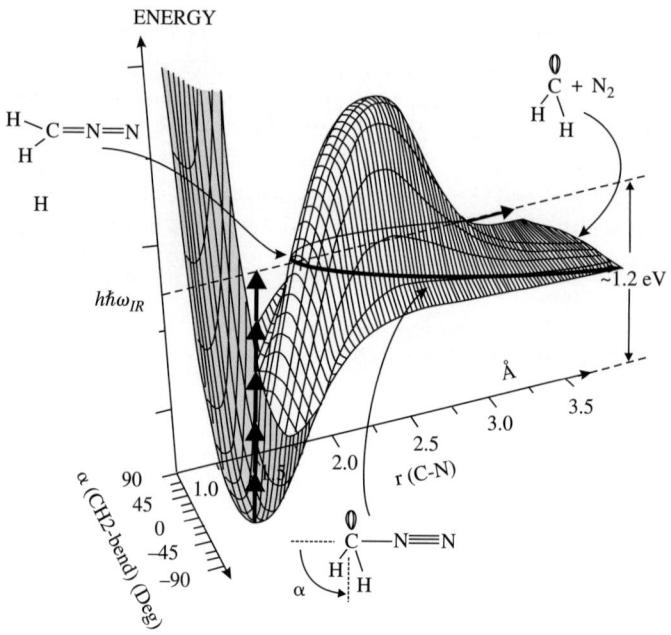

Fig. 11.15 Potential-energy surface of diazomethane, showing the proposed reaction pathway of the dissociation. The reaction coordinate R_{C-N} is the C–N bond length, given in Ångstroms, and α is the CH$_2$ out-of-plane bending angle in degrees. (Adapted and modified from Windhorn et al. 2003.)

The next step was to achieve a large population transfer to high-lying vibrational levels of the desired mode of the electronic ground state of the Cr(CO)$_6$ molecule by using negatively frequency-chirped femtosecond mid-IR pulses. This made it possible to optimize the climbing of the vibrational ladder by producing the necessary conditions for cascade excitation (Witte et al. 2003). The technique of frequency chirping of femtosecond pulses with time, developed primarily for the amplification of high-power visible femtosecond pulses (Mourou et al. 1998), became used for the "coherent quantum control" of quantum systems, considered in Chapter 12. This technique was then successfully used in the mid-IR region as well, to enhance dissociation by several orders of magnitude and make coherent ladder climbing possible by competing with the ultrafast IVR process.

In experiments with gaseous diazomethane (CH$_2$N$_2$) and mid-IR femtosecond pulses (Windhorn et al. 2003), IR MP-induced scission of the C–N bond on two distinct time scales, 480 fs and 36 ps was demonstrated. Windhorn et al. concluded that MPE of CH$_2$N$_2$ drove the molecule along the C–N stretch and brought it to a steep ridge on the potential-energy surface (Fig. 11.15). According to this experimental idea, the planar configuration corresponds to a maximum, rather than a minimum, in the CH$_2$ out-of plane bending coordinate. In this case, the energy is directly channeled into the bending coordinate as a result of fluctuations, so as to induce bond scission. This experiment clearly demonstrated that the coherent driving of unimolecular chemical reactions via MP vibrational excitation by femtosecond IR pulses was possible.

It would be very interesting to extend this method to condensed media, specifically biological molecules in a liquid. The pursuit of this goal is faced with difficulties, discussed at the 20th Solvay Conference on Chemistry (Letokhov 1997b), which are associated with the high peak intensity of the femtosecond pulses, causing the condensed medium to suffer optical breakdown, and the comparatively small vibrational-excitation cross section $\langle\sigma\rangle$. Nevertheless, a successful experiment has already been conducted on the femtosecond IR coherent excitation of molecules in a liquid phase (a $W(CO)_6$/n-hexane solution at room temperature), with a vibrational population distributed among the levels $v=0$ to $v=5$ (Witte *et al.* 2004). These results constitute a significant step toward the ultimate control of condensed phases with mid-IR femtosecond pulse-induced ground-state reactions relevant to practical synthetic transformations.

12
Coherent laser control of molecules

In Chapters 9–11, we have considered methods for *incoherent* control of atoms and molecules, whose realization relies on differences in their spectra. Of course, coherent effects inevitably manifest themselves in these methods, especially in the case of atoms, but with rare exceptions, they play no crucial role here. Noncoherent control was quite natural for the first stage of laser control, when the duration τ_p of the laser pulses used is much longer than the coherence time T_2 of the quantum system being controlled. A typical case is that of polyatomic molecules (Chapter 11) in highly excited vibrational states and molecules in condensed media, where the relaxation time of the coherence induced by a coherent laser pulse falls within the subpicosecond range. The progress in the development of nanosecond, picosecond, and femtosecond pulsed lasers has led to the corresponding development of new avenues of noncoherent and *coherent* laser control of molecules (Fig. 12.1). The first wave of successful applications of lasers for the purpose of effecting noncoherent laser control of polyatomic molecules was associated with the development in the early 1970s of the relatively simple CO_2 laser, generating high-power IR pulses in the range 9–11 μm with a duration of the order of 100 ns. The use of these lasers to excite vibrations in polyatomic molecules gave birth to multiphoton photoselective IR laser photochemistry, providing *intermolecular selectivity*, isotopic selectivity in particular (Chapter 11). At approximately that time the first picosecond pulses made their appearance, based on pulsed mode-locked Nd:glass and ruby lasers. These pulses were at once used to study the ultrafast relaxation of molecular vibrations in solutions (Laubereau and Kaiser 1975) and the ultrafast primary stages of photosynthesis (Rentzepis 1978). The advent of picosecond lasers led to the development of picosecond UV photochemistry of molecules in solution (Letokhov 1995a). This was, essentially, the second wave of applications of lasers to chemistry (Fig. 12.1). The development of femtosecond lasers (colliding-pulse-mode dye lasers (Shank and Ippen 1974) and Kerr mode-locked Ti-sapphire lasers (Spence *et al.* 1991)) led to rapid progress in studies of the primary stages of chemical reactions, that is, in experimental research into *molecular dynamics on the femtosecond timescale*, beginning with pioneering work on the photodissociation femtosecond dynamics of the ICN molecule (Rosker *et al.* 1988) and the Xe_2I molecule by use of two sequential femtosecond coherent laser pulses (Potter *et al.* 1992; Zewail 2000). Today this is the major avenue of investigation (the third wave in Fig. 12.1). This field was the subject matter of prestigious international conferences (Gaspard and Burghardt 1997; Sündström

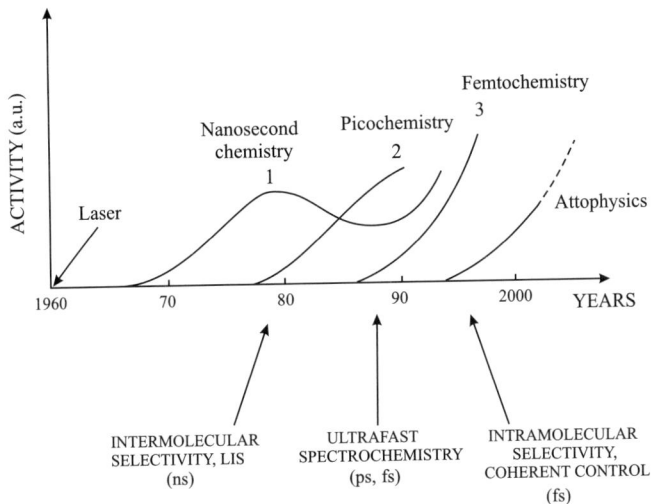

Fig. 12.1 Three waves of evolution of laser noncoherent and coherent control of molecules over a quarter of a century: from nanosecond to femtosecond laser pulses. The next wave is the development of attophysics. (Modified from Letokhov 1997a.) Integrated activity (funding, number of experiments, and number of publications) in arbitary units.

1997), and it was the achievements in this field that led to the development of coherent laser control of molecules.

Let us emphasize that noncoherent laser control of molecules depends on differences in the absorption spectrum between different molecular species, for example isotopic molecules. Coherent laser control is distinguished by its nontrivial manipulation of the phase coherence of excited states in molecules that can be similar in their absorption spectra but different in their phase coherence properties (Brixner et al. 2001b).

12.1 Introduction to coherent optimal control

Two main approaches to the control of molecules using wave interference in quantum systems have been proposed and developed in different "languages". The first approach (Tannor and Rice 1985; Tannor et al. 1986) uses pairs of ultrashort coherent pulses to manipulate quantum mechanical wave packets in excited electronic states of molecules. These laser pulses are shorter than the coherence lifetime and the inverse rate of the vibrational-energy redistribution in molecules. An ultrashort pulse excites vibrational wave packets, which evolve freely until the desired spacing of the excited molecular bond is reached at some specified instant of time on a subpicosecond timescale. The second approach is based on the wave properties of molecules as quantum systems and uses quantum interference between various photoexcitation pathways (Brumer and Shapiro 1986). Shaped laser pulses can be used to control this interference with a view to achieving the necessary final quantum state of the molecule. The probability of production of the necessary excited quantum state and the required final product depends, for example, on the phase difference between two CW lasers. Both these methods are based on the existence of multiple interfering pathways from the initial to the final (target) state of interest. Ultrashort, shaped laser pulses are effective owing to quantum interference between routes to the same final energy. This is contrary to the prevailing view that the main role of an ultrashort pulse is to beat the process of intramolecular vibrational redistribution (IVR) only (Shapiro and Brumer 2001).

To find optimal laser fields for controlling molecular systems requires a fairly accurate knowledge of the potential-energy surfaces and Hamiltonians of the molecules involved, and these are very complex and practically unknown. Of course, one can try to obtain this information by means of computer calculations, but so far this has been very hard to accomplish. Furthermore, it might prove very difficult and time-consuming to compute the optimal laser field for a given Hamiltonian. A way out of this difficult situation was suggested by Rabitz and coworkers (Judson and Rabitz 1992). They proposed using "trial and error" learning algorithms in computer simulations to find the necessary laser fields for the optimal control of quantum systems, complex molecular systems in particular. This method uses the quantum system to be controlled as an analog computer that tests every trial solution and guides the laser to generate the necessary optimal light field. Although the Hamiltonian of the desired complex molecule is unknown to us, the molecule itself contains all the information about its own Hamiltonian. Introducing feedback based on the required final quantum state or the yield of the product of a given chemical reaction makes it possible to find the optimal laser field through successive approximations made with the aid of "learning algorithms" (Section 12.4). A number of experiments have demonstrated the use of shaped pulses to control molecular systems with a learning algorithm. It was demonstrated that a learning algorithm could shape a femtosecond pulse with a positive chirp that was optimally suitable for controlling I_2 in the gas and the solid phase (Bardeen et al. 1997). Gerber and coworkers (Assion et al. 1998) demonstrated that molecular dissociation could be controlled through the use of laser pulses with a complex shape determined by a learning algorithm. All these new possibilities can be extended using very intense femtosecond laser pulses to substantially modify the energy potential or the potential-energy surfaces of the molecule under control and thus realize more complex coherent control (Levis et al. 2001).

A detailed consideration and bibliography of this rapidly developing and very interesting field of scientific exploration can be found in a monograph (Rice and Zhao 2000) and a number of reviews (Tannor and Rice 1988; Brixner et al. 2001a; Shapiro and Brumer 2003), a comprehensive bibliography of early work being provided by Manz (1997).

12.2 Coherent control using wave packets

When a molecule is excited by an ultrashort laser pulse with an appropriate center frequency, a localized wave packet can be created in the excited electronic state because of the excitation of a coherent superposition of many vibrational–rotational states. It follows from fundamental laws that the dynamics of molecular wave packets is governed by a time-dependent Schrödinger equation (eqn 2.29), where H is the relevant Hamiltonian of the given molecule. Because molecular potential-energy surfaces are anharmonic, this molecular wave packet tends to spread both in position (coordinates) and in momentum. However, in addition to expansion or defocusing, the wave packet also suffers delocalization at a certain instant of time. Coherent quantum

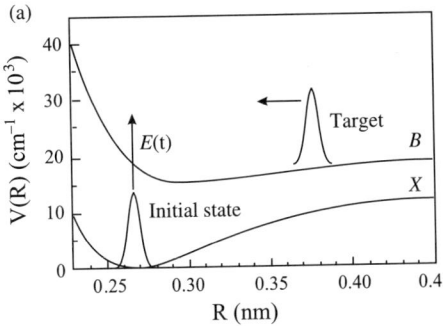 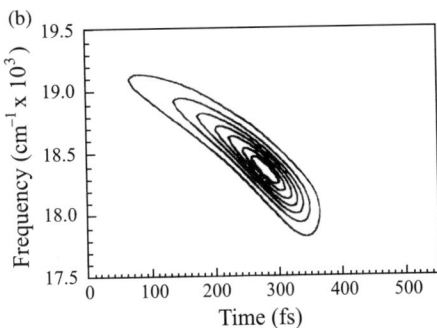

Fig. 12.2 Optimal pulsed excitation of a wave packet in an excited electronic state of the I_2 molecule: (a) simplified schematic energy diagram of I_2, for the ground state X and the excited state B; (b) optimal laser field in the weak-response (perturbative) limit for I_2 in the Wigner time–frequency representation. (Reprinted with courtesy and permission of the American Chemical Society from Kohler *et al.* 1995a.)

control uses the possibility of localization at a certain position and momentum. In other words, by making a coherent ultrashort laser pulse interact with an ensemble of molecules, the coherence of the light can be transferred to that of the molecular wave function, so that molecular wave packets evolve to the minimum size in time and space in the chosen target and at the chosen instant of time.

The idea of the experiment on controlling the vibrational dynamics of the I_2 molecule is schematically illustrated in Fig. 12.2(a) (Kohler *et al.* 1995a). The desired final state (or target) in this case is the minimum-uncertainty Gaussian distribution $\Delta p \, \Delta x = \hbar/2$ (where Δp is the momentum uncertainty and Δx is the coordinate uncertainty) in the electronically excited state B centered on the position $R_0 = 0.372$ nm, in which the iodine atoms are moving toward each other with a chosen mean velocity. It is necessary to find the laser pulse $E(t)$ that will excite the molecule from its ground state to form a vibrational wave packet with the best possible overlap with the final state (the target) at the chosen instant of time (550 fs in this case).

To find the optimal laser field to realize the wave packet motion illustrated in Fig. 12.3(a), one can use the Schrödinger equation (eqn 2.29) with the energy of interaction between the molecule and the electric field of the laser pulse (eqn 2.31), $V = -\mathbf{d} \cdot \mathbf{E}(t)$, where $\mathbf{d}$ is the dipole moment, added to the Hamiltonian. Wilson and coworkers (Kohler *et al.* 1995a, b) demonstrated this possibility using the density matrix method in Liouville space. Figure 12.2(b) presents the results of this calculation in the Wigner time–frequency representation. The optimal laser field is simple and smooth, and its main characteristic is a significant negative frequency chirp, in which the high-frequency components of the laser pulse arrive before their low-frequency counterparts. This is due to the anharmonicity of the vibrations in the electronically excited state B. This effect was verified by Shank and coworkers (Bardeen *et al.* 1995) in experiments with the I_2 molecule. Of course, all of this dynamics must take place in a time Δt much shorter than the time it takes for the molecular excitation

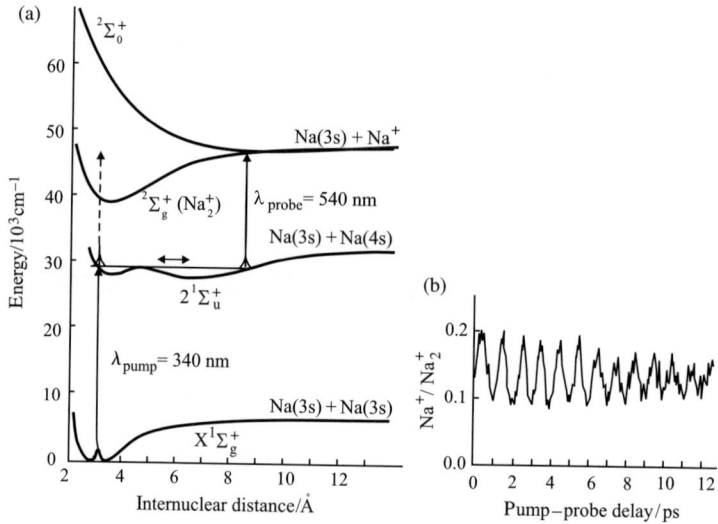

Fig. 12.3 (a) Potential-energy diagram of Na$_2$ and (b) Na$_2$/Na$_2^+$ ratio as a function of the delay of the 540 nm femtosecond laser pulse relative to the pumping 340 nm femtosecond pulse. (Reprinted with courtesy and permission of Academic Press from Brixner et al. 2001a.)

to lose its coherence, that is, the time it takes for the molecular wave packet to collapse.

It is clear from the excited-state wave packet dynamics of a molecule described above that the product yield of photochemical reactions can be controlled using temporal separation of at least two ultrashort pulses with a variable spacing, as is usually the case with the "pump–probe delay" method. This coherent-control approach is based on an idea advanced by Tannor, Kosloff, and Rice (Tannor and Rice 1985; Tannor et al. 1986) and relies on manipulation of the excited-state population. The first experimental demonstration of this method for coherent quantum control was performed by Gerber and coworkers (Baumert et al. 1991) using three-photon (or two-photon) ionization in competition with the fragmentation of molecular gas-phase Na$_2$. The potential-energy curves relevant to the two-photon ionization experiment are shown in Fig. 12.3(a). The first 100 fs pulse, of wavelength 340 nm, forms a wave packet on the $2^1\Sigma_u^+$ energy surface. The wave packet formed can propagate to its outer turning point at a large intermediate distance. The second, delayed femtosecond pulse of wavelength 540 nm transfers the wave packet to the repulsive ionic state, leading to the formation of Na(3s) and Na$^+$ fragments. On the other hand, the 540 nm femtosecond pulse can excite Na$_2$ from the inner turning point to the bound ionic state, leading to the formation of Na$_2^+$. The results of this experiment are presented in Fig. 12.3(b) as the Na$^+$/Na$_2^+$ ratio versus the time spacing of the two laser pulses. The observed strong modulation of the Na$^+$/Na$_2^+$ ratio corresponds to the propagation of the $2^1\Sigma_u^+$ wave packet between the inner and the outer turning point. So, by varying the delay time between the first and the second femtosecond pulse, one can control the breaking of the Na–Na bond.

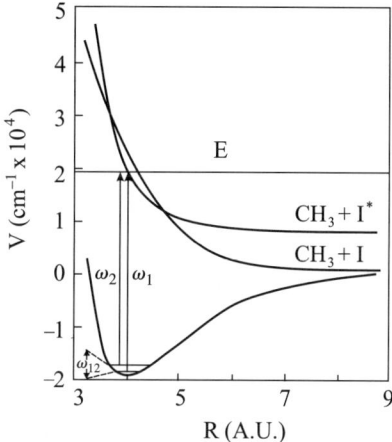

Fig. 12.4 A coherent-control scheme using one photon $\hbar\omega_{12}$ to prepare a superposition state on the ground-state energy surface of a molecule and then two additional photons $\hbar\omega_1$ and $\hbar\omega_2$ to excite the superposition state to an excited electronic-state surface. (Reprinted with courtesy and permission of Elsevier from Brumer and Shapiro 1986.)

12.3 Coherent control using quantum interference

The general principle of coherent control based on quantum interference between various photoexcitation pathways, including CW-laser weak excitation, is illustrated in Fig. 12.4. This quantum interference can be constructive or destructive, which allows control of the final state, that is, the control of a given reaction product. As an explicit example, Brumer and Shapiro (1986) have considered the process of photodissociation of methyl iodide, where the following two product channels are possible at an excitation energy of E:

$$CH_3I \begin{matrix} \nearrow CH_3 + I \\ \\ \searrow CH_3 + I^* \end{matrix} \qquad (12.1)$$

In order for these two channels to be coherent with respect to each other, that is, to be capable of interfering under CW-laser excitation conditions, it is necessary to use a coherent electromagnetic field of frequency ω_{12} to prepare a superposition state on the ground-state surface. Two additional laser fields of frequencies ω_1 and ω_2 are then used to raise the superposition state to the surfaces of the excited energetically degenerate state.[1]

---

[1] This scheme reminds me to some extent of a very early idea about the photoionization of atoms in coherent states (Letokhov 1966). The idea considered was the one-photon ionization of an atom in a state composed of a superposition of two states with close energies E_1 and E_2, prepared by the action of an electromagnetic field of frequency $\omega_{21} = (E_2 - E_1)/\hbar$. The coherent oscillations of the polarization at the frequency ω_{21} have the result that the probability of photoionization of

Experiments with HCl (Park *et al.* 1991) have confirmed the predictions of coherent-control theory, particularly the sinusoidal dependence of the ionization rate on the relative phases of the two exciting lasers, as well as the dependence of the degree of sinusoidal modulation of the ionization current on the relative laser field intensities. This technique was also used in experiments on controlling the product ratio in the photodissociation of HI (Zhu *et al.* 1995) and the branching ratio in the photodissociation of Na_2 (Shnitman *et al.* 1996).

Although the demonstration of this technique was a success, its efficiency is limited because CW lasers interact only with a small part of the thermal distribution of the molecules. The decay of the coherence of the molecules and radiation limits the amount of energy that can be used effectively for control purposes, because such coherence is a must for stable quantum wave interference. In this regard, the two-femtosecond-pulse approach (Section 12.2) seems to be more effective, especially when used in combination with optimally shaped electromagnetic fields. *Optimal control* of the shape of the laser pulses used can provide effective excitation of the desired final quantum mechanical state.

12.4 Optimal feedback control

The difficulty of calculating the optimal field for controlling a molecular system increases progressively with increasing complexity of the molecule. Rabitz and coworkers (Judson and Rabitz 1992) proposed an original approach to overcome this difficulty. They suggested that the experimental output should be involved in the optimization process. In this method, the molecules to be controlled are called upon to guide the search for the optimal field using a learning loop. A search algorithm evaluates the experimentally determined outcome of the laser-pulse–molecule interaction and then appropriately modifies the laser field $E(t)$. The learning algorithms employed to control molecular systems are typically of an evolutionary type (Schwefel 1995). These algorithms are so named because the idea underlying their design is based on the principles of biological evolution. Given the appropriate algorithm, automated cycling of the learning loop provides an effective means of finding optimal fields for the unknown Hamiltonian of the desired molecule. No prior knowledge of the potential-energy surface or any intermediate details of the chemical reaction is needed in this case.

A group of researchers headed by Gerber (Assion *et al.* 1998) experimentally realized the automated optimization of coherent control of independent chemical-reaction channels. The experimental setup (Fig. 12.5(a)) contained a computer-controlled femtosecond laser pulse shaper, complete with an evolutionary algorithm and feedback from the output of the femtosecond-laser-driven photoionization reaction. The femtosecond laser system was capable of producing pulses with a duration of 80 fs and an energy of 1 mJ, with a center wavelength of 800 nm and at a repetition frequency of 1 kHz. The laser pulses were focused into a high-vacuum chamber, where they interacted with a beam of the molecules of interest, which led to various multiphoton ionization and fragmentation processes. The ionic products were detected with a reflection time-of-flight mass spectrometer, from which the product yield data were

---

the atom also oscillates at the frequency ω_{21}. The oscillations of the photoionization probability are maintained in an atomic ensemble in a common field. In essence, owing to constructive or destructive interference, either enhancement or suppression of the photoionization yield takes place in this case.

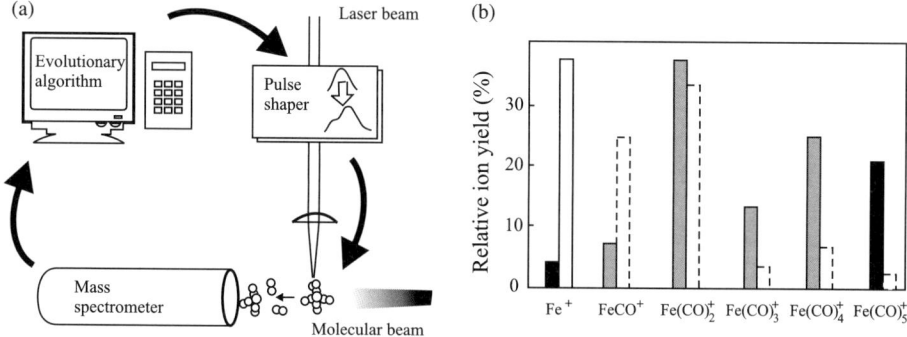

Fig. 12.5 Control of photodissociation of the Fe(CO)$_5$ molecule by feedback-optimized phase-shaped femtosecond laser pulses: (a) schematic diagram of the experimental setup; (b) relative product yields derived from the relative peak heights of the pertinent mass spectra. The Fe(CO)$_5^+$/Fe$^+$ ratio was both maximized (solid blocks) and minimized (open blocks) by the optimization algorithm, yielding significantly different Fe$^+$ and Fe(CO)$_5^+$ abundances in the two cases. (Reprinted with courtesy and permission of AAAS (USA) from Assion et al. 1998.)

fed into the controlling computer algorithm. In this key experiment, use was made of an organometallic molecule, specifically the well-studied Fe(CO)$_5$, which has a low dissociation energy and many channels of fragmentation and photoionization under the effect of intense femtosecond 800 nm laser radiation, depending on the pulse duration. Direct ionization of the parent molecule occurs efficiently under the action of ultrashort laser pulses with a duration less than 100 fs. A more complex process that occurs is a combined ionization and fragmentation, where the molecule loses not only an electron, but also from one to all of its five carbonyl groups. Complete dissociation of the molecule occurs under the actions of long laser pulses a few picoseconds in duration.

To test the self-learning coherent-control method, the branching ratio of two different product output channels, $\left[\text{Fe}(\text{CO})_5^+\right]$ and $\left[\text{Fe}^+\right]$, was used as a feedback signal in the optimization scheme. Maximization of this branching ratio was achieved within 30 generations of the evolutionary algorithm, yielding the product distribution shown in Fig. 12.5(b) (solid blocks). Minimization of the branching ratio was achieved within only a few generations (Fig. 12.5(b), open blocks) with longer, picosecond laser pulses. The results of this experiment with Fe(CO)$_5$ (and other complex molecules) demonstrated the automated coherent control of photodissociation reactions with tailored femtosecond laser pulses from a computer-controlled pulse shaper. Thus, a qualitative change from electronic population control to the direct control of various bond-breaking reaction channels by optimizing the final product yields was demonstrated without any a priori information about the molecular system of interest or about the experimental environment.

Of course, identifying the hidden mechanisms underlying the coherent control of a given chemical reaction involving a chosen molecule is of principal importance.

This fact was also successfully demonstrated in high-resolution two-femtosecond-pulse (pump–probe) experiments with the complex molecule CpMn(CO)$_5$ (Cp denotes cyclopentadienyl), along with *ab initio* quantum calculations and simulation of wave packet dynamics (Daniel et al. 2003).

12.5 Coherent optimal control by tailored strong-field laser pulses

When femtosecond laser pulses with an energy of the order of 1 mJ are focused, one may well attain intensities as high as 10^{13}–10^{15} W/cm^2. An intensity of 10^{14} W/cm^2 corresponds to an electric-field amplitude $E_0 \simeq 2$ V/Å, which is comparable to the electric field of the valence electrons. When the radiation intensity is this high, the Stark shift, polarization, and disturbance of the field-free electronic states become important. This leads to the formation of a quasi-continuum of new states in the molecule. The result of the action of such a strong electric field on a molecule is schematically illustrated in Fig. 12.6. Here a unidimensional electrostatic potential-energy surface of a diatomic molecule is distorted by a laser pulse with an electric-field strength of about 1 V/Å. During a femtosecond pulse, say 50 fs in duration, the magnitude and sense of the electric field $E(t) = E_0(t) \sin(\omega t + \varphi)$ oscillate a few tens of times. Therefore, the perturbed molecular potential in Fig. 12.6 also oscillates in direction and amplitude.

As before, the essence of coherent control by a strong laser field consists in transferring the molecular system at hand from the initial state $|\Psi\rangle$ into the desired final state $|\Psi_f\rangle$. This should be achieved by means of a tailored electric field, $E(t)$, that enters into the Schrödinger equation

$$i\hbar \frac{\partial}{\partial t} |\Psi\rangle = [H_0 - \mathbf{d} \cdot \mathbf{E}(t)] |\Psi\rangle \qquad (12.2)$$

through the dipole **d**. The optimal tailored laser pulse should produce the maximum constructive interference in the state $|\Psi_f\rangle$, while simultaneously achieving the maximum destructive interference in all the other intermediate states $|\Psi_{\text{int}}\rangle$ at the desired target instant of time.

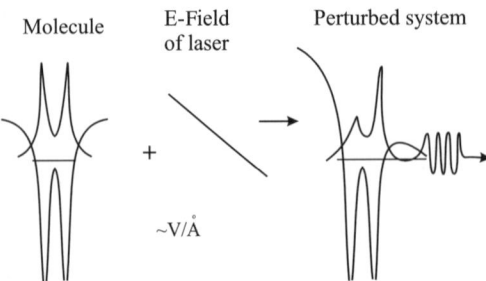

Fig. 12.6 Schematic diagram of a diatomic molecule interacting with an instantaneous strong electric-field. In this case, the electric-field strength is of the order of that of the H$_2$ molecule at a distance of 1 Å from a nucleus. The process shown represents the tunneling of an electron into the continuum. (Reprinted with courtesy and permission of the American Chemical Society from Levis and Rabitz 2002.)

This requirement for the optimization of quantum interference with a view to attaining the maximum yield of the desired product makes it necessary to appropriately control the laser field $E(t)$, including manipulation of the phases and amplitudes of all the pathways coupling the initial and the final state (Fig. 12.2). To this end, use is made of the spatial light modulation technique. But the hope of computing the necessary time-dependent laser field in order to achieve the desired reaction seems unrealistic. This is due to the lack of knowledge about the Hamiltonians of polyatomic molecules. Though the field-free Hamiltonians for simple molecules are well known, the highly nonlinear strong-field interaction and excitation make it practically impossible to calculate the appropriate pulse shape even for the simplest of them. The only possibility remaining thus far is to use a closed-loop control technique for laser-induced processes.

The closed-loop learning procedure for "teaching" lasers to control quantum systems under strong-field conditions has been demonstrated in numerous experiments, including the generation of higher vacuum ultraviolet (VUV) femtosecond-pulse harmonics in Ar gas (Bartels *et al.* 2000). Specially shaped strong-field laser pulses were employed to enable the control of photodissociation processes in various organic molecules (Levis *et al.* 2001). To control organic chemistry with optical fields has been a long-sought-after goal. This objective goes far beyond traditional photochemistry, where a variety of products are obtained by tuning the wavelength of monochromatic radiation into the absorption bands of electronic transitions to excited states of molecules. However, a tunable tailored laser pulse can guide the distribution of products to a chosen channel characterized by dissociation or rearrangement etc. Dissociative ionization is most convenient channel to study, thanks to the simplicity of the recording of the product distribution by means of mass spectrometry.

As an example of simple and successful verification of the coherent control of molecular dissociative ionization under strong-field conditions, we shall present some results of experiments (Fig. 12.7) with the acetone molecule, CH_3–CO–CH_3 (Levis and Rabitz 2002). There are a number of mass-spectral peaks corresponding to various photoreaction channels. The first channel corresponds to the simple removal of an electron from the molecule to produce the acetone radical cation, with mass $M = 58$. The second channel corresponds to the cleavage of one methyl group, CH_3, and the third pathway corresponds to the removal of two methyl groups. Only one of the product species in each channel is shown, with a positive charge. It was the second pathway that was optimized in this experiment. The pertinent mass spectra are presented in Fig. 12.7(a) as a function of the serial number of the pulse generation, the learning algorithm being aimed at increasing the intensity of the methyl carbonyl ion at $M = 43$. The intensity of this ion increases by an order of magnitude by the fifth pulse generation in comparison with the initial, randomly generated pulses (Fig. 12.7(b)). This experiment demonstrated two important features of the closed-loop control technique. The first is that a learning algorithm can find suitable solutions within a reasonable laboratory time (10 min in the above case). The second is that the resulting appropriately shaped strong-field pulses can dramatically change the relative ion yields.

The next important question is whether applications exist where the coherent laser control of molecular reactions offers special advantages (e.g. new products or

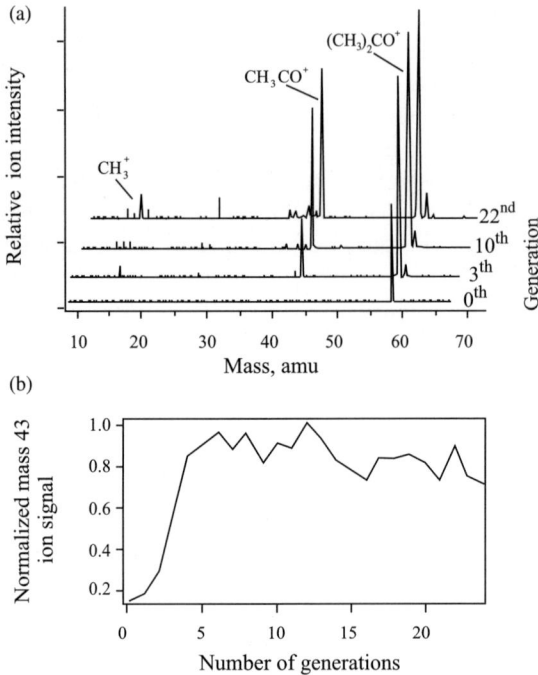

Fig. 12.7 (a) Representative strong-laser-field-induced mass spectra of acetone for the initial(0th), 3rd, 10th, and 22nd generation of a laboratory learning process aimed at maximizing the production of CH_3CO^+; (b) average signal for the members of the population at each generation. (Reprinted with courtesy and permission of the American Chemical Society from Levis and Rabitz 2002.)

better performance) over existing techniques that operate under noncoherent kinetic conditions (Chapters 10 and 11).

12.6 Coherent control of large molecules in liquids

To achieve quantum control of simple molecules is a problem that is relatively simple to solve. Can we hope to control the quantum dynamics of large molecules, specifically macromolecules such as proteins and nucleic acids? We should distinguish between the following two, rather different cases: the gas phase or a molecular beam (Sections 12.4 and 12.5), and a condensed (liquid) medium.

First, the obvious difficulty encountered in attaining quantum control of large molecules is the lack of adequate knowledge about their detailed properties and Hamiltonians. Second, even if the Hamiltonians and the necessary properties of large molecules with many degrees of freedom were known accurately enough, calculating the optimal laser fields to control them would seem to be problematic. So, optimal quantum control based on a learning algorithm and closed-loop feedback has been the only suggested solution to the problem. This approach has been successfully demonstrated in experiments with polyatomic molecular beams (Section 12.4 and 12.5). In these experiments, use could even be made of strong laser fields with intensities of the order of 10^{14} W/cm^2, which allowed the excitation of high-lying electronic states and, correspondingly, opened up many ionization, dissociation, and fragmentation channels.

In the case of large molecules in a liquid, particularly biomolecules in water, the situation can become more involved because of the problem of photodamage (optical breakdown and multiphoton absorption) of the solvent itself. To realize coherent control of a chemical reaction with a view to increasing the yield of the desired product, it is necessary to photoexcite the molecule with high probability. For a typical excitation cross section $\sigma_{\text{exc}} \simeq 10^{-17} - 10^{-18}$ cm^2, use should be made of a femtosecond laser pulse with an energy fluence of $\Phi = \hbar\omega/\sigma_{\text{exc}} \simeq 0.03$–$0.3$ J/cm^2, at any rate. For coherent interaction, the laser pulse duration τ_p must be shorter than the molecular decoherence time T_2 or the intramolecular vibrational relaxation time τ_{IVR} (Fig. 12.8). This means that the laser pulse intensity $I = \Phi/\tau_p$ will be over 10^{12} W/cm^2. This intensity is quite acceptable for gas-phase or molecular-beam experiments, but can be less suitable for condensed-medium experiments because of various nonlinear (damage) effects such as self-focusing and optical breakdown. We should bear in mind these effects when we are discussing, for example, the potential applications of femtosecond coherent control to biomolecules and the photochemical synthesis of pharmaceutical molecules in solution. Nevertheless, there is an "experimental window" for the coherent control of photochemical reactions in liquids (Fig. 12.8).

The fact that such an "experimental window" for coherent control in liquids does actually exist was verified in experiments on the selective multiphoton excitation of two distinct electronically and structurally complex dye molecules in solution (Brixner et al. 2001(b)). In these experiments, despite the failure of single-parameter variation (wavelength, intensity, or linear chirp control), adaptive femtosecond pulse shaping revealed that complex laser fields could achieve chemically selective molecular excitation. These results prove, first, that the phase coherence of complex molecules persists for more than 100 fs in a solvent environment. Second, this is direct proof that it is the nontrivial coherent manipulation of the excited state and not of the frequency-dependent two-photon cross sections that is responsible for the coherent control of the population of the excited molecular state.

12.7 Perspectives

New concepts of coherent femtosecond control, quantum learning, and feedback control of molecules have been proposed and demonstrated successfully in several laboratories for the past two decades. Moreover, significant advances have been made toward establishing broad foundations and laboratory implementation of control over quantum phenomena in various quantum systems, from atoms to complex molecules.

One of the most important goals in the coherent control of complex molecules is the realization of the control of stereoselective processes. The difference in the absorption spectrum between molecular enantiomers is very small and becomes manifest only with circularly polarized light. For this reason, the demonstration of the coherent control of left/right molecules by Shapiro et al. (2000) seems important for future

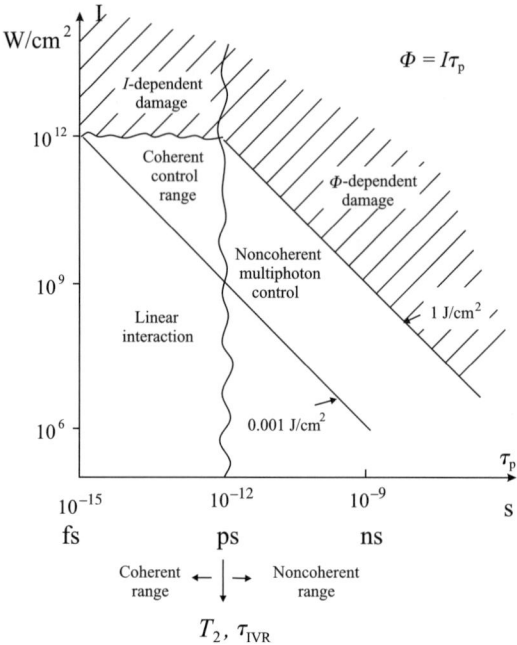

Fig. 12.8 Diagram of radiation intensity I (W/cm^2) versus laser pulse duration τ_p (s), with the various laser-pulse–condensed-medium interaction regimes being indicated very qualitatively. At high radiation intensities I and energy fluences $\Phi = \tau_p I$, the medium suffers optical damage. Coherent interaction takes place with subpicosecond pulses that have a duration of $\tau_p \leq T_2, \tau_{\text{IVR}}$. At low energy fluences ($\Phi < 0.001$ J/cm^2), the efficiency of the laser excitation of molecules is very low (linear interaction region). As a result, the experimental window for coherent control occupies a restricted area of this approximate diagram with flexible borderlines. (From Letokhov 1997b.)

applications, especially to biomolecules. The first step toward controlled stereoselectivity in photochemistry was taken by Gerber and coworkers (Vogt et al. 2005) in experiments on the optimal control of the *trans–cis* photoisomerization of a complex molecule (3,3-diethyl-2,2-thiacyanine iodide) dissolved in methanol. By optimizing the *trans/cis* isomer ratios in multiparameter optical-control experiments, these authors demonstrated that the isomerization efficiencies could be either enhanced or reduced. The results of the optimization obtained show that adaptive femtosecond pulse shaping can be applied to many challenges in research where isomerization reactions are of importance.

It is therefore quite natural to put the following question: Whither the future of controlling quantum phenomena in the broad context? (Rabitz et al. 2000). Even today, one can expect with confidence that these new concepts will lead to the extraction and deciphering of microscopic information from coherent molecular dynamics. This will be an important qualitative jump from the preceding period of studying and using only spectral information about molecules. The concept of optimal control has already been successfully used to control the phase of an electron wave packet with shaped attosecond laser pulses, to optimize the generation of perfect spatially coherent higher harmonics in the XUV region, and so on. One can completely agree with the statement that "the synergism of femtosecond laser pulse-shaping capabilities, laboratory closed-loop learning algorithms, and control theory concepts now provides the basis to fulfill the promise of coherent control" (Rabitz *et al.* 2000).

13
Related topics: laser control of microparticles and free electrons

The concept of the laser control of atomic motion, discussed in Chapters 5 and 6, was developed simultaneously and successfully for neutral microparticles, specifically with a view to creating optical tweezers. There are prospects for using the nonresonance gradient force produced by strong femtosecond laser pulses to control to some extent the motion of electrons and thus develop laser-induced electron optics in the future. It therefore seems quite natural to discuss, in conclusion, these problems related to the main subject matter of this book.

13.1   Laser trapping of microparticles

13.1.1   Optical forces on a microsphere

To trap a microparticle, it is necessary that the optical force acting on it should be strong enough to compensate for the gravitational force. A simple estimate shows that this condition is easy to satisfy by means of a focused CW laser beam. Let us consider as an example a laser beam with a power of 1 W focused onto a 1 µm sphere around 1 g/cm^3 in density and let the sphere scatter all of the incident light isotropically. The acceleration of the particle under the action of such a beam is of the order of

$$a = \frac{F}{M} = \frac{P/c}{M} \simeq 10^6 g, \tag{13.1}$$

where g is the acceleration due to gravity, P is the power of the laser beam, and M is the mass of the particle. Such a "back of the envelope" estimate led Ashkin to the idea of the optical trap (Ashkin 1970, 2000).

Like an atom, a particle in a light beam is acted upon by two forces—the scattering force and the gradient force. The former is associated with various types of scattering, whereas the latter is due to the polarizability of the particle in the nonuniform light field. Depending on the relation between the radius r of a sphere and the wavelength λ of light incident upon it, there can occur different conditions for scattering of this light by the sphere. When $r \ll \lambda$, the scattering is of symmetrical character (Rayleigh scattering), and the scattering force in that case is defined by the following expression (Ashkin et al. 1986):

$$F_{\text{scat}} = \frac{P_{\text{scat}}}{c} = \frac{I_0}{c} \frac{128\pi^5 r^6}{3\lambda^4} \left(\frac{n^2 - 1}{n^2 + 2}\right), \tag{13.2}$$

where P_{scat} is the Rayleigh scattering power, I_0 is the intensity of the incident light, and n is the refractive index of the sphere. If a sphere is located in a beam with an intensity gradient, it is additionally acted upon by the gradient force

$$F_{grad} = -\tfrac{1}{2} \alpha \nabla E^2, \tag{13.3}$$

where α is the polarizability of the sphere, given by

$$\alpha = r^3 \left(\frac{n^2 - 1}{n^2 + 2} \right). \tag{13.4}$$

When $n > 1$, the polarizability α is positive, and the gradient force pulls particles into the high-intensity regions of the beam.

The above two forces govern the behavior of a particle with $n > 1$ at a given point P on the axis of a single-beam trap (Fig. 13.1). The radial stability of the particle is ensured by the pulling effect of the gradient force, its axial stability being also maintained by the axial gradient force, the pushing effect of the scattering force notwithstanding. Estimates show that the minimum size of a silica particle that can be trapped in water by a 1.5 W argon-laser beam with a wavelength of $\lambda = 5145\,\text{Å}$ is around 20 nm. Under Mie scattering conditions, diffraction effects for particles with a radius of $r \gg \lambda$ are less important, and the scattering force on such particles can be estimated by considering the transfer of photon momentum upon refraction of light by a sphere. This is illustrated in Fig. 13.2 for the case of single-beam gradient trap formed by a strongly convergent laser beam focused by a microscope lens to the focal point f. The figure shows displacements of the sphere in two axial directions (Figs. 13.2(a) and (b)) and in the radial direction (Fig. 13.2(c)). From this figure one can see, even without making any calculations, that in all cases the net force $\mathbf{F} = \mathbf{F}_a + \mathbf{F}_b$ acting on the sphere brings it back to the focal point f.

Ashkin and Dziedzic (1987) gave an impressive demonstration of the optical trapping of dielectric particles in experiments with a biological particle (tobacco mosaic virus) in water by means of a single-beam gradient trap formed by an argon laser 0.1–0.3 W in power. The rodlike tobacco mosaic virus is 3000 nm long and 200 nm in diameter and has a refractive index of 1.57. In this experiment, Ashkin and Dziedzic

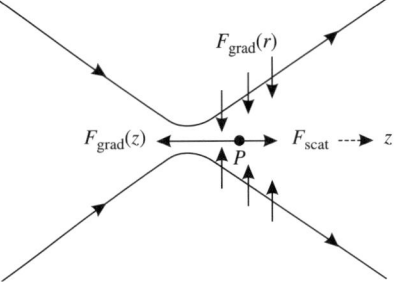

Fig. 13.1 Force components on a submicron particle placed at a point P in a single-beam gradient laser trap.

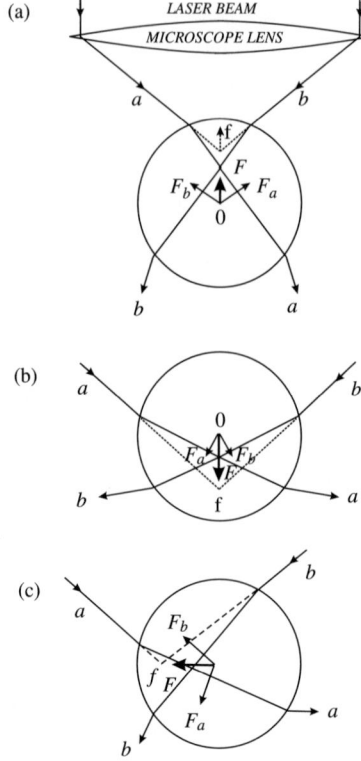

Fig. 13.2 Forces on a Mie-sized particle responsible for its axial (a, b) and transverse (c) displacement away from the focus f_0 in a single-beam gradient trap. The refraction of the rays a and b gives rise to the radiative pressure forces F_a and F_b. The net force F on the sphere is in all cases a restoring force back to the focus. (Reprinted with courtesy and permission of IEEE (USA) from Ashkin 2000.)

clearly observed a strong Brownian motion of the virus about its equilibrium position. To prevent biological particles from suffering photodamage, it proved expedient to use the near-infrared light at $\lambda = 1.06\,\mu$m from a CW Nd:YAG laser.

13.1.2 Optical tweezers

The possibility of controlling the position of the focal point of a laser beam, and hence the position of the optical trap formed by it, in an optical microscope makes the latter a powerful tool—optical tweezers—for microparticles and, what is exceptionally important, biological objects such as various types of cells, bacteria, and biological macromolecules (DNA, RNA, and others). In essence, optical tweezers have proved to be a powerful nonresonance, contactless control instrument for manipulating biological macromolecules and biostructures (Ashkin and Dziedzic 1987; Svoboda and Block 1994). Figure 13.3 presents a schematic diagram of an optical-tweezers trap for separating a single desired bacterium from a mixture of bacteria. This instrument helped researchers to trap one or more bacteria and manipulate them into the core of

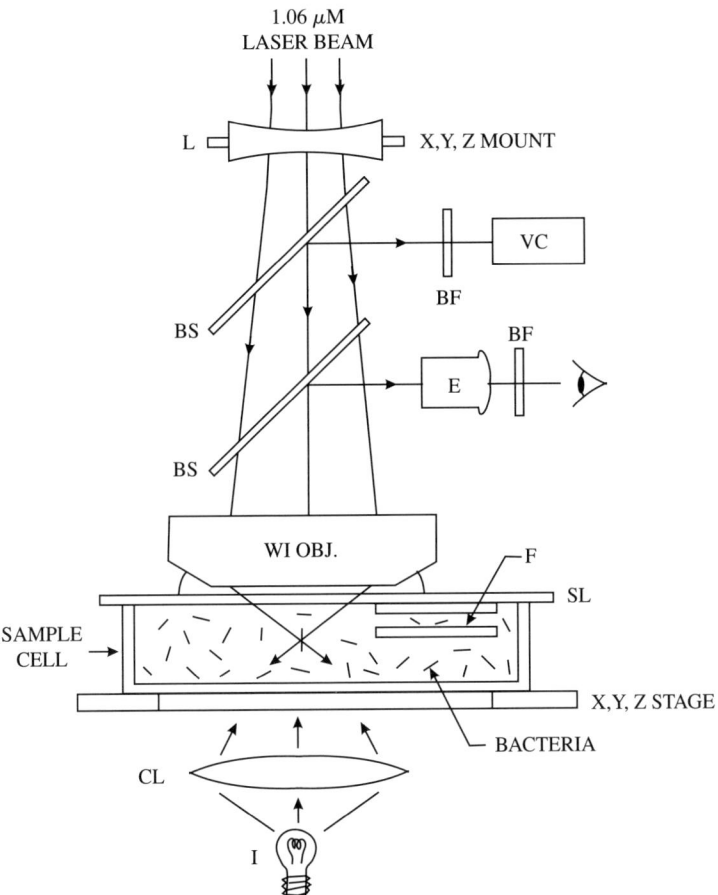

Fig. 13.3 Combined high-resolution optical microscope and 1.06 μm infrared laser trap for observing, manipulating, and separating bacteria and other organisms (abbreviations: VC, video camera; BS, beam splitter; BF, beam filter; WI OBJ, objective; F, flow; CL, condenser lens; I, illumination lamp). (Reprinted with courtesy and permission of IEEE (USA) from Ashkin 2000.)

the hollow fiber F, with an inside diameter of 15 μm. Despite the significant optical distortions at the input of the fiber, it proved possible to maneuver the bacteria into the fiber core without losing them. The ability of optical tweezers to separate and manipulate bacteria and other biological cells, including macromolecules, is a relatively simple but important property that can find numerous applications.

Optical tweezers have also been used to transfer various types of photon angular momentum to macroparticles, as was done long ago with the photon spin and atoms (Chapter 4). It is not out of place here to note that the photon angular momentum is, first, associated with the spin for circularly polarized light and has a magnitude of $\hbar = h/2\pi$ per photon. Second, there is an orbital angular momentum that is associated with the inclination of the wavefront an the laser beam (Allen *et al.* 1999). Unlike the plane wavefront of the ordinary laser beam, inclined wavefronts can be found in the "helical" light beams (optical vortices) of special Laguerre–Gaussian modes. The orbital angular momentum of light can therefore exceed $\hbar$ per photon by many times. The first observation of the transfer of orbital angular momentum to an absorbing particle trapped on the axis of a Laguerre–Gaussian laser beam, which caused the particle to rotate, was made by He et al. (1995), a holographic technique being used to form the beam. Strictly speaking, the spin angular momentum of a photon causes a particle to spin around its axis, whereas the orbital angular momentum should cause a particle to orbit about the beam axis (O'Neil and Padgett 2000). Both these effects—the rotation of a microparticle as a result of the transfer to it of the spin angular momentum of photons and its orbiting on account of the transfer of the orbital angular momentum in a multiringed laser beam—were experimentally observed by Garus-Chavez et al. (2003). The second effect can be observed in the case of off-axis displacement of the particle. Thus, an optical trap for microparticles is an efficient tool for direct observation of fine quantum effects exerted on matter by light.

13.1.3 Biomedical applications of optical tweezers

The widest application field of optical tweezers is in biomedical research, starting with the seminal paper by Ashkin and Dziedzic (1987) (a detailed bibliography on laser-based optical tweezers can be found in the review by Lang and Block 2003). The successful applications of optical tweezers at the microscopic level in the biomedical field are based on their capabilities for ultrafine noninvasive positioning, measurement, and control. Optical forces of up to 200 pN or thereabouts can be applied with sub-pN resolution to biological objects whose characteristic size is comparable to the wavelength of light. Optical tweezers are useful for a variety of bioobjects, such as biopolymers (microtubules and DNA molecules), lipid membranes, cells, and single biological macromolecules. There are many fields of research in biophysics, for example the mechanical unfolding and refolding of proteins and nucleic acids, the strength of receptor–ligand bonding interactions, and the nanoscale features of biological motors, that have been studied with optical tweezers. As an illustration of these surprisingly effective biomedical applications of optical tweezers, let us present a few examples.

1. *Single-molecule DNA trapping and stretching*. To understand the physics that keeps DNA tightly packed in the cell, the forces that enzymes exert to uncoil DNA, and the mechanics behind gene expression, it is necessary to study single DNA molecules. Optical tweezers are an ideal instrument for this purpose, because they can provide and measure forces in the 0.1–100 pN range—the range that is the scale of most primary biological processes on the molecular level. Optical traps exert just the right magnitude of force by focusing a laser beam into a micron-sized spot through a microscope objective, as shown in Fig. 13.3. A dielectric bead attached to a DNA molecule can be trapped at the center of the spot. To exert force on the DNA molecule, its other end is stabilized on a surface. As the DNA molecule is pulled away from the trap, the bead moves slightly out of the focus. The amount of this displacement can be measured and converted into the magnitude of the force being applied. The same kind of setup can be used to study molecular motors and other mechanical processes in the cell.

2. *Molecular motors acting on single DNA molecules.* Extensive work has been done using optical tweezers to study the interaction of the *E. coli* RNA polymerase enzyme with DNA. RNA polymerases are enzymes that copy the DNA sequence to construct single-stranded messenger RNA (mRNA) in a process called *transcription.* The mRNA is then used by a ribosome to construct a specific protein in a process called *translation.* It takes energy for the mRNA molecule to move along the DNA molecule and read its sequence. RNA polymerases are molecular nanomotors because they use energy to induce motion and generate forces in the cell. The forces generated by this motor have been measured directly using optical tweezers. Figure 13.4 shows how optical tweezers have been used for this purpose. The most effective technique consists in attaching a single polystyrene bead to a single DNA molecule and optically trapping the dielectric bead. For example, in the experiment conducted by Davenport *et al.* (2000), the RNA polymerase molecule was attached to another polystyrene bead that was held fast at the end of a glass micropipette. In order to stretch the DNA molecule and thus exert on it a force opposing transcription, the micropipette was moved to a specified position or was moved until a specified force was applied to the bead in the optical trap. Thus, experiments of this type help to explain the role of sequence-dependent pausing in transcriptional regulation.

3. *Medical applications.* One can cite numerous examples of the successful use of optical tweezers for biological-research purposes. The relative simplicity and accessibility of this technique make it very promising. Weak optical forces will probably find not only use for the manipulation of biological structures as a whole, but also direct medical applications. An experiment on the guiding of neuronal growth with laser light (Ehrlicher *et al.* 2002) confirms this possibility. It is well known that control over neuronal growth is a fundamental objective in neuroscience, cell biology, and biomedicine, and it is particularly important for the formation of neuronal circuits *in vitro*, as well as nerve regeneration *in vivo*. Ehrlicher *et al.* demonstrated experimentally that they could use weak optical forces to guide the direction taken by the leading edge, the growth cone, of a nerve cell. In actively extending growth cones, the focal spot of a laser beam was placed in front of a specified area of the nerve's leading

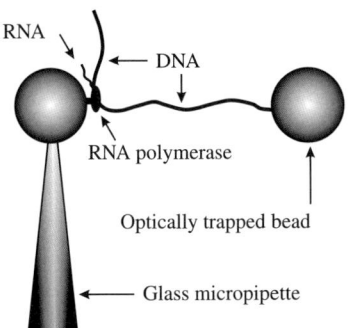

Fig. 13.4 Schematic diagram of an optical-tweezers experiment to measure the transcriptional forces generated by *E. coli* RNA polymerase. (Reprinted with permission of AAAS (USA) from Davenport *et al.* 2000.)

edge, which increased growth into the beam focus and resulted in guided neuronal turns, as well as enhanced growth. The laser power was chosen so that the resulting gradient force was sufficiently strong to bias the polymerization-driven extension of the lamellipodia, but too weak to hold and move the growth cone. Thus, a laser-induced optical force could control a natural biological process at the low light power used for optical tweezers. These results therefore open a new avenue for controlling neuronal growth *in vitro* and *in vivo* with weak optical gradient forces.

13.2 Laser control of free-electron motion

The effective laser control of atoms is based on resonant interaction, whose cross section at maximum can reach a huge magnitude, $\sigma_{res} \simeq \lambda^2 \simeq 10^{-8}\,\text{cm}^2$. In the case of microparticles, their interaction with light is of nonresonant character, but the efficiency of this light-scattering interaction depends on the size of the particles. In the case of free electrons, their interaction with light is due to nonresonant scattering (the Thomson effect), whose cross section $\sigma_T \simeq 6.65 \times 10^{-25}\,\text{cm}^2$ is associated with the classical electron radius $r_e = e^2/mc^2 = 2.8 \times 10^{-13}\,\text{cm}$. Therefore, laser light can be used to control the motion of free electrons only if its intensity is approximately ten orders of magnitude higher than in the case of atoms and microparticles. Such intensities are quite achievable today with femtosecond laser pulses.

Leaving aside the important problem of interaction between ultrahigh-intensity femtosecond laser pulses and relativistic electrons, we shall consider below only the effects involved in the control of nonrelativistic electrons, such as coherent diffraction, deflection, focusing, and reflection. The diffraction of an electron beam by a standing light wave (the Kapitza–Dirac effect, Kapitza and Dirac 1933) is essentially the earliest proposal for the control of matter by light.

13.2.1 Free-electron–laser-field interaction

Consider for the sake of simplicity the nonrelativistic motion of an electron, disregarding the electron spin, in a laser field. In this approximation, the interaction between an electron and a laser field may be described by the Schrödinger equation with the ordinary nonrelativistic Hamiltonian

$$H = \frac{1}{2m}\left(\mathbf{P} - \frac{e}{c}\mathbf{A}\right)^2, \tag{13.5}$$

where $\mathbf{P} = -i\hbar\nabla$ is the canonical momentum, $\mathbf{A}$ is the vector potential of the laser field, and e and m are the charge and mass, respectively, of the electron. The original Hamiltonian (eqn 13.5) can be simplified substantially by taking into account the high-frequency oscillations of the laser field. In the oscillating electromagnetic field described by the vector potential $\mathbf{A}$, an electron executes both fast oscillations and a slow motion. By applying a standard procedure of averaging the motion of the electron over its fast oscillations, one can obtain from eqn (13.5) a Hamiltonian describing only the slow electron motion in the light field:

$$H = \frac{1}{2m}\mathbf{P}^2 + \frac{e^2}{2mc^2}\langle\mathbf{A}^2\rangle, \tag{13.6}$$

where the brackets denote the time-averaging operation.

In the simple case of a monochromatic light beam, the electric field can be taken to be $\mathbf{E} = \mathbf{E}_0(\mathbf{r})\cos(\mathbf{kr} - \omega t)$, where $\mathbf{k}$ is the wave vector and ω is the frequency of the field. Accordingly, the vector potential is $\mathbf{A} = \mathbf{A}_0(\mathbf{r})\sin(\mathbf{kr} - \omega t)$, where $\mathbf{A}_0 = (c/\omega)\mathbf{E}_0$, and so the Hamiltonian becomes

$$H = \frac{1}{2m}\mathbf{P}^2 + \frac{e^2}{4m\omega^2}\mathbf{E}_0^2. \tag{13.7}$$

The second term in the above equation describes the effective, or ponderomotive, potential of the electron in the light field. Noting that the intensity of the light beam is $I = (c/8\pi)E_0^2$, the ponderomotive potential can be written as

$$U_{\text{eff}}(\mathbf{r}) = \frac{2\pi\alpha\hbar}{m\omega^2}I(\mathbf{r}), \tag{13.8}$$

where $\alpha = e^2/\hbar c$ is the fine-structure constant. The ponderomotive potential can be considered responsible for the ponderomotive, or gradient, force acting on the electron in the light field:

$$\mathbf{F}(\mathbf{r}) = -\nabla U_{\text{eff}} = -\frac{2\pi\alpha\hbar}{m\omega^2}\nabla I(\mathbf{r}). \tag{13.9}$$

The magnitude of the potential produced by a laser field can be estimated as follows. For a laser wavelength $\lambda = 500\,\text{nm}$ and an intensity I of about $10^{14}\,\text{W/cm}^2$, the magnitude of the potential U_{eff} is of the order of $1\,\text{eV}$.

The polarizability of a free electron in an optical field of frequency ω is described by

$$\alpha_{\text{pol}}(\omega) = -\frac{e^2}{m\omega^2}. \tag{13.10}$$

According to eqns (13.9) and (13.10), the electron is expelled from a high-intensity region into a low-intensity region of the field. This expulsion can be used to change the trajectory of the electron.

One can also treat the "light medium" (at the optical wavelength λ_{opt}) as a medium of reduced refractive index $n_{\text{el}}(\omega)$ for nonrelativistic electrons ($v \ll c$), given by the expression

$$n_{\text{el}}(\omega) = \left[1 - \left(\frac{\mu c}{v}\right)^2\right]^{1/2}, \tag{13.11}$$

where μ^2 is a dimensionless intensity parameter given by the expression

$$\mu^2 = \frac{2}{\pi}\frac{r_e}{mc^3}\lambda^2 I(r), \tag{13.12}$$

where r_e is the classical electron radius, m is the electron rest mass, λ is the optical wavelength, and I is the intensity of the light. Figure 13.5 shows n_{el} as a function of the intensity of laser radiation of various wavelengths for electrons with an energy of $E_{\text{ph.el.}} \approx 10\,\text{eV}$. At an intensity in the range 10^{13}–$10^{14}\,\text{W/cm}^2$, n_{el} drops perceptibly.

This means that for electrons with a kinetic energy of $E_{\text{kin}} = 10\,\text{eV}$, such a "photon medium" at a laser intensity as high as $10^{14}\,\text{W/cm}^2$ and $\lambda_{\text{opt}} \simeq 1\,\mu\text{m}$ has a refractive index differing substantially from unity. Electrons will be pushed out of such a medium.

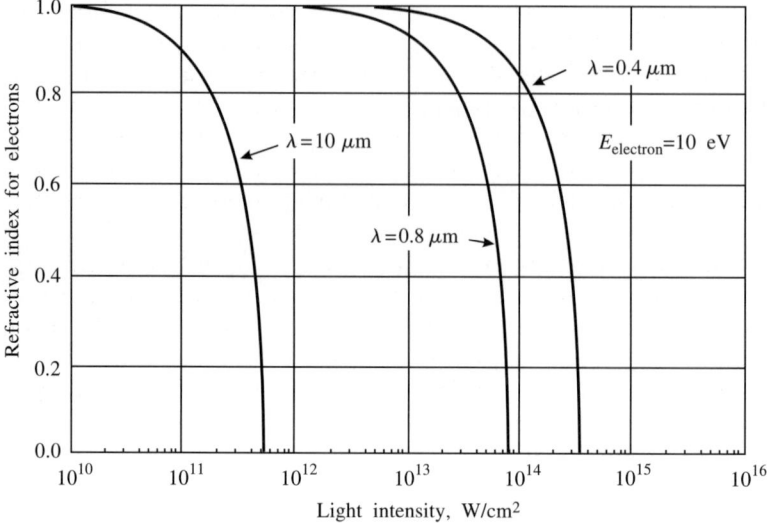

Fig. 13.5 Variation of the effective refractive index n_e for electrons with an energy of $E_{el} = 10$ eV in the presence of a laser wave with an intensity of I (W/cm^2). (From Balykin et al. 1996.)

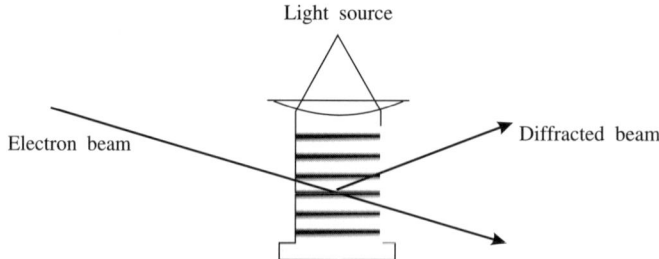

Fig. 13.6 General idea of an experiment to observe the Kapitza–Dirac effect.

And it is exactly femtosecond laser pulses that can help one obtain the necessary high laser intensities. In essence, by using high-intensity ultrashort laser pulses in a vacuum, one can form regions of a desired shape with a reduced refractive index for an ultrashort period of time. It is precisely this fact that provides the basis for developing ultrafast laser-induced electron optics using specific features of laser light, namely, high intensity, short pulse duration, and a high concentration of energy in a small region of space.

13.2.2 Diffraction of free electrons by a standing laser wave

In 1933, Kapitza and Dirac suggested that a well-collimated electron beam could be diffracted by a standing light wave. Figure 13.6 illustrates the idea of the experiment that they suggested. They estimated that the intensity of the diffracted beam would be 10^{-14} relative to that of the undiffracted beam using light from a mercury arc lamp.

Obviously, such experiment became possible only with the advent of the laser. Following early controversial experiments, Freimond et al. (2001) succeeded in observing the Kapitza–Dirac effect, with the diffraction pattern featuring well-resolved diffraction orders. Because the diffracted electron beams are coherent with respect to each other, the Kapitza–Dirac effect essentially provides a coherent electron beam splitter based on stimulated Compton scattering in the particular case of electromagnetic radiation in the form of a standing light wave. The relationship between the stimulated Compton scattering and the ponderomotive forces was considered by Fedorov et al. (1997).

In the case of a very intense standing wave produced by a femtosecond laser pulse, there can take place the stimulated Compton scattering of many photons, n_m, by each electron. In that case, the diffraction pattern contains many diffraction maxima, corresponding to changes of the initial electron momentum by amounts of $\pm 2\hbar \mathbf{k}, \pm 4\hbar \mathbf{k}, \ldots, \pm 2n_m \hbar \mathbf{k}$. The electric field of a standing light wave produced by two counterpropagating waves is $\mathbf{E} = 2\mathbf{E}_0 \cos \omega t \cos kz$. The periodic ponderomotive potential can be expressed as

$$U_{\text{eff}} = 2U_0 \cos^2 kz, \tag{13.13}$$

where the amplitude of the potential is written in a form including the intensity $I_0 = (c/8\pi)E_0^2$ of a single traveling wave:

$$U_0 = \frac{\alpha \hbar}{\pi} \frac{I_0 \lambda^2}{mc^2}. \tag{13.14}$$

In a very strong standing light wave, an initially broad electron wave packet will be split into two parts, each being the envelope of many Bragg maxima, with the angle 2Θ between the split beams given by

$$\Theta_{\pm} = 2n_m \frac{v_r}{v_0}, \quad n_m = \frac{U_0 \tau_0}{\hbar}, \tag{13.15}$$

where $v_r = \hbar k/m$ is the recoil velocity of the electron and v_0 is its velocity. At a laser pulse intensity of the order of 10^{13} W/cm^2 and a duration $\tau_0 \simeq 10^{-12}$ s, the angle Θ can reach 10–20° (Minogin et al. 1997). With a small laser beam diameter $w_0 \simeq 100\,\mu$m (the waist of a Gaussian beam), the duration of the diffracted electron-beam pulse can fall within the picosecond range, which can be of interest in producing ultrashort electron pulses.

13.2.3 Focusing of an electron beam by a laser beam

According to eqn (13.9), an electron moving in a light beam is pulled into the low-intensity region of the light beam. This circumstance can be used to develop a technique for focusing an electron beam by use of a copropagating laser beam produced by the transverse laser mode TEM$_{01}^*$ (Fig. 13.7). If such a beam propagates along the z-axis, its transverse intensity distribution profile is given by

$$I(r) = 8I_0 \frac{w_0^2 r^2}{w^4(z)} \exp\left(\frac{-2r^2}{w^2(z)}\right), \tag{13.16}$$

where r is the transverse coordinate, w_0 is the beam-waist radius at $z = 0$, and $w(z) = w_0^2(1 + z_R)^{1/2}$ is the beam waist at the coordinate z, $z_R = (\pi/\lambda)w_0^2$ being the

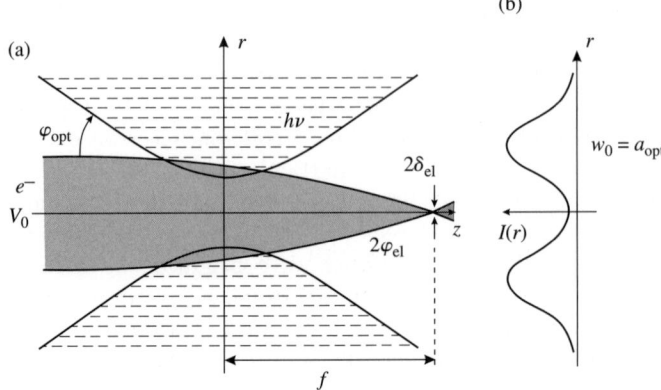

Fig. 13.7 Focusing of an electron beam by the TEM$_{01}^*$ laser mode: (a) propagation geometry of the electron beam inside the laser beam; (b) transverse intensity profile of the laser mode. (From Letokhov 1995b.)

Rayleigh length. Considering the potential well (eqn 13.8) produced by a laser beam with the intensity profile in eqn (13.16), one can evaluate the focusing properties of the "electron lens" formed by the laser beam.

The diameter $2a_{el}$ of the spot into which the electron beam is focused is related to the focusing diameter $2a_{opt} = 2w_0$ of the laser beam by the simple relation

$$a_{el} = a_{opt} \left[\frac{\lambda_{dBr}}{\lambda} \right] \frac{1}{\delta n_{el}}, \quad \delta n_{el} = \frac{1}{2} \left(\frac{\mu c}{v} \right)^2 \ll 1, \tag{13.17}$$

where $\lambda_{dBr} = 2\pi\hbar/mv$ is the de Broglie wavelength of the electron, $E_{kin} = mv^2/2$ is its kinetic energy, and δn_{el} is the change of the effective refractive index. Let us consider the focusing of electrons with an energy of $E_{kin} = 100\,\text{eV}$ ($v = 5.9 \times 10^8$ cm/s, $\lambda_{dB} = 1.2$ Å) by electromagnetic radiation at $\lambda = 10\,\mu\text{m}$ with an intensity of $I_0 = 0.5 \times 10^{12}\,\text{W/cm}^2$ in the focus. In this case, the coefficient δn_{el} in eqn (13.17) is δn_{el} equal to 0.12, and with $\sin\varphi_{opt} = 0.3, \ldots, 0.5$, for example, the electron beam can be focused into a spot with a radius of $a_{el} \simeq 10\text{–}15$ Å. The transit time of the electron through the focusing region is $\tau_{int} \simeq 5$ ps, and so focusing can be achieved even with picosecond pulses with a length of 100 ps and an energy of a mere $E_{las} \simeq \lambda^2 I_0 \tau_p \simeq 10^{-4}$ J.

The laser lens for electrons suffers from spherical aberration, but this has a negligible effect on the estimate in eqn (13.17), and also from chromatic aberration, which is a more serious matter. For the particular numerical example we are discussing here, the electrons should be monochromatic within a factor on the order of 0.01%. The same stringent requirement is also imposed on the stability of the laser pulse intensity. We thus need a square pulse with minimal rise and fall times ($\tau_{fr} \ll \tau_p$), since the position of the focal point varies at the leading and trailing edges of the pulse. On the other hand, the rapid reciprocating motion of the electron focal point along the z-axis, induced by the varying laser pulse intensity, might be utilized to produce femtosecond electron pulses.

On the basis of these estimates, one can imagine the possibility of developing a scanning electron microscopy technique with picosecond time resolution using low-energy electrons and laser focusing. This modification of laser-controlled electron microscopy would be of interest because low-energy electrons are sensitive to the atomic and molecular structure of the surface of the specimen.

13.2.4 Reflection of an electron beam by an evanescent laser wave

The idea of reflecting an electron beam by an evanescent light wave is illustrated in Fig. 13.8. The internal reflection of a laser beam at a dielectric–vacuum interface produces an evanescent light wave with an intensity distribution given by

$$I(z) = I_0 \exp\left(-\frac{z}{z_0}\right), \tag{13.18}$$

where z_0 is the characteristic decay length of the light wave along the z-direction normal to the dielectric–vacuum interface. The intense evanescent light wave formed by the total internal reflection of a femtosecond light pulse penetrates into the vacuum to a depth of the order of the wavelength of the light (see Section 7.2.1). The spatially nonuniform intensity distribution produces a ponderomotive force (eqn 13.9) that pushes the electrons out of the evanescent field region. The corresponding ponderomotive potential (eqn 13.8) is thus responsible for reflection of the electron beam from the evanescent light field (Balykin *et al.* 1996).

The change in the electron velocity can be described in terms of the effective index of refraction ($n_{el} < 1$) determined by eqn (13.11). Simple estimates show that an electron beam with an energy of 100 eV crossing an evanescent field produced by femtosecond laser pulses can be reflected through an angle of about 0.1–1 rad. This angle is sufficiently large to allow separation of the reflected beam from the original electron beam. Moreover, the reflected electron beam may have a very short duration, of about 100 femtoseconds.

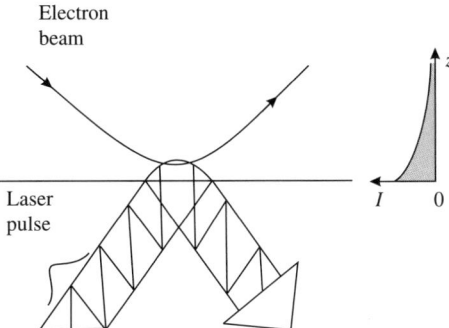

Fig. 13.8 Idea of reflection of an electron beam by an evanescent light wave produced by total internal reflection of a femtosecond laser pulse at a dielectric–vacuum interface. (From Balykin *et al.* 1996.)

By using a curved evanescent wave, one can, in principle, obtain focusing of an electron beam simultaneously with reflection. Of course, all these potentialities of "laser-induced" reflective electron optics should be the subject matter of future studies, specifically into the damage threshold of the materials used for the formation of the high-intensity femtosecond evanescent laser light waves.

14
Concluding comments

When I was writing this book, the names of many remarkable scientists from various countries, pioneers of new avenues of scientific exploration, whom I have already mentioned in this book, crossed my mind. I have had occasion to meet many of these scientists, who have worked during the past fifty years, in various countries and at various conferences and to receive them at the Institute of Spectroscopy of the Russian Academy of Sciences in the city of Troitsk and at my home in the village of Puchkovo, both near Moscow. I have kept many thousands of pertinent photographs and have decided to publish some of them in this concluding section, along with very brief commentaries.

In 1960, while a student at the Moscow Physical-Technical Institute, I did my graduate work at the P. N. Lebedev Physical Institute, where I had a good chance to regularly see at seminars Professors Nikolai Basov and Alexander Prokhorov, Nobel Prize winners to be. I had no camera at the time, so I present photographs of them from subsequent years.

Between 1963 and 1970 I was a graduate and postgraduate student of Professor Basov. During that period I was engaged, almost simultaneously, in several, rather different, problems: (1) nonlinear amplification of high-power nanosecond lasers, (2) methods to obtain ultranarrow spectral resonances for the purposes of optical frequency standards and Doppler-free laser spectroscopy, and (3) evaluation of the parameters of high-power explosively pumped photodissociation lasers (classified research at the time). The first problem was studied with an experimental setup built by V. Zuev and P. Kryukov in the form of a multistage ruby laser amplifier, which led to the discovery of the "shift of the leading edge" of the laser pulse with a superluminal velocity of some 9c. At that time, the Department of Quantum Radiophysics, headed by Professor Basov, was visited by Professor C. H. Townes, who shared the Nobel Prize with Professor Basov and Prokhorov. I have kept an amateur photograph (Fig. 14.1) of a discussion of this effect with Professor Townes.

The problem of production of ultranarrow resonances was also being worked on independently by Professor Veniamin P. Chebotayev (who became my close friend) and Dr. J. Hall (Boulder, USA), and we all managed to suggest in 1967 using the effect of saturation of CH_4 for this purpose, in a cell placed inside the cavity of a $3.39\,\mu m$ He–Ne laser. We met all together in Warsaw in 1968 at an URSI conference (Fig. 14.2). To avoid parallelism in our work, we came to an agreement with Professor Chebotayev that I should concentrate on other problems, mainly the laser control of atoms and molecules, while he should continue with the study of ultranarrow

Fig. 14.1 Discussion of the effect of the superluminal velocity of laser pulses in a nonlinear amplifier with Professor C. H. Townes (*left*) during his visit to the USSR (P. N. Lebedev Physical Institute) in the middle 1960s (the author is on the *right*).

Fig. 14.2 Conference on laser measurements in Warsaw, Poland, September, 1968. On the *left* of the fifth row Dr. John Hall; the third from the *right* in the same row is Professor Veniamin Chebotayev. The author is on the *right* of the last row.

saturation resonances and ultrastable-frequency lasers relying on them. And he was doing so very successfully until his premature death in 1992. During our collaboration period we wrote our joint monograph *Nonlinear Laser Spectroscopy Without Doppler Broadening* in Russian, and in English for Springer-Verlag, and received the Lenin Prize for Science and Technology for this in 1978. The photograph in Fig. 14.3 shows a discussion of this problem at a seminar.

In the autumn of 1970 I spent three months in the USA, within the framework of an exchange of scientists between the Academies of Science of the USSR and the USA, as a visiting professor at the laboratory headed by Professor Ali Javan at the Massachusetts

Fig. 14.3 Professor V. P. Chebotayev (*left*) and the author (*right*) discussing the problem of ultranarrow resonances in optics in a seminar at Moscow State University (1978).

Fig. 14.4 Professor Ali Javan with his first He–Ne laser at his office at MIT. On the *right* is Dr. N. Kurnit, and the author is on the *left*. October 1970, MIT, USA.

Institute of Technology (Fig. 14.4). During the course of this visit I went to a number of laboratories in the USA (Boulder, Berkeley University, Stanford University, Bells Labs, IBM at Yorktown Heights, and others), where I met with many outstanding scientists who made a strong impact on me. Having returned to the USSR, at the Institute of Spectroscopy of the USSR Academy of Sciences, just established in a small town called Akademgorodok (later named the city of Troitsk) near Moscow, I tried, in many respects, to use my American experience. I learned in particular that proposals were a good and important thing, but the main thing was the experimental results obtained. After we had succeeded with our first experiments on laser isotope separation and discovered the isotope-selective multiple-photon dissociation of polyatomic molecules, many prominent scientists started visiting our Institute of Spectroscopy.

One of the first visitors was Professor A. M. Prokhorov, Chairman of the General Physics and Astronomy Department of the USSR Academy of Sciences at that time, the department that our institute belonged to (Fig. 14.5).

254 *Laser control of atoms and molecules*

Fig. 14.5 Discussion of the results of the first experiments on the laser separation of isotopes by way of multiple-photon dissociation under the effect of high-power CO_2-laser pulses with Professor A. M. Prokhorov (*left*) during his visit to the Institute of Spectroscopy in October 1974 (the author is on the *right*).

Late in the 1960s, thanks to the efforts of Professor N. G. Basov, friendly contacts were established between scientists working in the USSR and the GDR in the field of quantum electronics. I frequently had to visit the Central Institute of Optics and Spectroscopy in Adlershof (Berlin) and Jena University, where central part in this field was played by Professors B. Wilhelmi and M. Schubert (Fig. 14.6).

One of the first foreign scientists who visited the newly established Institute of Spectroscopy was Professor Peter Franken from the Optical Center of the University of Arizona Fig. 14.7). He was always very well-disposed towards our institute, and later on (till his unexpected death in 1998) I frequently visited him in Tucson. Peter was a very versatile and talented man, especially as regards design, a good amateur jeweler in particular (Fig. 14.8). Thereafter, early in the 1990s (years of severe depression in the financing of science in Russia), Professor Franken argued very strongly for the financial support of several institutes in Russia on the part of the Department of Defense of the USA.

Fig. 14.6 My friendly meeting with Professor B. Wilhelmi (*left*) and Professor M. Schubert (*right*) during my visit to Jena University to give a lecture in the early 1970s.

Fig. 14.7 Discussion of experimental results at the Laboratory of Laser Spectroscopy with Professor Peter Franken (*left*) during his visit to the USSR in 1975. In the *center* is Professor R. V. Ambartzumian and on the *right*, the author.

In the middle 1970s, Vavilov's Conferences on Nonlinear Optics were held every two years in Novosibirsk in the USSR, with foreign scientists participating. When they stopped in Moscow on their way to Novosibirsk, they usually visited the Institute of Spectroscopy. The photograph in Fig. 14.9 shows Dr. Mooradian, a future Nobel prize winner of 2005, Dr. John Hall, the present author, and Professor R. V. Ambartzumian at the Institute during one such visit, and the photograph in Fig. 14.10 shows Professor M. Feld, Professor V. P. Chebotayev, Professor J.-C. Lehmann, and the present author at Vavilov's conference held in the town of Akademgorodok near Novosibirsk in 1975.

256 *Laser control of atoms and molecules*

Fig. 14.8 Professor Peter Franken at his private amateur workshop (in the middle 1970s).

Fig. 14.9 Dr. A. Mooradian, Dr. J. Hall, the author, and Professor R. V. Ambartzumian (from *left* to *right*) at the Institute of Spectroscopy of the USSR Academy of Sciences in the city of Troitsk (1975).

Professor Herbert Walther from the Max Planck Institute of Quantum Optics (Fig. 14.11) was a frequent guest at the Institute of Spectroscopy, who was greatly interested in everything that was new there. He argued very much for the establishment of scientific contacts between West German and Soviet scientists during the "Cold War" period. Later on he was elected a foreign Member of the Russian Academy of Sciences.

There were also some remarkable meetings abroad during various scientific schools and conferences. I recall with pleasure my talks with Professor Ali Javan at the 1975 Les Houches School on Laser Spectroscopy (Fig. 14.12), and with Professor A. Siegman during my visit to Stanford University (Fig. 14.13), where we discussed animatedly the effect of isotope-selective multiple-photon dissociation of polyatomic molecules by IR laser pulses. At regular international conferences on laser spectroscopy and atomic

Fig. 14.10 Some of the participants in V. Vavilov's Conference on Nonlinear Optics in the town of Akademgorodok near Novosibirsk. From *right* to *left*: Professor M. Feld, Professor V. Chebotayev, Professor S. Haroche, Professor J.-C. Lehmann, and the author.

Fig. 14.11 Professor H. Walther discussing the problem of the resonance multiple-photon excitation of polyatomic molecules with Dr. I. N. Knyazev (*left*) at the Laboratory of Laser Spectroscopy of the Institute of Spectroscopy in the late 1970s.

physics, one could meet many scientists at the same time and discuss many problems with them (Figs. 14.14 and 14.15).

In 1985, a conference was held in Ringbergschloss (Bavaria), devoted to the 25th anniversary of the discovery of tunable dye lasers, which played a very important role in the development of methods for laser control of atoms and molecules. This small conference gathered a remarkable number of scientists, some of whom can be seen in Fig. 14.16.

Fig. 14.12 Professor Ali Javan at the Les Houches Summer School on Laser Spectroscopy in the French Alps in 1975.

Fig. 14.13 Professor Anthony Siegman discussing with the author the mechanism of isotope-selective multiple-photon dissociation of SF_6 molecules under the effect of high-power IR-laser pulses at his office at Stanford University (in the late 1970s).

Among international undertakings, the annual Soviet–German seminars on laser spectroscopy held since 1978 alternately in the USSR and Germany were very important to Soviet scientists. These seminars allowed young scientists from both countries to take part in international contacts that were very difficult to maintain at that time. One such seminar was held in Samarkand (in the Uzbek Soviet Socialist Republic). Some of the participants in this seminar were photographed during a country stroll (Fig. 14.17) and thereafter during their visit to my home in Puchkovo (near Troitsk) (Fig. 14.18).

Many prominent scientists, including Nobel Prize winners, took part in the 13th International Conference on Atomic Physics held in Göteborg, Sweden, in 1982, at

Fig. 14.14 Professor Herbert Walther, Professor Theodore Hansch, Professor Christos Flytzanis, and Professor Steven Harris at an international conference on laser spectroscopy in Germany in the 1980s.

Fig. 14.15 Professor M. Feld, Professor N. Bloembergen, and Professor V. Chebotayev (from *left* to *right*) during a break between sessions at an international conference on laser spectroscopy in the middle 1980s.

which the most pressing problems of laser applications in atomic physics were discussed (Fig. 14.19).

The central problems that were being investigated at the Laser Spectroscopy Department of the Institute of Spectroscopy at the time were (1) control of the motion of atoms, their cooling, atom optics, etc., and (2) resonance multiple-photon excitation and dissociation of polyatomic molecules by high-power IR laser radiation. Scientists from many countries visited the institute in order to get acquainted with the latest results obtained in these fields. In particular, Professor S. Stenholm visited our institute more than once (Fig. 14.20). Among the outstanding scientists who visited

Fig. 14.16 Professor Y. Lee, Professor F. P. Schafer, Dr. P. Sorokin, Professor A. Schawlow, and Professor K. Siegbahn (from *left* to *right*)—participants in the conference "Dye Lasers: 25 Years"—on the terrace of Ringberg Castle (Bavaria).

Fig. 14.17 Open-air discussions during the third Soviet–German seminar in the suburbs of Samarkand. From *left* to *right*: Professor F. Schafer, Professor K. Kompa, Mrs. H. zu Putlitz, Professor G. zu Putlitz, Mrs. I. Wolfrum, and Professor J. Wolfrum.

the Institute of Spectroscopy was Professor Willis Lamb, who at that time was interested in the computer simulation of the classical model of multiple-photon excitation of vibrations in the SF_6 molecule (Fig. 14.21).

Close contacts in the field of laser fusion existed between the Department of Quantum Radiophysics of the P. N. Lebedev Physical Institute, headed by Professor N. G. Basov, and Lawrence Livermore National Laboratory. A delegation from LLNL also visited the Institute of Spectroscopy late in the 1980s (Fig. 14.22). After visiting the institute, Professor Basov, together with the American delegation, usually

Fig. 14.18 Some of the participants in a Soviet–West German seminar on laser spectroscopy on a visit to the present author's home in the village of Puchkovo (near the city of Troitsk). From *right* to *left*: Professor E. Otten, Professor G. zu Putlitz, Mrs. H. zu Putlitz, and the author with his wife, Professor T. Karu.

Fig. 14.19 Winners of the Nobel Prize in physics for various years, participants in the 13th International Conference on Atomic Physics in Göteborg, Sweden, on the way to their hotel after a session. From *left* to *right*: Professor A. L. Schawlow, Professor K. Siegbahn, Professor A. M. Prokhorov, Professor I. Rabi, and Professor N. Bloembergen.

visited the author's home, the distance to be traveled being not very long (8 km). During one such visit, Professor Basov and the present author went to the nearby ruined Trinity-Kazan church (Fig. 14.23). During the years of "perestroika" the reconstruction of this church was started on the author's initiative, and today it is functioning fully.

262 *Laser control of atoms and molecules*

Fig. 14.20 Professor S. Stenholm discussing the problems of the laser cooling and trapping of atoms on a visit to the author's home. From *right* to *left*: Professor V. G. Minogin, Professor S. Stenholm, Professor T. Karu, and Professor V. I. Balykin.

Fig. 14.21 Professor W. Lamb (*left*) at the Institute of Spectroscopy in 1986. On the *right* is Professor E. Ryabov, and in the *center*, the author.

In 1987, an international symposium on the physics of trapped low-energy particles was held in Stockholm, at which the problems of laser cooling of trapped charged and neutral particles were discussed. Such researchers of international renown in this field as Professor W. Paul, Professor H. Dehmelt, Professor N. Ramsey, and others took part in this symposium (Fig. 14.24). Within the framework of this symposium, the problems associated with Bose–Einstein condensation were discussed, with hundreds of scientists from various countries participating. After this symposium, Professor S. Svanberg organized the eighth regular International Conference on Laser Spectroscopy in the small town of Åre in the northern part of Sweden (Figs. 14.25 and 14.26). The number of participants proved greater than there were seats in the conference hall. But they could take part in the conference by TV while remaining in their rooms.

Fig. 14.22 Delegation from Lawrence Livermore National Laboratory and Professor N. Basov during a visit to the Institute of Spectroscopy: Dr. W. Krupke, Dr. J. Emmett, (?), Professor P. G. Kruykov, Dr. I. Zubarev, the author, and Professor N. G. Basov.

Fig. 14.23 Professor N. Basov and the author beside the ruined Trinity-Kazan church near the author's home in the village of Puchkovo (near the city of Troitsk) in the late 1980s.

Fig. 14.24 Professor W. Paul and Professor H. Dehmelt (*at the top left*) during discussions at the International Symposium on the Physics of Trapped Low-Energy Particles (Stockholm, 1987). Professor N. Ramsey (*top right*), author (*bottom left*), Professor W. Paul and the author (*bottom right*).

Fig. 14.25 At an excursion in the suburbs of Åre during the 8th International Conference on Laser Spectroscopy. From *left* to *right*: in the first row, Professor H. Walther and Professor H. Dehmelt; in the second row, Professor M. Scully and the author; in the third row, Professor R. Glauber; and in the last row, Professor I. Lindgren and Professor D. Pritchard.

Fig. 14.26 Professor Sune Svanberg (*right*) and the author.

Fig. 14.27 Professor M. El-Sayed, Professor A. M. Prokhorov, Professor A. Zewail, and the present author at the office of Professor Prokhorov (from *right* to *left*).

Late in the 1980s, during the "perestroika" period in the USSR, it proved possible to organize Soviet–American seminars on laser chemistry. The first such seminar was held in Santa Barbara, USA, in March 1988, and the leading scientists in this field from the two countries took part in it. Before the seminar, Professor M. El-Sayed and Professor A. Zewail came to the USSR, where they met with Professor A. M. Prokhorov (Fig. 14.27). The seminar in Santa Barbara was a great success (Fig. 14.28), and the next seminar was held in the USSR.

During the "perestroika" period it proved possible to establish scientific contacts between Soviet and Israeli researchers. The present author visited Israel at that time to read the James Frank Lectures and take part in scientific discussions at the Israel Academy of Sciences presided over by Professor Joshua Jortner (Fig. 14.29).

In the 1990s, Professor Ahmed Zewail initiated international conferences on femtochemistry. In 1995, the Solvay Congress on Chemistry in Belgium and the Nobel Symposium in Sweden were held on the same problem. These conferences played an

Fig. 14.28 Some of the participants at the first Soviet–American seminar on laser chemistry held in March 1988 in Santa Barbara: Professor C. B. Moore, Professor R. Marcus, and Professor M. El-Sayed after a session.

Fig. 14.29 Professor J. Jortner at his office during the author's visit to Israel in December 1989.

important role in the formation of this novel line of research on the science of ultrafast processes and the allied technologies (Figs. 14.30 and 14.31).

In 1995, Professor C. H. Townes came to the International Conference on Coherent Optics in St. Petersburg, where he was given a warm reception by hundreds of the participants. He himself also read an interesting paper on radio astronomy and astronomical lasers and met with his old friend Sasha (Professor A. M. Prokhorov) (Fig. 14.32)

Fig. 14.30 Some of the participants in the 20th Solvay Congress on Chemistry, "Chemical Reactions and their Control on the Femtosecond Time Scale" (November 1995, Brussels). From *left* to *right*: the author, Professor J. Troe, Professor R. Marcus, Professor A. Zewail, and Professor R. Levine.

Fig. 14.31 Professor C. V. Shank, one of the creators of CW femtosecond dye lasers (*right*), Professor A. Zewail, the founder of laser femtochemistry (*center*), and the author.

In 1996, a conference on laser physics was held in Moscow on the occasion of Professor Prokhorov's 80th birthday, and Professor Townes and his wife Frances took part in it. They accepted my invitation to visit my home in Puchkovo (Fig. 14.33). Charlie and Frances are remarkably agile people. They stayed in Moscow to attend Prokhorov's conference during the course of their round-the-world trip, with only two small suitcases as baggage.

Fig. 14.32 Professor A. M. Prokhorov and Professor C. H. Townes at the International Conference on Coherent and Nonlinear Optics held in St. Petersburg in 1995.

Fig. 14.33 Professor C. H. Townes (*extreme right*), Mrs. Frances Townes (*center*), Professor A. M. Prokhorov (*extreme left*), the author (second from the *right*), and Professor T. Karu's daughter Inga (second from the *left*) at the author's home in the village of Puchkovo during Professor Townes's visit to Moscow to attend the conference devoted to Professor Prokhorov's 80th birthday in July 1996.

In the early to middle 1990s, I spent a few months every year in France, working with Professor M. Ducloy in the field of nanooptics at Université Paris-Nord. I happened to meet there with many noted French scientists (Fig. 14.34).

One cannot but mention numerous laser-related conferences held in the United Kingdom (Oxford, Edinburgh, Glasgow, York etc.). What has especially stuck in my memory is my latest visit to the UK in 2002, by invitation of Professor Colin Webb,

Fig. 14.34 Professor C. Borde, Professor C. Cohen-Tannoudji, the author, and Professor M. Himbert at the Université Paris-Nord after the Honoris Causa presentation ceremony in 1996.

Fig. 14.35 Professor Colin Webb (Oxford University) and Professor Anthony Siegman (Stanford University) at the Athenaeum Club in London in May 2002.

who also argued for the appearance of this book. It is always pleasant to meet with friends after a lapse of many years (Fig. 14.35).

One of the latest highlight events was the 19th International Conference on Atomic Physics held in Rio de Janeiro in 2004. Some topics intensively discussed at this conference were the problems of collective effects in quantum gases, and laser-cooled and trapped atoms and molecules. One could meet there with the pioneers of this new area of research, as well as with atomic physicists engaged in other fields (Fig. 14.36).

In recent years I have happened to attend many 60th, 65th, and 70th birthday celebrations of prominent scientists. The latest was the 65th birthday of my old friend Professor J. Wolfrum, celebrated at Heidelberg University, and it was this outstanding scientist who helped me in establishing contacts between Soviet and West German scientists in the form of laser seminars. The photograph in Fig. 14.37 shows him in

Fig. 14.36 Professor Daniel Kleppner (*left*) and Professor Ingvar Lindgren from Göteborg having a discussion during a break between sessions at the 19th International Conference on Atomic Physics held in Rio de Janeiro in July 2004.

Fig. 14.37 Professor J. Wolfrum (*right*) and Professor W. Ketterle at Professor Wolfrum's 65th birthday celebration at Heidelberg University in November 2004.

company with his postdoc W. Ketterle, Nobel Prize winner, who also participated in these seminars.

I have picked only 37 photographs out of ten thousand, which were taken during the course of approximately 35 years at various international undertakings associated with the problem of "laser control of atoms and molecules."

References

Abragam, A. (1961). *The principles of nuclear magnetism*. Clarendon Press, Oxford.

Ackerhalt, J. R., and Eberly, J. H. (1976). Coherence versus incoherence in stepwise laser excitation of atoms and molecules. *Physical Review A*, **14**, 1705–1710.

Adams, C. S., Sigel, M., and Mlynek, J. (1994). Atom optics. *Physics Reports*, **240**, 143–210.

Albert, M. S., Cates, G. D., Driehuys, B., Happer, W., Saam, B., Springer, C. S. Jr., and Wishnia, A. (1994). Biological magnetic resonance imaging using laser-polarized ^{129}Xe. *Nature*, **370**, 199–201.

Alexandrov, E. B., Bonch-Bruevich, A. M., and Khodovoi, B. A. (1967). Possibility of weak magnetic field measurements by optical orientation of atoms. *Optics and Spectroscopy* (Russia), **23**, 282-286 [*Optics and Spectroscopy*, **23**, 151–154].

Alimpiev, S. S., Bagratashvili, V. N., Karlov, N. V., Letokhov, V. S., Lobko, V. V., Makarov, A. A., Sartakov, B. G., and Khokhlov, E. M. (1977). Effect of depletion of many rotational states upon vibrational excitation of molecules in a strong IR field. *Journal of Experimental and Theoretical Physics Letters*, **25**, 547–550.

Alimpiev, S. S., Karlov, N. V., Sartakov, B. G., and Khokhlov, E. M. (1978). The spectral structure of the low level excitation in the process of the collisionless dissociation of polyatomic molecules. *Optics Communications*, **26**, 45–49.

Alimpiev, S. S., Karlov, N. V., Nikiforov, S. M., Prokhorov, A. M., Sartakov, B. G., Khokhlov, E. M., and Shtarkov, A. L. (1979). Spectral characteristics of the SF$_6$ molecules excitation by a strong IR laser field at continuously tuned radiation frequency. *Optics Communications*, **31**, 309–312.

Alkhazov, G. D., Letokhov, V. S., Mishin, V. I., Panteleyev, V. N., Romanov, V. I., Sekatskii, S. K., and Fedoseyev, V. N. (1989). Highly effective Z-selective photoionization of atoms in hot metal cavity, followed by electrostatic confinement of ions. *Journal of Technical Physics Letters*, **15**, 63–67.

Allen, L., and Eberly, J. H. (1975). *Optical resonance and two-level atoms*. Wiley Interscience, New York.

Allen, L., Padgett, M. J., and Babiker, M. (1999). The orbital angular momentum of light. In *Progress in Optics*, vol. 39 (ed. E. Wolf), pp. 291–371. Elsevier, Amsterdam.

Alvarez, L. W., Alvarez, W., Azaro, F., and Mitchell, H. V. (1980). Extraterrestial cause for the Cretaceous–Tertiary extension. *Science*, **208**, 1095–1108.

Alzetta, G., Gozzini, A., Moi, L., and Orriols, G. (1976). An experimental method for the observation of RF transitions and laser beat resonances in oriented Na vapor. *Nuovo Cimento B*, **36**, 5–20.

Ambartzumian, R. V., and Letokhov, V. S. (1972). Selective two-step (STS) photoionization of atoms and photodissociation of molecules by laser radiation. *Applied Optics*, **11**, 354–358.

Ambartzumian, R. V., and Letokhov, V. S. (1977). Multiple infrared photon laser photochemistry. In *Chemical and biochemical applications of lasers*, vol. 3 (ed. C. B Moore), pp. 166–316. Academic Press, Orlando.

Ambartzumian, R. V., Kalinin, V. P., and Letokhov, V. S. (1971). Selective two-step photoionization of rubidium atoms by laser radiation. *Journal of Experimental and Theoretical Physics Letters*, **13**, 217–219.

Ambartzumian, R. V., Letokhov, V. S., Makarov, G. N., and Puretzkii, A. A. (1973). Laser separation of nitrogen isotopes. *Journal of Experimental and Theoretical Physics Letters*, **17**, 63–65.

Ambartzumian, R. V., Letokhov, V. S., Ryabov, E. A., and Chekalin, N. V. (1974). Isotope-selective chemical reaction of BCl_3 molecules in a strong IR laser field. *Journal of Experimental and Theoretical Physics Letters*, **20**, 273–274.

Ambartzumian, R. V., Bekov, G. I., Letokhov, V. S., and Mishin, V. I. (1975a). Excitation of high-lying states of Na atoms by means of tunable laser radiation and autoionization of these states in an electric field. *Journal of Experimental and Theoretical Physics Letters*, **21**, 279–281.

Ambartzumian, R. V., Gorokhov, Yu. A., Letokhov, V. S., and Makarov, G. N. (1975b). Separation of sulfur isotopes with an enrichment ratio higher than 10^3 by acting on the SF_6 molecule with a CO_2 laser radiation. *Journal of Experimental and Theoretical Physics Letters*, **21**, 171–174.

Ambartzumian, R. V., Furzikov, N. P., Gorokhov, Yu. A., Letokhov, V. S., Makarov, G. N., and Puretzky, A. A. (1976). Selectivity dissociation of SF_6 molecules in a two frequency IR laser field. *Optics Communications*, **18**, 517–521.

Aminoff, C. G., Steane, A. M, Bouyer, P., Desbiolles, P., Dalibard, J., and Cohen-Tannoudji, C. (1993). Cesium atoms bouncing in a stable gravitational cavity. *Physical Review Letters*, **71**, 3083–3086.

Anderson, A., Haroche, S., Hinds, E. A., Jhe, W., Meschede, D., and Moi, L. (1986). Reflection of thermal Cs atoms grazing a polished glass surface. *Physical Review A*, **34**, 3513–3516.

Anderson, M. H., Ensher, J. R., Matthews, M. R., Wieman, C. E., and Cornell, E. A. (1995). Observation of Bose–Einstein condensation in a dilute atomic vapor. *Science*, **269**, 198–201.

Andrews, M. R., Townsend, C. G., Miesner, H.-J., Durfee, D. S., Kurn, D. M., and Ketterle, W. (1997). Observation of interference between two Bose condensates. *Science*, **275**, 637–641.

Andreyev, S. V., Antonov, V. S., Knyazev, I. N., Letokhov, V. S., and Movshev, V. G. (1975). Photoionization of molecules by VUV laser radiation. *Physics Letters*, **54A**, 91–92.

Andreyev, S. V., Balykin, V. I., Letokhov, V. S., and Minogin, V. G. (1981). Radiative deceleration to 1.5 K and monochromatization of a beam of sodium atoms in counter-propagating laser beam. *Journal of Experimental and Theoretical Physics Letters*, **34**, 442–445.

Andreyev, S. V., Balykin, V. I., Letokhov, V. S., and Minogin, V. G. (1982). Radiative slowing down and monochromatization of a beam of sodium atoms in a counterpropagating laser beam. *Journal of Experimenal and Theoretical Physics*, **55**, 828–834.

Andreyev, S. V., Letokhov, V. S., and Mishin, V. I. (1987). Laser resonance photoionization spectroscopy of Rydberg levels in Fr. *Physical Review Letters*, **59**, 1274–1276.

Andreyev, S. V., Mishin, V. I., and Letokhov, V. S. (1988). Rydberg levels and ionization potential of francium measured by laser-resonance ionization in a hot cavity. *Journal of the Optical Society of America B*, **5**, 2190–2198.

Anton, K.-R., Kaufman, W., Moruzzi, G., Neugart, R., Otten, E.-W., and Schinzler, E. (1978). Collinear laser spectroscopy on fast atomic beams. *Physical Review Letters*, **40**, 642–645.

Antonov, V. S., Knyazev, I. N., Letokhov, V. S., Matyuk, V. M., Movshev, V. G., and Potapov, V. K. (1977). Mass-spectrometer with selective stepwise photoionization of molecules with laser radiation. *Journal of Technical Physics Letters*, **3**(23), 1287–1291 (in Russian).

Antonov, V. S., Knyazev, I. N., Letokhov, V. S., Matyuk, V. M., Movshev, V. G., and Potapov, V. K. (1978). Stepwise laser photoionization of molecules in mass-spectrometer: a new method of probing and detection of polyatomic molecules. *Optics Letters*, **3**(2), 37–39.

Antonov, V. S., Letokhov, V. S., and Shibanov, A. N. (1980a). Formation of molecular ions upon irradiation of molecular crystal with a UV laser radiation. *Journal of Experimental and Theoretical Physics Letters*, **31**, 441–444.

Antonov, V. S., Letokhov, V. S., and Shibanov, A. N. (1980b). Photoionization mass spectrometry of benzene and benzaldehyde with an excimer laser. *Applied Physics*, **22**, 293–297.

Antonov, V. S., Letokhov, V. S., and Shibanov, A. N. (1981a). Laser photoionization and mass-spectral detection of single naphthalene molecules. *Optics Communications*, **38**(3), 182–184.

Antonov, V. S., Letokhov, V. S., and Shibanov, A. N. (1981b). Nonthermal desorption of molecular ions of polyatomic molecules induced by UV laser radiation. *Applied Laser Physics*, **25**, 71–76.

Antonov, V. S., Yegorov, S. E., Letokhov, V. S., Matveets, Yu. A., and Shibanov, A. N. (1982). Detachment of chromophore ions from a complex molecule on a surface by picosecond UV laser pulses. *Journal of Experimental and Theoretical Physics Letters*, **36**, 33–35.

Antonov, V. S., Yegorov, S. E., Letokhov, V. S., and Shibanov, A. N. (1983). Laser photoionization detection of submonomolecular surface layers. *Journal of Experimental and Theoretical Physics Letters*, **37**, 185–187.

Apatin, V. M., Krivtzun, V. M., Kuritzyn, Yu. A., Makarov, G. N., and Pak, I. (1983). Diode laser study of IR multiphoton-induced depletion of rotational sublevels of the ground vibrational state of SF_6 molecules cooled in a pulsed free jet. *Optics Communications*, **47**, 251–256.

Arimondo, E. (1996). Coherent population trapping in laser spectroscopy. In *Progress in optics* (ed. E. Wolf), vol. 39, pp. 257–354. Elsevier, Amsterdam.

Arndt, M., Nairz, O., and Zeilinger, A. (2002). Interferometry with macromolecules: quantum paradigms tested in mesoscopic world. In *Quantum (un)speakables* (ed. R. Bertlmann and A. Zeilinger), pp. 333–351. Springer, New York.

Aseyev, S. A., Kudriavtsev, Yu. A., Letokhov, V. S., Petrunin, V. V. (1991). Laser detection of the rare isotopes ^{3}He at concentration as low as 10^{-9}. *Optics Letters*, **16**(7), 514–516.

Ashfold, M. N. R., and Howe, J. P. (1994). Multiphoton spectroscopy of molecular species. *Annual Review of Physical Chemistry*, **45**, 57–82.

Ashkin, A. (1970). Acceleration and trapping of particles by radiation pressure. *Physical Review Letters*, **24**, 156–159.

Ashkin, A. (2000). History of optical trapping and manipulation of small-neutral particles, atoms and molecules. *IEEE Journal of Selected Topics in Quantum Electronics*, **6**, 841–856.

Ashkin, A., and Dziedzic, J. M. (1987). Optical trapping and manipulation of viruses and bacteria. *Science*, **235**, 1517–1520.

Ashkin, A., Dziedzic, J. M., Bjorkholm, J. E., and Chu, S. (1986). Observation of a single-beam gradient force optical trap for dielectric particles. *Optics Letters*, **11**, 288–291.

Aspect, A., Arimondo, E., Kaiser, R., Vansteenkiste, N., and Cohen-Tannoudji, C. (1988). Laser cooling below the one-photon recoil energy by velocity-selective coherent population trapping. *Physical Review Letters*, **61**, 826–829.

Aspect, A., Arimondo, E., Kaiser, R., Vansteenkiste, N., and Cohen-Tannoudji, C. (1989). Laser cooling below the one-photon recoil energy by velocity-selective coherent population trapping: theoretical analysis. *Journal of the Optical Society of America B*, **6**, 2112–2124.

Assion, A., Baumert, T., Bergt, M., Bixner, T., Kiefer, B., Seyfried, V., Strehle, M., and Gerber, G. (1998). Control of chemical reactions by feedback optimized phase shaped femtosecond laser pulse. *Science*, **282**, 919–922.

Autler, S. H., and Townes, C. H. (1955). Stark effect in rapidly varying field. *Physical Review*, **100**, 703–722.

Backe, H., Dretzke, A., Hies, M., Kube, G., Kunz, H., Lauth, W., Sewtz, M., Trautman, N., Repnow, R., and Majer, H. J. (2000). Isotope shift measurement at ^{242}Am. *Hyperfine Interactions*, **127**, 35–39.

Bagayev, S. N., Chebotayev, V. P., Dmitriev, A. K., Om, A. E., Nekrasov, Yu. V., and Skvortsov, B. N. (1991). Second-order Doppler-free spectroscopy. *Applied Physics B*, **52**, 63–66.

Bagratashvili, V. N., Burimov, V. I., Deyev, D. E., Kudriavtsev, Yu. A., Kuz'min, M. V., Letokhov, V. S., and Sviridov, A. P. (1982). Observation of electronic predissociation of vibrationally overexcited polyatomic molecules. *Journal of Experimental and Theoretical Physics Letters*, **35**, 189–192.

Bagratashvili, V. N., Kuzmin, M. V., Letokhov, V. S., and Shibanov, V. N. (1983). Observation of the proton and electron detachment from the anthracene molecule during strong IR multiple-photon superexcitation. *Journal of Experimental and Theoretical Physics Letters*, **37**, 112–116.

Bagratashvili, V. N., Kuzmin, M. V., Letokhov, V. S., and Stuchebrukhov, A. A. (1985a). Theory of multiple-photon IR excitation of polyatomic molecules in the model of active and passive modes of vibrational reservoir. *Chemical Physics*, **97**, 13–29.

Bagratashvili, V. N., Letokhov, V. S., Makarov, A. A., and Ryabov, E. A. (1985b). *Multiple photon infrared laser photophysics and photochemistry*. Harwood Academic, Chur.

Bagratashvili, V. N., Ionov, S. I., and Makarov, G. N. (1989). Laser IR spectroscopy of polyatomic molecules near and above the dissociation limit. In *Laser spectroscopy of highly vibrationally excited molecules* (ed. V. S. Letokhov), pp. 265–328. Adam Hilger, Bristol.

Bahns, J. T., Gould, P. L., and Stwalley, W. C. (2000). Formation of cold ($T \gtrsim 1$ K) molecules. In *Advances in atomic, molecular and optical physics* (eds. B. Bederson and H. Walther), vol. 42, pp. 171–224. Academic Press, San Diego.

Baklanov, E. V., Dubetsky, B. Ya., and Chebotayev, V. P. (1976a). Nonlinear Ramsey resonance in the optical range. *Applied Physics*, **9**, 171–173.

Baklanov, E. V., Chebotayev, V. P., and Dubetsky, B. Ya. (1976b). The resonance of two-photon absorption in separated optical fields. *Applied Physics*, **9**, 201–202.

Baldwin, K. G., Hajnal, J. V., and Fisk, P. T. H. (1990). Optics for neutral atomic beams: a reflection and diffraction of sodium atoms by evanescent laser light fields. *Journal of Modern Optics*, **37**, 1839–1848.

Balykin, V. I. (1999). Atomic waveguides. In *Advances in atomic, molecular and optical physics* (eds. B. Bederson and H. Walther), vol. 41, pp. 181–260. Academic Press, San Diego.

Balykin, V. I., and Letokhov, V. S. (1987). The possibility of deep laser focusing of an atomic beam into Å-region. *Optics Communications*, **64**, 151–156.

Balykin, V. I., and Letokhov, V. S. (1988). Deep focusing of an atomic beam into Å-size region by means of laser radiation. *Journal of Experimental and Theoretical Physics*, **67**, 78–83.

Balykin, V. I., and Letokhov, V. S. (1995). *Atom optics with laser light*. Harwood Academic, Chur.

Balykin, V. I., Letokhov, V. S., and Mishin, V. I. (1979). Observation of the cooling of free sodium atoms in a resonance laser field with a scanning frequency. *Journal of Experimental and Theoretical Physics Letters*, **29**, 560–569.

Balykin, V.I., Letokhov, V.S., and Minogin, V.G. (1981). Radiative redistribution of the velocities of free sodium atoms by resonance laser radiation. *Journal of Experimental and Theoretical Physics*, **53**(5), 919–924.

Balykin, V. I., Letokhov, V. S., and Sidorov, A. I. (1984). Radiative collimation of an atomic beam by two-dimensional cooling by a laser beam. *Journal of Experimental and Theoretical Physics*, **40**(6), 1026–1029.

Balykin, V., Letokhov, V., Ovchinnikov, Yu., and Sidorov, A. (1988a). Quantum-state selective mirror reflection of atoms by laser light. *Physical Review Letters*, **60**, 2137–2140.

Balykin, V. I., Letokhov, V. S., Ovchinnikov, Yu. B., Sidorov, A. I., and Shul'ga, S. V. (1988b). Channeling of atoms in standing spherical light wave. *Optics Letters*, **13**, 958–960.

Balykin, V. I., Letokhov, V. S., Sidorov, A. I., and Ovchinnikov, Yu. B. (1988c). Focusing of an atomic beam and "imaging" of atomic source by means of "laser lens" based on resonance radiation pressure. *Journal of Modern Optics*, **35**, 17–34.

Balykin, V. I., Klimov, V. V., and Letokhov, V. S. (1994). Laser near field lens for atoms. *Journal de Physique II, France*, **4**, 1981–1997.

Balykin, V. I., Subbotin, M. V., and Letokhov, V. S. (1996). Reflection of an electron beam by femtosecond light waves. *Optics Communications*, **129**, 177–183.

Balykin, V. I., Minogin, V. G., and Letokhov, V. S. (2000). Electromagnetic trapping of cold atoms. *Reports on Progress in Physics*, **63**, 1429–1510.

Balykin, V. I., Klimov, V. V., and Letokhov, V. S. (2003). Atomic nanooptics based on "photon dots" and "photon holes". *Journal of Experimental and Theoretical Physics Letters*, **78**(1), 8–12.

Band, Y. B., and Julienne, P. S. (1995). Ultracold molecule production by laser-cooled atom photoassociation. *Physical Review A*, **51**, R4317–R4320.

Bardeen, C., Wang, Q., and Shank, C. (1995). Selective excitation of vibrational wave packet motion using chirped pulses. *Physical Review Letters*, **75**, 3410–3413.

Bardeen, C., Yakovlev, V., Wilson, K., Carpenter, S., Weber, P., and Warren, W. (1997). Feedback quantum control of molecular electronic population transfer. *Chemical Physics Letters*, **280**, 151–158.

Barger, R. L., and Hall, J. L. (1969). Pressure shift and pressure broadening of methane line at 3.39 μm studied by laser-saturated molecular absorption. *Physical Review Letters*, **22**, 4–8.

Barnett, W. L. (2000). Spontaneous emission and energy transfer in the microcavity. *Contemporary Physics*, **41**(5), 287–300.

Barnett, S. M., and Loudon, R. (1996). Sum rule for modified spontaneous emission rates. *Physical Review Letters*, **77**, 2444–2446.

Bartels, R., Backus, S., Zeek, E., Misogutt, L., Vdovin, G., Christov, L. P., Murnane, M. M., and Kapteyan, H. C. (2000). Shaped-pulse optimization of coherent emission of high-harmonic soft X-rays. *Nature*, **4–6**, 164–166.

Basov, N. G., and Letokhov, V.S. (1968). Optical frequency standards. *Soviet Physics—Uspekhi*, **96**(4), 585–631.

Basov, N. G., and Prokhorov, A. M. (1954). Applications of molecular beam for radiospectroscopy of a rotational spectra of molecules. *Journal of Experimental and Theoretical Physics*, **27**, 431–438 (in Russian).

Basov, N. G., Kompanets, I. N., Kompanets, O. N., Letokhov, V. S., and Nikitin, V. V. (1969). Narrow resonance upon saturation of SF_6 with CO_2-laser radiation. *Journal of Experimental and Theoretical Physics*, **9**, 568–571 (in Russian).

Bast, R., and Schwerdtfeger, P. (2003). Parity-violation effects in the C–F stretching mode of heavy-atom methyl fluorides. *Physical Review Letters*, **91**(2), 023001.

Baudon, J., Mathevet, R., and Robert, J. (1999). Atomic interferometry. *Journal of Physics B: Atomic, Molecular and Optical Physics*, **32**, R173–R195.

Baumert, T., Grosser, M., Thalweiser, R., and Gerber, G. (1991). Femtosecond time-resolved molecular multiphoton ionization: the Na_2 system. *Physical Review Letters*, **67**, 3753–3756.

Behr, J. A., Gorelov, A., swanson, T., Häusser, O., Jackson, K. P., Trinczek, M., et al. (1997). Magneto-optic trapping of β-decaying $^{38}K^m$, ^{37}K from an on-line isotope separator. *Physical Review Letters*, **79**, 375–378.

Bekov, G. I., Letokhov, V. S., Matveyev, O. I., and Mishin, V. I. (1978). Detection of a long-lived autoionization state in the spectrum of gadolinium. *Journal of Experimental and Theoretical Physics Letters*, **28**, 283–285.

Bekov, G. I., Letokhov, V. S., Radayev, V. N., Baturin, G. N., Egorov, A. S., Kursky, A. N., and Narseyev, V. A. (1984). Ruthenium in the ocean. *Nature*, **312**, 748–750.

Bekov, G. I., Letokhov, V. S., Radayev, V. N., Badykov, D. D., and Nazarov, M. A. (1988). Rhodium distribution at the Cretaceous/Tertiary boundary analyzed by ultrasensitive laser photoionization. *Nature*, **332**, 146–148.

Bennett, W. R. Jr. (1962). Hole burning effects in He–Ne maser. *Physical Review*, **126**, 580–593.

Bergeman, T., Erez, G., and Metcalf, H. (1987). Magnetostatic trapping fields for neutral atoms. *Physical Review A*, **35**, 1535–1546.

Bergmann, K., Theur, H., and Shore, B. W. (1998). Coherent population transfer among quantum states of atoms and molecules. *Reviews of Modern Physics*, **70**, 1003–1025.

Berman, P. R. (ed.) (1997). *Atom interferometry*. Academic Press, New York.

Bethe, H. A., and Salpeter, E. A. (1957). *Quantum mechanics of one- and two-electron atoms*. Academic Press, New York.

Binnewies, T., Wilpers, G., Sterr, U., Riehle, F., Helmeke, J., Mehlstäubler, T.E., Rasel, E.M., and Ertmer, W. (2001). Doppler cooling and trapping on forbidden transitions. *Physical Review Letters*, **87**, 123002.

Bjorkholm, J. E., Freeman, R. E., Ashkin, A. A., and Pearson, D. B. (1978). Observation of focusing of neutral atoms by the dipole forces of resonance-radiation pressure. *Physical Review Letters*, **41**, 1361–1364.

Bjorkholm, J. E., Freeman, R. E., Ashkin, A. A., and Pearson, D. B. (1980). Experimental observation of the influence fluctuations of resonance-radiation pressure. *Optics Letters*, **5**, 111–113.

Blaum, K., Beck, D., Bollen, G., Delahaye, P., Guenaut, C., Heufurth, F., Kellerbauer, A., Kluge, H.-J., Lunney, D., Schwartz, S., Schweikhard, L., and Yazidjian, C. (2004). Population inversion of nuclear states by a Penning trap mass-spectrometer. *Europhysics Letters*, **67**(4), 586–592.

Bloch, F. (1946). Nuclear induction. *Physical Review*, **70**, 460–474.

Bloembergen, N. and Yablonovich, E. (1978). Infrared-laser-induced unimolecular reactions. *Physics Today*, **31**(5), 23–30.

Bloembergen, N., and Zewail, A. H. (1984). Energy distribution in isolated molecules and the question of mode-selective laser chemistry revisited. *Journal of Physical Chemistry*, **88**, 5459–5465.

Bluhm, R., Kostelecky, V. A., and Russuel, N. (1999). CPT and Lorentz tests in hydrogen and antihydrogen. *Physical Review Letters*, **82**, 2254–2257.

Boesl, U., Neusser, H. J., and Schlag, E. W. (1978). Two-photon ionization of polyatomic molecules in a mass-spectrometer. *Zeitschrift Naturforschung*, **33a**, 1546–1548.

Boesl, U., Neusser, H. J., and Schlag, E. W. (1980). Visible and UV multiphoton ionization and fragmentation of polyatomic molecules. *Journal of Chemical Physics*, **72**, 4327–4333.

Borde, C. J. (1989). Atomic interferometry with internal state labelling. *Physics Letters*, **140**, 10–12.

Borde, C., Camy, G., Decomps, D., and Pottier, L. (1973). Mise en evidence experimentale de phenomene di dispersion saturee daus l'iode a 5145. *Comptus Rendus Academie des Sciences* (Paris) *B*, **277**, 381–383.

Born, M., and Wolf, E. (1984). *Principles of optics*. Pergamon Press, Oxford.

Bouchiat, M. A., Carver, T. R., and Varnum, C. M. (1960). Nuclear polarization in He^3 gas induced by optical pumping and dipolar exchange. *Physical Review Letters*, **5**, 373–375.

Bouwkamp, C. J. (1954). Diffraction theory. *Reports on Progress in Physics*, **7**, 35–100.

Boyarkin, O. V., Kovalczyk, M., and Rizzo, T. R. (2003). Collisionally enhanced isotopical selectivity in multiphoton dissociation of vibrationally excited CF_3H. *Journal of Chemical Physics*, **118**, 93–103.

Bradley, C. C., Sackett, C. A., Tollet, J. J., and Hulet, R. G. (1995). Evidence of Bose–Einstein condensation in atomic gas with attractive interactions. *Physical Review Letters*, **75**, 1687–1690.

Brillouin, L. (1960). *Wave propagation and group velocity*. Academic Press, New York.

Brixner, T., Damrauer, N. H., and Gerber, G. (2001a). Femtosecond quantum control. *Advances in Atomic, Molecular and Optical Physics*, **46**, 1–54.

Brixner, T., Damrauer, N. H., Niklaus, P., and Gerber, G. (2001b). Photoselective adaptive femtosecond quantum control in the liquid phase. *Nature*, **414**, 57–60.

Brossel, J., and Bitter, F. (1952). A new "double resonance" method for investigating atomic energy levels. Application to Hg $^3P_1$. *Physical Review*, **86**, 308–316.

Brumer, P., and Shapiro, M. (1986). Control of unimolecular reactions using coherent light. *Chemical Physics Letters*, **126**, 541–564.

Budker, D., Kimball, D. F., Rochester, S. M., Yashchuk, V. V., and Zolotarev, M. (2000). Sensitive magnetometry based on nonlinear magneto-optical rotation. *Physical Review A*, **62**, 043403.

Burnett, K., Juliene, P. S., Lett, P. D., Tresing, E., and Williams, C. J. (2002). Quantum encounters of the cold kind. *Nature*, **416**, 205–232.

Bushaw, B. A., Nortershäuser, W., Blaum, K., and Wendt, K. (2003). Studies of narrow autoionizating resonances in gadolinium. *Spectrochimica Acta, Part B*, **58**, 1083–1095.

Bykov, V. P. (1975). Spontaneous emission from medium with a band spectrum. *Soviet Journal of Quantum Electronics*, **4**, 861–871.

Cantrell, C. D., Letokhov, V. S., and Makarov, A. A. (1980). Coherent excitation of multilevel quantum system by laser light. In *Coherent nonlinear optics* (ed. M. S. Feld and V. S. Letokhov), Topics in Current Physics, vol. 21, pp. 165–269. Springer, Berlin.

Carnal, O., and Mlynek, J. (1991). Young's double-slit experiment with atoms: a simple atom interferometer. *Physical Review Letters*, **66**, 2689–2692.

Cesar, C., Fried, D., Killian, T., Polcyn, A., Sandberg, J., Yu, I., Greytak, T., Kleppner, D., and Doyle, J. (1996). Two-photon spectroscopy of trapped atomic hydrogen. *Physical Review Letters*, **77**, 255–258.

Chang, S., and Minogin, V. G. (2002). Density matrix approach to dynamics of multilevel atoms in the laser field. *Physics Reports*, **365**, 65–143.

Chapman, M. S., Ekstrom, C. R., Hammond, T. D., Schmiedmayer, J., Tannian, E., Wehinger, S., and Pritchard, D. E. (1995). Near-field imaging of atom diffraction gratings: the atomic Talbot-effect. *Physical Review A*, **51**, R14–R17.

Chardonnet, C., Guernet, F., Charton, G., and Borde, C. J. (1994). Ultra-high resolution saturation spectroscopy using slow molecules in an external cell. *Applied Physics B*, **59**, 333–343.

Chebotayev, V. P. (1985). Superhigh resolution spectroscopy. In *Laser handbook* (ed. M. Bass and M. L. Stitch), pp. 291–404. Elsevier Science, Amsterdam.

Chebotayev, V. P., Dubetsky, B. Ya., Kasantsev, A. P., and Yakovlev, V. P. (1985). Interference of atoms in separated optical fields. *Journal of the Optical Society of America B*, **2**, 1791–1798.

Chekalin, S. V., Golovlev, V. V., Kozlov, A. A., Matveetz, Yu. A., Yartsev, A. P, and Letokhov, V. S. (1988). Femtosecond laser photoionization mass spectrometry of molecules on surfaces. In *Ultrafast phenomena VI* (ed. T. Yajima, K. Yoshihara, C. B. Harris, and S. Shionoya), pp. 414–419. Springer, Berlin. Also published in *Journal of Physical Chemistry*, **92**, 6585–6588.

Chen, C. Y., Li, Y. M., Bailey, K., O'Connor, T. P., Young, L., and Lu, Z. T. (1999). Ultrasensitive isotope trace analysis with magneto-optical trap. *Science*, **286**, 1139–1141.

Chikkatur, A. P., Shin, Y., Leanhardt, A. E., Kielpinski, D., Tsikata, E., Gustafson, T. L., Pritchard, D. E., and Ketterle, W. (2002). A continuous source of Bose–Einstein condensed atoms. *Science*, **296**, 2193–2195.

Chin, C., Bartenstein, M., Altmeyer, A., Riedl, S., Jochim, S., Denschlag, J. H., and Grimm, R. (2004). Observation of the pairing gap in strongly interacting Fermi gas. *Science*, **305**, 1128–1130.

Chu, S. (2002). Cold atoms and quantum control. *Nature*, **416**, 206–210.

Chu, S., Hollberg, L., Bjorkholm, J., Cable, A., and Ashkin, A. (1985). Three-dimensional viscous confinement and cooling of atoms by resonance radiation pressure. *Physical Review Letters*, **5**, 48–51.

Chu, S., Bjorkholm, J., Ashkin, A., and Cable, A. (1986). Experimental observation of optically trapped atoms. *Physical Review Letters*, **57**, 314–317.

Chupp, T., and Swanson, S. (2000). Medical imaging with laser-polarized noble gases. In *Advances in atomic, molecular and optical physics* (ed. B. Bederson and H. Walther), vol. 45, pp. 42–97. Academic Press, San Diego.

Clairon, A., Laurent, P., Santarelli, G., Ghezali, S., Lea, S. N., and Bahoura, M. (1995). A cesium fountain frequency standard: recent results. *IEEE Transactions on Instrumentation and Measurement*, **44**(2), 128–131.

Cohen-Tannoudji, C. (1961). Observation d'un deplacement de raie de resonance magnetique cause par l'excitation optique. *Comptus Rendus Academie des Sciences* (Paris), **252**, 394–396.

Cohen-Tannoudji, C. (1962). Theorie quantique du cycle de pompage optique. Verification expérimentale des nonveaux effets prévus. *Annuales de Physique* (Paris), 13e série, **7**, 423–425.

Cohen-Tannoudji, C. (1992). Laser cooling and trapping of neutral atoms theory. *Physics Reports*, **219**, 153–164.

Cohen-Tannoudji, C., and Dupont-Roc, J. (1972). Experimental study of Zeeman light shifts in weak magnetic fields. *Physical Review A*, **5**, 968–984.

Cohen-Tannoudji, C., and Kastler, A. (1966). Optical pumping. In *Progress in optics* (ed. E. Wolf), pp. 1-81. North-Holland, Amsterdam.

Cohen-Tannoudji, C., and Phillips, W.D. (1990). New mechanisms for laser cooling. *Physics Today*, **43**, October, 33–40.

Cohen-Tannoudji, C., Dupont-Roc, J., and Grynberg, G. (1992). *Atom–photon interactions*. Wiley, New York.

Cook, R. J., and Hill, R. K. (1982). An electromagnetic mirror for neutral atoms. *Optics Communications*, **43**, 258–260.

Cornell, E. A., and Wieman, C. E. (2002). Nobel lecture: Bose–Einstein condensation in a dilute gas, first 70 years and some recent experiments. *Reviews of Modern Physics*, **74**, 875–893.

Cornish, S. L., Claussen, N. P., Roberts, J. L., Cornell, E. A., and Wieman, C. E. (2000). Stable ^{85}Rb Bose–Einstein condensates with widely tunable interactions. *Physical Review Letters*, **85**, 1795–1798.

Courteille, P., Feeland, R. S., Heinzen, D. J., van Abeelen, F. A., and Verhaar, B. J. (1998). Observation of a Feshbach resonance in cold atom scattering. *Physical Review Letters*, **81**, 69–72.

Courteille, P. W., Bagnato, V. S., and Yukalov, V. I. (2001). Bose–Einstein condensations of trapped atomic gases. *Laser Physics*, **11**, 659–800.

Crim, F. F. (1993). Vibrationally mediated photodissociation: exploring excited-state surfaces and controlling decomposition pathways. *Annual Review of Physical Chemistry*, **44**, 397–428.

Dalfovo, F., Giorgini, S., Pitaevskii, L. P., and Stringari, S. (1999). Theory of Bose–Einstein condensation of trapped gases. *Reviews of Modern Physics*, **71**, 463–512.

Dalibard, J., and Cohen-Tannoudji, C. (1989). Laser cooling below the Doppler limit by polarization gradients—simple theoretical models. *Journal of the Optical Society of America B*, **2**, 2023–2045.

Daniel, C., Full, J., Gonzalez, L., Lupulescu, C., Manz, J., Merli, A., Vajda, S., and Wöste, L. (2003). Deciphering the reaction dynamics underlying optimal control laser fields. *Science*, **299**, 536–539.

Daussy, C., Marrel, T., Any-Klein, A., Nguyen, C. T., Borde, C. J., and Chardonnet, C. (1999). Limit of the parity nonconservating energy difference between the enantiomers of a chiral molecule by laser spectroscopy. *Physical Review Letters*, **83**(8), 1554–1557.

Davenport, J. R., Wuite, G. J., Landick, R., and Bustamante, C. (2000). Single-molecule study of transcriptional pausing and arrest by *E. coli* RNA polymerase. *Science*, **287**, 2497–2500.

Davis, K. B., Mewes, M.-O., Andrews, M. R., van Druten, N. J., Durfee, D. S., and Kurn, D. S., Ketterle, W. (1995). Bose–Einstein condensation in the gas of Na atoms. *Physical Review Letters*, **75**, 3969–3972.

Dehmelt, H. G. (1957). Slow spin relaxation of optically polarized sodium atoms. *Physical Review*, **105**, 1487–1489.

Dehmelt, H. G. (1958). Spin resonance of free electrons polarized by exchange collisions. *Physical Review*, **109**, 381–385.

Dehmelt, H. (1990). Experiments with isolated subatomic particle at rest. *Reviews of Modern Physics*, **62**, 525–530.

DeMarco, B. and Jin, D. S. (1999). Onset of Fermi degeneracy in a trapped atomic gas. *Science*, **285**, 1703–1706.

DeMarco, B., Papp, S. B., and Jin, D. S. (2001). Pauli blocking of collisions in quantum degenerate atomic Fermi gas. *Physical Review Letters*, **86**, 5409–5412.

Dicke, R. (1953). The effect of collisions upon the Doppler width of spectral lines. *Physical Review*, **89**, 472–473.

Diddams, S. A., Bergquist, J. C., Jefferts, S. R., and Oates, C. W. (2004). Standards of time and frequency at the outset of the 21st century. *Science*, **306**, 1318–1324.

Dirac, P. A. M. (1958). *The principles of quantum mechanics*. Clarendon Press, Oxford.

Doherty, A. C., Lynn, T. W., Hood, C. J., and Kimble, H. J. (2000). Trapping of single atoms with single photons in cavity QED. *Physical Review A*, **63**, 013401.

Doljikov, V. S., Letokhov, V. S., Makarov, A. A., Malinovsky, A. L., and Ryabov, E. A. (1986). Raman probing of overtone and combinations band to study vibrational energy distribution produced by multi-photon excitation. *Chemical Physics Letters*, **124**, 304–308.

Dowling, J. P., and Gea-Banacloche, J. (1996). Evanescent light-wave atom mirrors, resonators wave-guides, and traps. In *Advances in atomic, molecular and optical physics* (ed. B. Bederson and H. Walther), vol. 37, pp. 1–94. Academic Press, San Diego.

Dubetskii, B. Yu., Kazantsev, A. P., Chebotayev, V. P., and Yakovlev, V. P. (1985). Interference of atoms in spatially-separated optical fields. *Journal of Experimental and Theoretical Physics*, **89**, 1190–1204.

Ducas, T. W., Littman, M. G., Freeman, R. R., and Kleppner, D. (1975). Stark ionization of high-lying states of sodium. *Physical Review Letters*, **35**, 366–369.

Dupont-Roc, J., Haroche, S., and Cohen-Tannoudji, C. (1969). Detection of very weak magnetic fields (10^{-9} Gauss) by ^{87}Rb zero-field level crossing resonances. *Physics Letters*, **28A**, 107–108.

Dyson, F. (1998). *The imaged worlds*. Oxford University Press, Oxford.

Egorov, S. E., Letokhov, V. S., and Shibanov, A. N. (1984a). Mechanism of ion formation by irradiation of molecular crystal surface with laser pulses. *Soviet Journal of Quantum Electronics*, **14**, 940–946.

Egorov, S. E., Letokhov, V. S., and Shibanov, A. N. (1984b) Laser multiphotonionization detection of molecules adsorbed on a surface. *Chemical Physics*, **85**, 349–353.

Ehrlicher, A., Betz, T., Stuhrmann, B., Koch, D., Milner, V., Raisen, M. G., and Käs, J. (2002). Guiding neuronal growth with light. *Proceedings of the National Academy of Sciences*, **99**, 16024–16028.

Einstein, A. (1916). Strahlungs-Emission und -Absorption nach der Quantentheorie. *Verhandl. Dtsch. Phys. Ges.*, **18**, 318–323.

Esslinger, T., Weidemuller, M., Hemmerich, A., and Hansch, T. (1993). Surface-plasmon mirror for atoms. *Optical Letters*, **18**, 450–452.

Esslinger, T., Bloch, I., and Hansch, T. W. (1998). Bose–Einstein condensations in a quadrupole-Ioffe-configuration trap. *Physical Review A*, **58**, R2664–2667.

Evseev, A. V., Letokhov, V. S., and Puretzky, A. A. (1985). Highly selective and efficient multiphoton dissociation of polyatomic molecules. *Applied Physics B*, **39**, 93–103.

Fano, U. (1961). Effects of configuration interaction on intensities and phase shifts. *Physical Review*, **124**, 1866–1878.

Fedichev, P. O., Kagan, Yu., Shlyapnikov, G. V., and Walvaren, J. T. M. (1996). Influence of nearly resonant light on the scattering length in low-temperature atomic gases. *Physical Review Letters*, **77**, 2913–2916.

Fedorov, M. V., Goreslavsky, S. P., and Letokhov, V. S. (1997). Pondermotive forces and stimulated Compton scattering of free electrons in a laser-field. *Physical Review E*, **55**, 1015–1027.

Fedoseyev, V. N., Fedorov, D. V., Horn, R., Huber, G., Köster, U., Lassen, J., Mishin, V. I., Seliverstov, M. D., Weisman, L., Wendt, K., and the ISOLDE Collaboration (2003). Atomic spectroscopy studies of short-lived isotopes and nuclear isomer separation with ISOLDE RILIS. *Nuclear Instruments and Methods in Physical Research B*, **204**, 353–358.

Feron, S., Reinhardt, J., Le Boiteux, S., Gorciex, O., Baudon, J., Ducloy, M., Robert, J., Miniatura, C., Nic-Chormaic, S., Haberland, H., and Lorent, V. (1993). Reflections of metastable neon atoms by a surface plasmon wave. *Optics Communications*, **102**, 83–88.

Feynman, R. (1992). There's plenty of room at the bottom. In *Nanotechnology: research and perspectives* (ed. B.C. Crandall and J. Lewis), p. 360. MIT Press, Cambridge, MA.

Feynman, R. P., Vernon, F. L., and Hellwarth, R. W. (1957). Geometrical representation of the Schrödinger equation for solving maser problems. *Journal of Applied Physics*, **28**, 49–52.

Foot, C. J. (2004). *Atomic physics*. Oxford University Press, Oxford.

Flugge, S. (1971). *Practical quantum mechanics*, vol. 1. Springer, Berlin.

Freimond, D. L., Aflatooni, K., and Batelaan, H. (2001). Observation of the Kapitza–Dirac effect. *Nature*, **413**, 142–143.

Friebel, S., D'Andrea, C., Walz, J., Weitz, M., and Hänsch, T. W. (1998). CO_2-laser optical lattice with cold rubidium atoms. *Physical Review A*, **57**, R20–R23.

Fried, D. G., Killian, T. C., Willmann, L., Landhuis, D., Moss, S. C., Kleppner, D., and Greytak, T. J. (1998). Bose–Einstein condensation of atomic hydrogen. *Physical Review Letters*, **81**, 3811–3814.

Fujita, J., Morinaga, M., Kishimoto, T., Yasuda, M., Matsui, M., and Shimizu, F. (1996). Manipulation of an atomic beam by a computer-generated hologram. *Nature*, **380**, 691–694.

Gallagher, T. F., Edelstein, S. A., and Hill, R. M. (1975). Radiative lifetimes of the S and D Rydberg levels of Na. *Physical Review*, **11**, 1504–1506.

Gallatin, G., and Gould, P. J. (1991). Laser focusing of atomic beams. *Journal of the Optical Society of America B*, **8**, 502–519.

Garus-Chavez, V., McGloin, D., Padgett, M. J., Dulotz, W., Schmitzer, H., and Dholakia, K. (2003). Observation of the transfer of the local angular momentum density of a multiringed light beam to an optically trapped particle. *Physical Review Letters*, **91**, 093602.

Gaspard, P., and Burghardt, I. (ed.) (1997). *Chemical reactions and their control on the femtosecond time scale*, Proceedings of the 20th Conference on Chemistry. Wiley, New York.

Gauthier, M, Cureton, C. G., Hackett, P. A., and Willis, C. (1982). Efficient production of $^{13}C_2F_4$ in the infrared laser photolysis of $CHClF_2$. *Applied Physics B*, **28**, 43–50.

Gennini, G., Ritt, G., Geckeler, C., and Weitz, M. (2003). All-optical realization of an atom laser. *Physical Review Letters*, **91**, 240408.

Glauber, R. J. (1963a). Quantum theory of optical coherence. *Physical Review*, **130**, 2529–2539.

Glauber, R. J. (1963b). Coherent and incoherent states of the radiation field. *Physical Review*, **131**, 2766–2788.

Goldenberg, H. M., Kleppner, D., and Ramsey, N. F. (1960). Atomic hydrogen maser. *Physical Review Letters*, **5**, 361–362.

Gordon, J. P., and Ashkin, A. (1980). Motion of atom in a radiation trap. *Physical Review A*, **21**, 1606–1617.

Gordon, J. P., Zeiger, H.J., and Townes, C. H. (1954). Molecular microwave oscillator and new hyperfine structure in the microwave spectrum of NH_3. *Physical Review*, **95**, 282–284.

Gould, P. I., Ruff, G.A., and Pritchard, D. E. (1986). Diffraction of atoms by light: the near-resonant Kapitza–Dirac effect. *Physical Review Letters*, **56**, 827–830.

Goy, P., Raimond, J. M., Gross, M., and Haroche, S. (1983). Observation of cavity-enhanced single-atom spontaneous emission. *Physical Review Letters*, **50**, 1903–1906.

Greiner, M., Regal, C., and Jin, D. S. (2003). Emergence of a molecular Bose–Einstein condensate from Fermi gas. *Nature*, **426**, 537–540.

Grimm, R., Weidemüller, M., and Ovchinnikov, Yu. B. (2000). Optical dipole traps for neutral atoms. In *Advances in atomic, molecular, and optical physics* (ed. B. Bederson and H. Walther), vol. 42, pp. 95–170. Academic Press, San Diego.

Grynberg, G., and Courtois, J. (1994). Proposal for a magneto-optical lattice for trapping atoms in nearly-dark states. *Europhysics Letters*, **27**, 41–46.

Gustavson, T. L., Boyer, P., and Kasevich, M. A. (1997). Precision rotation measurements with an atom interferometer gyroscope. *Physical Review Letters*, **78**, 2046–2049.

Hajnal, J. V., and Opat, G. I. (1989). Diffraction of atoms by standing light wave—a reflection grating for atoms. *Optics Communications*, **71**, 119–124.

Haken, H. (1981). *Light*, vols. 1 and 2. North-Holland, Amsterdam.

Hall, J. L., Ye, J., Diddams, S. A., Ma, L.-S., Cundiff, S. T., and Jones, D. J. (2001). Ultrasensitive spectroscopy, the ultrastable lasers, the ultrafast lasers, and the seriously nonlinear fiber: a new alliance for physics and metrology. *IEEE Journal of Quantum Electronics*, **37**(12), 1482–1492.

Han, D. J., Courteille, P. W., and Wynar, R. H. (1998). Bose–Einstein condensation of large numbers of atoms in a magnetic time-averaged orbital potential trap. *Physical Review A*, **57**, R4114–4117.

Hansch, T. W. (1989). High resolution spectroscopy of hydrogen. In *The Hydrogen Atom* (ed. C. F. Bassani, M. Inguscio, and T. W. Hansch), pp. 93–102. Springer, Berlin.

Hänsch, T., and Schawlow, A. (1975). Cooling of gases by laser radiation. *Optics Communications*, **13**, 68–71.

Harris, D. J., and Savage, C. M. (1995). Atomic gravitational cavities from hollow optical fibers. *Physical Review A*, **51**, 3967–3971.

Harris, S. (1997). Electromagnetically induced transparency. *Physics Today*, **50**(7), 36–42.

Hau, L. V., Harriss, S., Dutton, Z., and Behroozi, C. H. (1999). Light speed reduction to 17 meters per second in an ultracold atomic gas. *Nature*, **397**, 594–598.

He, H., Friese, M. E. J., Heckenberg, N. R., and Rubinstein-Dunlop, H. (1995). Direct observation of transfer of angular-momentum to absorptive particles from a laser beam with phase singularities. *Physical Review Letters*, **75**, 826–829.

Heindenrich, R. D., (1964). *Fundamentals of transmission electron microscopy*. Interscience, New York.

Hemmerich, A., Weidemüller, M., Esslinger, T., Zimmermann, C., and Hänsch, T. (1995). Trapping atoms in a dark optical lattice. *Physical Review Letters*, **75**, 37–40.

Henkel, C., Molmer, K., Kaiser, R., Vansteenkiste, N., Westbrook, C. I., and Aspect, A. (1997). Diffuse atomic reflection at a rough mirror. *Physical Review*, **55**, 1160–1178.

Henkel, C., Wallis, H., Westbrook, N., Westbrook, C. I., Aspect, A., Sengstock, K., and Ertmer, W. (1999). Theory of atomic diffraction from evanescent waves. *Applied Physics B*, **69**(4), 277–289.

Henon, M., and Heiles, C. (1964). The applicability of the third integral of numerical experiments. *Astronomical Journal*, **69**, 73–79.

Hensley, J., Wicht, A., Young, B., and Chu, S. (2000). Progess towards a measurment of t/M_{cs}. In *Proceedings of 17th International Conference on Atomic Physics* (ed. E. Arimondo, P. DeNatale, and M. Inguscio), pp. 43–57. American Institute of Physics, New York.

Herbig, J., Kraemer, T., Mark, M., Weber, T., Chin, C., Nägerl, H.-C., and Grimm, R. (2003). Preparation of a pure molecular quantum gas. *Science*, **301**, 1510–1513.

Hess, H. F. (1986). Evaporative cooling of magnetically-trapped and compressed spin-polarized hydrogen. *Physical Review B*, **34**, 3476–3479.

Hillenkamp, F., Karas, M., Beavis, R. C., and Chait, B. T. (1991). Matrix-assisted laser desorption/ionization mass spectrometry of biopolymers. *Analytical Chemistry*, **63**, 1193A–1203A.

Hippler, M., Quack, M., Schwartz, R., Seyfang, G., Matt, S., and Märk, T. (1997). Infrared multiphoton excitation, dissociation and ionization of C_{60}. *Chemical Physics Letters*, **278**, 111–120.

Hollberg, L., Oates, C. W., Curtis, E. A., Ivanov, E. N., Diddams, S. A., Udem, T., Robinson, H. G., Bergquist, J. C., Rafac, R. J., Itano, W. M., Drullinger, R. E., and Wineland, D. J. (2001). Optical frequency standards and measurements. *IEEE Journal of Quantum Electronics*, **37**(12), 1502–1513.

Honda, K., Takahashi, Y., Kuwamoto, T., Fujimoto, M., Toyoda, K. (1999). Magneto-optical trapping of Yb atoms and a limit of the branching ratio of the $^1P_1$ state. *Physical Review A*, **59**, R934–937.

Horsley, J. A., Cox, D. M., Hall, R. B., Kaldor, A., Maas, E. T. Jr., Priestly, E. B., and Kramer, G. M. (1980). Isotopical selectivity in the laser induced dissociation of molecules with overlapping absorption bands. *Journal of Chemical Physics*, **73**, 3660–3665.

Hulet, R. G., Hilfer, E., and Kleppner, D. (1985). Inhibited spontaneous emission by a Rydberg atom. *Physical Review Letters*, **55**, 2137–2140.

Hurst, G. S., and Payne, M. G. (1988). *Principles and applications of resonance ionization spectroscopy*. Adam Hilger, Bristol.

Hurst, G. S., Nayfeh, M. H., and Young, J. P. (1977a). A demonstration of one-atom detection. *Applied Physics Letters*, **30**, 229–231.

Hurst, G. S., Nayfeh, M. H., and Young, J. P. (1977b). One-atom detection using resonance ionization spectroscopy. *Physical Review A*, **15**, 2283–2292.

Inoye, S., Andrews, M. R., Stenger, J., Miesner, H.-J., Stamper-Kurn, D. M., and Ketterle, W. (1998). Observation of Feshbach resonances in a Bose–Einstein condensate. *Nature*, **392**, 151–154.

Ivanov, L. N., and Letokhov, V. S. (1975). Selective ionization of atoms by light and electric field. *Kvantovaya Elektronika* (Russia), **2**, 585–590. [Erratum: 1977, **5**, 877].

Ivanov, L. N., and Letokhov, V. S. (1976). On the possibility of discharge of metastable nuclei upon a negative muon capture. *Journal of Experimental and Theoretical Physics*, **43**, 9–14.

Javan, A., Bennett, W. R. Jr., and Herriott, D. R. (1961). Population inversion and continuous optical maser oscillation in gas discharge containing a He–Ne mixture. *Physical Review Letters*, **6**, 106–110.

Jessen, P. S., Gerz, C., Lett, P. D., Phillips, W. D., Rolston, W. D., Spreeuw, R. J. C., and Westbrook, C. I. (1992). Observation of quantized motion of Rb atoms in an optical field. *Physical Review Letters*, **69**, 49–52.

Jochim, S., Bertenstein, M., Altmeyer, A., Hendi, G., Chin, C., Denschlag, J. H., and Grimm, R. (2003). Pure gas of optically trapped molecules created from fermionic atoms. *Physical Review Letters*, **91**, 240402.

Jortner, J., Levine, R. D., and Rice, S. A. (ed.) (1981). *Photoselective chemistry.* Wiley, New York.

Judson, R., and Rabitz, H. (1992). Teaching lasers to control molecules. *Physical Review Letters*, **68**, 1500–1504.

Kaiser, R., Levy, Y., Vansteenkiste, N., Aspect, A., Seifert, W., Leipold, D., and Mlynek, J. (1994). Resonant enhancement of evanescent waves with a thin dielectric waveguide. *Optics Communications*, **104**, 234–240.

Kapitza, P., and Dirac, P. A. M. (1933). Reflection of electrons by standing light waves. *Proceedings of Cambridge Philosophical Society*, **29**, 297–300.

Karas, M., and Hillenkamp, F. (1988). Laser desorption ionization of proteins with molecular masses exceeding 10 000 Daltons. *Analytical Chemistry*, **60**, 2299–2301.

Karas, M., Bachman, D., Bahr, U., and Hillenkamp, F. (1987). Matrix-assisted ultraviolet laser desorption of non-volatile compounds. *International Journal of Mass Spectrometry and Ion Processes*, **78**, 53–68.

Karplus, R., and Schwinger, J. (1948). A note on saturation in microwave spectroscopy. *Physical Review*, **73**, 1020–1034.

Kasevich, M. (2002). Coherence with atoms. *Science*, **298**, 1363–1368.

Kasevich, M., and Chu, S. (1992). Laser cooling below a photon recoil with 3-level atoms. *Physical Review Letters*, **69**, 1741–1744.

Kasevich, M., Weiss, D., and Chu, S. (1990). Normal-incidence reflection of slow atoms from an optical evanescent wave. *Optics Letters*, **15**, 607–609.

Kash, M. M., Sautenkov, V. A., Zibrov, A. S., Hollberg, L., Welch, G. K., Lukin, M. D., Rostovtsev, Yu., Fly, E. S., and Scully, M. O. (1999). Ultraslow group velocity and enhanced nonlinear optical effects in a coherently driven hot atomic gas. *Physical Review Letters*, **82**, 5229–5232.

Kastler, A. (1950). Quelques suggestions concernant la production optique et la détection optique d'une integalité de population des niveaux de quantification spatiale des atomes: application à l'epérrence de Stern et Gerlach et à la résonance magnétique. *Journal de Physique et le Radium*, **11**, 255–265.

Kastler, A. (1966). Optical methods for studying Hertzian resonances. In: *Nobel Lecturers, Physics 1963–1970*, pp. 186–204. Elsevier, Amsterdam (1972).

Katori, H., Ido, T., Isoya, Y., and Kuwata-Gonokami, M. (1999). Magneto-optical trapping and cooling of strontium atoms down to the photon recoil effect. *Physical Review Letters*, **82**, 1116–1119.

Kazantsev, A., Surdutovich, G., and Yakovlev, V. (1990). *Mechanical action of light on atoms*. World Scientific, Singapore.

Keith, D. W., Ekstrom, C. R., Turchette, Q. A., and Pritchard, D. E. (1991). An interferometer for atoms. *Physical Review Letters*, **66**, 2693–2696.

Ketterle, W. (2002). Nobel lecture: When atoms behave as waves: Bose–Einstein condensation and the atom laser. *Reviews of Modern Physics*, **74**, 1131–1149.

Ketterle, W., and van Drutten, N. J. (1996). Evaporative cooling of trapped atoms. In *Advances in atomic, molecular and optical physics* (ed. B. Bederson and H. Walther), vol. 37, pp. 181–236. Academic Press, San Diego.

Ketterle, W., Rendall, B. D., Jotte, M. A., Marun, A., and Pritchard, D. (1993). High densities of cold atoms in a dark spontaneous-force optical trap. *Physical Review Letters*, **70**, 2233–56.

Klimov, V. V., and Letokhov, V. S. (1995). New atom trap configuration in the near field of laser radiation. *Optics Communications*, **121**, 130–136.

Klimov, V. V., and Letokhov, V. S. (2003). Laser focusing of cold atoms: analytical solution of the problem. *Laser Physics*, **13**(3), 339–349.

Knappe, S., Wynands, R., Kitching, J., Robinson, H.G., and Hollberg, L. (2001). Characterization of coherent population-trapping resonances as atomic frequency references. *Journal of the Optical Society of America*, **18**, 1545–1553.

Knight, P. L., and Allen, L. (1983). *Concepts of quantum optics*. Pergamon Press, Oxford.

Knyazev, I. N., Kudriavtsev, Yu. A., Kuzmina, N. P., Letokhov, V. S., and Sarkisian, A. S. (1978). Laser isotope separation of carbon by multiple IR photon and subsequent UV excitation of CF_3I molecules. *Applied Physics*, **17**, 427–429.

Kocharovskaya, O. (1992). Amplification and lasing without inversion. *Physics Reports*, **219**, (3–6), 175–190.

Kohler, B., Krause, J. L., Raksl, F., Wilson, K. R., Yakovlev, V., Whitenell, R. M., and Yan, Y. (1995a). Controlling the future of matter. *Accounts of Chemical Research*, **28**, 133–140.

Kohler, B., Yakovlev, V., Che, J., Krause, J., Messina, M., Wilson, K., Schwentner, R., Whitnell, R., and Yan, Y. (1995b). Quantum control of wave packet evolution with tailored femtosecond pulses. *Physical Review Letters*, **74**, 3360–3363.

Köhler, S., Diesenberger, R., Eberhardt, K., Erdman, N., Hermann, G., Huber, G., Kratz, J. V., Nunnemann, M., Passler, G., Rao, P. M., Riegel, J., Trautman, N., and Wendt, K. (1997). Determination of the first ionization potential of actinide elements by resonance ionization mass spectroscopy. *Spectrochimica Acta B*, **52**, 717–726.

Komins, I. K., Kornack, T. W., Allred, J. C., and Romalis, M. V. (2003). A subfemtotesla multichannel atomic magnetometer. *Nature*, **422**, 596–599.

Kompanets, O. N., Kukudzhanov, A. R., Letokhov, V. S., and Gervits, L. L. (1976). Narrow resonances of saturated absorption of dissymetrical molecules CHFClBr and the possibility of weak current detection in molecular physics. *Optics Communications*, **19**(3), 414–416.

Köster, U., Cathrell, R., Fedoseyev, V. N., Franchoo, S., Georg, H., Huyse, M., Kruglov, K., Lettry, J., Mishin, V. I., Oinonen, M., Ravn, H., Seliverstov, M. D., Simon, H., van Dupper, P., van Roosbroeck, J., Weismann, L., the IS365 Collaboration, and the ISOLDE Collaboration (2000). Isomer separation and measurement of nuclear moments with the ISOLDE RILIS. *Hyperfine Interactions*, **127**, 417–420.

Kramer, S. D., Bemis, C. E. Jr., Young, J. P., and Hurst, G. S. (1978). One-atom detection in individual ionization tracks. *Optics Letters*, **3**, 16–18.

Kreutzmann, H., Poulsen, U. V., Lewenstein, M., Dumke, R., Ertmer, W., Birkl, G., and Sanpera, A. (2004). Coherence properties of guided interferometers. *Physical Review Letters*, **92**, 163201.

Kudriavtsev, Yu. A., and Letokhov, V. S. (1982). Method of highly selective detection of rare radioactive isotopes through multistep photoionization of accelerated atoms. *Applied Physics B*, **29**, 219–221.

Kuhn, T. S. (1996). *The structure of scientific revolutions*. 3rd edn. University of Chicago Press, Chicago.

Kuwamoto, T., Honda, R., Takahashi, Y., and Yabuzaki, T. (1999). Magneto-optical trapping of Yb atoms using an intercombination transition. *Physical Review A*, **60**, R745–R748.

Kuz'min, M. V., and Stuchebrukhov, A. A. (1989). Dynamical chaos and intramolecular vibrational relaxation in polyatomic molecules. In *Laser spectroscopy of highly vibrationally excited molecules* (ed. V. S. Letokhov), pp. 178–264. Adam Hilger, Bristol.

Lang, M. J., and Block, S. M. (2003). Laser-based optical tweezers. *American Journal of Physics*, **71**, 201–215.

Lamb, W. Jr. (1964). Theory of optical maser. *Physical Review*, **134**, A1429–A1450.

Lamb, W. E. Jr. (1975). Physical concepts in the development of the maser and laser. In *Impact of basic research and technology* (ed. B. Kursunogly and A. Perlmutter). Plenum Press. 59–111.

Laubereau, A., and Kaiser, W. (1975). Picosecond spectroscopy of molecular dynamics in liquids. *Annual Review of Physical Chemistry*, **26**, 83–99.

Lee, P. H., and Skolnick, M. L. (1967). Saturated neon absorption inside a 6328 Å laser. *Applied Physics Letters*, **10**, 303–305.

Leggett, A. L. (2001). Bose–Einstein condensation in the alkali gases: some fundamental concepts. *Reviews of Modern Physics*, **73**, 307–356.

Lehmann, J.-C., and Cohen-Tannoudji, C. (1964). Pompage optique en champ magnetique faible. *Comptus Rendus Academe des Science* (Paris), **258**, 4463–4466.

Letokhov, V. S. (1966). Photoionization of atoms in coherent states. *Optics and Spectroscopy*, **20**(2), 349–350 (in Russian).

Letokhov, V. S. (1967). Laser frequency autostabilizaiton by nonlinear absorbing gas cell. *Journal of Experimental and Theoretical Physics Letters*, **6**, 101–104.

Letokhov, V. S. (1968). Narrowing of the Doppler width in a standing light wave. *Journal of Theoretical and Experimental Physics Letters*, **7**(9), 272–275.

Letokhov, V. S. (1969). On the possibility of isotope separation by resonant atomic photoionization and molecular photodissociation with laser radiation. Report, of Lebedev Physical Institute, Nov. 1969. (Published in preprint No. 1 (1979) of the Institute of Spectroscopy, USSR Academy of Sciences, pp. 1–54).

Letokhov, V. S. (1973a). Use of lasers to control selective chemical reactions. *Science*, **180**, 451–458.

Letokhov, V. S. (1973b). Possibility of the optical separation of the isomeric nuclei by laser radiation. *Optics Communications*, **7**, 59–60.

Letokhov, V. S. (1973c). New possibilities for the spectroscopy inside the Doppler line in the optical and γ-ranges. In *Methods of spectroscopy without Doppler broadening of excited levels of simple molecules*, Proceedings of the conference, Aussois, France, May 23–26, 1973, pp. 127–138. Also published in *Laser and Unconventional Optics*, **46**, 3–27 (1973).

Letokhov, V. S. (1973d). On the problem of the nuclear transition γ-laser. *Journal of Experimental and Theoretical Physics*, **37**, 787–793.

Letokhov, V. S. (1975a). On difference of energy levels of left and right molecules due to weak interactions. *Physics Letters*, **53A**, 275–276.

Letokhov, V. S. (1975b). Nonlinear high resolution laser spectroscopy. *Science*, **190**(4212), 344–351.

Letokhov, V. S. (1976). Future applications of selective laser photophysics and photochemistry. In *Tunable lasers and their applications*, Proceedings of the Conference, Loen, Norway, June 8–11, 1976, (ed. A. Mooradian, T. Jaeger, and P. Stokseth), pp. 122–139. Springer, Berlin.

Letokhov, V. S. (1983). *Nonlinear laser chemistry. Multiple-photon excitation.* Springer, Berlin.

Letokhov, V. S. (1987). *Laser photoionization spectroscopy*, pp. 1-357. Academic Press, Orlando.

Letokhov, V. S. (ed.) (1989). *Laser spectroscopy of highly-vibrational excited molecules.* Adam Hilger, Bristol.

Letokhov, V. S. (1995a). Multiphoton photochemistry and photobiochemistry with ultrashort laser pulses. In *Ultrafast processes in chemistry and photobiology* (ed. M. El-Sayed, I. Tanaka, and Yu. Molin), pp. 195-214. Blackwell Science, Oxford.

Letokhov, V. S. (1995b). Electron-beam focusing by the dipole force of an ultrashort laser pulse. *Journal of Experimental and Theoretical Physics Letters*, **61**, 805–808.

Letokhov, V. S. (1997a). Ultrafast processes: from the past to the future. In *Femtochemistry and femtobiology: ultrafast reaction dynamics at atomic-scale resolution* (ed. V. Sündström), pp. 755–763. Imperial College Press, London.

Letokhov, V. S. (1997b). Concluding remark on XXth Solvay Conference on Chemistry. In *Chemical reactions and their control on the femtosecond time scale* (ed. P. Gaspard and I. Burghardt), Advances in Chemical Physics, vol. 101, pp. 873–887. Wiley, New York.

Letokhov, V. S., and Chebotayev V. P. (1977). *Nonlinear laser spectroscopy.* Springer, Berlin.

Letokhov, V. S., and Makarov, A. A. (1973). Kinetics of vibrational excitation of molecules by IR radiation. *Soviet Physics Journal of Experimental and Theoretical Physics*, **36**, 1091–1096.

Letokhov, V. S., and Minogin, V. G. (1981a). Laser radiation pressure of free atoms. *Physics Reports*, **73**(1), 1–65.

Letokhov, V. S., and Minogin, V. G. (1981b). Laser cooling of atoms and its application in frequency standards. *Journal de Physique, Colloque CB*, **42**(12), Suppl. CB-347–CB-355.

Letokhov, V. S., and Mishin, V. I. (1979). Highly selective multistep ionization of atoms by laser radiation. *Optics Communications*, **29**, 168–171.

Letokhov, V. S., and Ryabov, E. A. (2004). Laser infrared multiphoton noncoherent control of intermolecular (isotope) selectivity for polyatomic molecules on a practical scale. *Israel Journal of Chemistry*, **44**, 1–7.

Letokhov, V. S., Minogin, V. G., and Pavlik, B. D. (1976). Cooling and trapping of atoms and molecules by a resonant laser field. *Optics Communications*, **19**(1), 72–75.

Letokhov, V. S., Minogin, V. G., and Pavlik, B. D. (1977). Cooling and trapping of atoms by a resonant laser field. *Journal of Experimental and Theoretical Physics*, **45**(4), 698–705 (1977).

Lett, P., Watts, R., Westbrook, C., Phillips, W., Gould, P., and Metcalf, H. (1988). Observation of atoms laser cooled below the Doppler limit. *Physical Review Letters*, **61**, 169–172.

Lett, P. D., Julienne, P. S., and Phillips, W. D. (1995). Photoassociative spectroscopy of laser-cooled atoms. *Annual Review of Physical Chemistry*, **46**, 423–452.

Levis, P. J., and Rabitz, H. A. (2002). Closing the loop on bond selective chemistry using tailored strong field laser pulse. *Journal of Physical Chemistry*, **106**, 6427–6444.

Levis, P. J., Menkir, M., and Rabitz, H. (2001). Selective bond dissociation and rearrangement with optimally tailored, strong field laser pulses. *Science*, **292**, 709–713.

Levy, D. H. (1980). Laser spectroscopy of cold gas-phase molecules. *Annual Review of Physical Chemistry*, **31**, 197–225.

Lisitsyn, V. N., and Chebotayev, V. P. (1968). Absorption saturation effects in a gas laser. *Journal of Experimental and Theoretical Physics*, **54**(2), 419–423 (in Russian).

Lodahl, P., van Driel, F., Nikolaev, I. S., Irman, A., Overgaay, K., Vanmaekelberg, D., and Vos, W. L. (2004). Controlling the dynamics of spontaneous emission from quantum dots by photonic crystals. *Nature*, **430**, 654–657.

Lokhman, V. N., Makarov, A. A., Petrova, I. Yu., Ryabov, E. A., and Letokhov, V. S. (1999). Transition spectra in the vibrational quasicontinuum of polyatomic molecules. IR multiple-photon absorption in SF_6. II. Theoretical simulations and comparison with experiment. *Journal of Physical Chemistry*, **103**, 11299–11309.

Loudon, R. (1973). *The quantum theory of light*. Oxford University Press, Oxford.

Lu, Z. T., Bowers, C. J., Freedman, S. J., Fujikawa, B. K., Mortara, J. L., and Shang, S.-Q. (1994). Laser trapping of short-lived radioactive isotopes. *Physical Review Letters*, **72**, 3791–3794.

Lu, Z. T., Corwin, K. L., Vogel, K. R., Wieman, C. E., Dinneen, T. P., Maddi, J., and Gould, H. (1997). Efficient collection of ^{221}Fr into a vapor cell magneto-optical trap. *Physical Review Letters*, **79**, 994–997.

Lubman, D. M. (ed.) (1990). *Lasers and mass-spectrometry*. Oxford University Press, New York.

Lubman, D. M., and Li, Liang (1990). Resonant two-photon ionization spectroscopy of biological molecules in supersonic jets volatilized by pulsed laser desorption. In *Lasers and Mass-Spectrometry* (ed. D. M. Lubman), pp. 352-382. Oxford University Press, New York.

Lyman, J. L., Jensen, R. J., Rink J., Robinson, P., and Rockwood, S. D. (1975). Isotopical enrichment of SF_6 in S^{34} by multiple absorption of CO_2-laser radiation. *Applied Physics Letters*, **27**, 87–89.

McClelland, J. J., and Scheinfein, M. P. (1991). Laser focusing of atoms: a particle-optics approach. *Journal of the Optical Society of America B*, **8**, 1974–1986.

McClelland, J. J., Scholten, R. E., Palm, E. C., and Cellota, R. J. (1993). Laser-focused atomic deposition. *Science*, **262**, 877–880.

McFarlane, R. A., Bennett, W. R. Jr., and Lamb, E. W. Jr. (1963). Single mode tuning dip in the power output of He–Ne optical maser. *Applied Physics Letters*, **2**, 189–190.

McIlroy, A.,and Nesbitt, D. J. (1990). Vibrational mode mixing in terminal acetylenes: high-resolution infrared laser study of isolated J states. *Journal of Chemical Physics*, **92**, 2229–2243.

Maiman, T. (1960). Stimulated optical radiation in ruby. *Nature*, **187**, 493–494.

Malinovsky, A. L., Makarov, A. A., and Ryabov, E. A. (2004). Real-time observation of the dynamics of vibrational-energy redistribution within an isolated polyatomic molecule by spontaneous Raman spectroscopy. *Journal of Experimental and Theoretical Physics Letters*, **80**, 532–534.

Manz, J. (1997). Molecular wave packet dynamics: theory for experiments 1926–1996. In *Femtochemistry and femtobiology: ultrafast reaction dynamics at atomic-scale resolution* (ed. V. Sündström), pp. 80–318. Imperial College Press, London.

Martin, P. J., Oldaker, B. G., Miklich, A. H., and Pritchard, D. E. (1988). Bragg scattering of atoms from a standing light wave. *Physical Review Letters*, **60**, 515–518.

Masnou-Seeuws, F., and Pillet, P. (2001). Formation of ultracold molecules ($T \gtrsim 200\,\mu K$) via photoassociation in a gas of laser-cooled atoms. *Advances in Atomic, Molecular and Optical Physics*, **47**, 54–127.

Meschede, D., and Metcalf, H. (2003). Atomic nanofabrication: atomic deposition and lithography by laser and magnetic forces. *Journal of Physics D*, **36**, R17–R38.

Metcalf, H. J., and van der Straten, P. (1999). *Laser cooling and trapping*. Springer, New York.

Mewes, M.-O., Andrews, M. P., Kurn, D. M., Durfee, D. S., Townsend, C. G., and Ketterle, W. (1997). Output coupler for Bose–Einstein condensed atoms. *Physical Review Letters*, **78**, 582–585.

Meystre, P. (2001). *Atom optics*. Springer, Berlin.

Meystre, P., and Sargent III, M. (1990). *Elements of quantum optics*. Springer, Berlin.

Miesner, H.-J., Stamper-Kurn, D. M., Andrews, M. R., Durfee, D. S., Innouye, S., and Ketterle, W. (1998). Bosonic stimulation in the formation of a Bose–Einstein condensate. *Science*, **279**, 1005–1007.

Migdal, A. L., Prodan, J. V., Phillips, W. D., Bergeman, T. N., and Metcalf, H. J. (1985). First observation of magnetically trapped neutral atoms. *Physical Review Letters*, **54**, 2596–2599.

Minogin, V. G. (1980). Deceleration and monochromatization of atomic beams by laser radiation pressure. *Optics Communications*, **34**(2), 265-268.

Minogin, V. G., and Letokhov, V. S. (1987). *Laser light pressure on atoms*. Gordon and Breach, New York.

Minogin, V. G., Fedorov, M. V., and Letokhov, V. S. (1997). Formation of ultrashort electron pulses on scattering of an electron beam by a standing laser wave of ultrashort duration. *Optics Communications*, **140**, 250-254.

Mishin, V. I., Sekatskii, S. K., Fedoseyev, V. N., Buyanov, N. B., Letokhov, V. S., Barzakh, V. S., Denisov, V. I., Dernyatin, A. G., Ivanov, V. S., Chubukov, I. Ya., and Alkhazov, G. D. (1987). Resonance photoionization spectroscopy and laser separation of ^{141}Sm and ^{164}Tm nuclear isomers. *Optics Communications*, **61**, 383–386.

Mishin, V. I., Fedoseyev, V. N., Kluge, H.-J., Letokhov, V. S., Ravn, H. L., Scheerer, F., Shirakabe, Y., Sundell, S., Tegnblad, O., and the ISOLDE Collaboration chemically selective laser ion-source for the CERN-ISOLDE on-line mass separator facility. (1993). *Nuclear Instruments and Methods in Physical Research B*, **73**, 550–560.

Mollow, B. R. (1969). Stimulated emission and absorption near resonance for driven systems. *Physical Review A*, **5**, 2217–2222.

Monroe, C., Swann, W., Robinson, H., and Wieman, C. E. (1990). Very cold trapped atoms in a vapor cell. *Physical Review Letters*, **65**, 1571–1574.

Mourou, G. A., Barty, P. J., and Perry, M. D. (1998). Ultrahigh-intensity lasers: physics of the extreme on a tabletop. *Physics Today*, January, 22–28.

Nagourney, W., Sandberg, J., and Dehmelt, H. (1986). Shelved optical electron amplifier: observation of quantum jumps. *Physical Review Letters*, **56**(26), 2797–2799.

Nairz, O., Brezger, B., Arndt, M., and Zeilinger, A. (2001). Diffraction of complex molecules by structure made of light. *Physical Review Letters*, **87**, 160401.

Nesbitt, D. J., and Field, R. W. (1996). Vibrational energy flow in highly excited molecules: role of intramolecular vibrational redistribution. *Journal of Physical Chemistry*, **100**, 12735–12756.

Neuhauser, W., Hohenstatt, M., Toschek, P., and Dehmelt, H. (1978). Optical sideband cooling of visible atom cloud confined in particle well. *Physical Review Letters*, **41**, 233–236.

Nikolov, A. N., Eyler, E. E., Wang, X., Wang, H., Li, J., Stwalley, W. C., and Gould, P. L. (1999). Observation of ultracold ground-state potassium molecules. *Physical Review Letters*, **82**, 703–706.

Nogues, G., Rauschenbeutel, A., Osnaghi, S., Brune, M., Raimond, J.M., and Haroche, S. (1999). Seeing a single photon without destroying it. *Nature*, **400**, 239–242.

Notkin, G. E., Rautian, S. G., and Feoktistov, A. A. (1967). On the theory of spontaneous emission of atoms in external electromagnetic field. *Journal of Experimental and Theoretical Physics*, **52**(6), 1673–1687 (in Russian).

O'Hara, K. M., Granado, S. R., Gehm, M. E., Savard, T. A., Bali, I., Ficed, C., and Thomas, J. E. (1999). Ultrastable CO_2 laser trapping of lithium fermions. *Physical Review Letters*, **82**, 4204–4207.

O'Neil, A. T., and Padgett, M. J. (2000). Three-dimensional optical confinement of micron-sized metal particles and the decoupling of the spin and orbital angular momentum within an optical spanner. *Optics Communications*, **185**, 139–143.

Ohtsu, M. (ed.) (1998). *Near field nano/atom optics and technology*. Springer, Berlin.

Ol'shanii, M. A., Letokhov, V. S., and Minogin, V. G. (1992). Role of interaction time in light pressure force on atoms. *Nonlinear Optics*, **3**, 283–294.

Ol'shanii, M. A., Ovchinnikov, Yu. B., and Letokhov, V. S. (1993). Laser guiding of atoms in hollow optical fiber. *Optics Communications*, **98**, 77–79.

Ovchinnikov, Yu. B., Soding, J. and Grimm, R. (1995). Cooling atoms in dark gravitational laser trap. *Journal of Experimental and Theoretical Physics Letters*, **61**, 21–23.

Ovchinnikov, Yu. B., Manek, I., and Grimm, R. (1997). Surface trap for Cs atoms based on evanescent-wave cooling. *Physical Review Letters*, **79**, 2225–2228.

Paisner, J. A. (1988). Atomic vapor laser isotope separation. *Applied Physics B*, **46**, 253–260.

Park, S. M., Lu, S.-P., and Gordon, R. J. (1991). Coherent laser control of the resonance-enhanced multiphoton ionization of HCl. *Journal of Chemical Physics*, **94**, 822–824.

Patorski, K. (1989). The self-imaging phenomenon and its applications. In *Progress in optics* (ed. E. Wolf), vol. 27, pp. 1-108. Elsevier Science, Amsterdam.

Paul, W. (1990). Electromagnetic traps for charged and neutral particles. *Reviews of Modern Physics*, **62**, 531–540.

Peters, A., Chung, K. Y., and Chu, S. (2001). High precision gravity measurements using atom interferometry. *Metrology*, **38**, 25–61.

Petrich, W., Anderson, M. H., Ensher, J. R., and Cornell, E. A. (1995). Stably, tightly confined magnetic trap for evaporation cooling of neutral atoms. *Physical Review Letters*, **74**, 3352–3355.

Petrov, A. K., Chesnokov, E. N., Gorelik, S. R., Straub, K. D., Szarmes, E. B., and Madey, J. M. J. (1997). Multiphoton isotope-selective dissociation of formic acid molecules under action of a free electron laser. *Journal of Physical Chemistry*, **101**, 7200–7207.

Phillips, W., and Metcalf, H. (1982). Laser deceleration of an atomic beam. *Physical Review Letters*, **48**, 596–599.

Popova, T. Ya., Popov, A. K., Rautian, S. G. and Sokolovskii, R. I. (1969). Nonlinear interference effects in spectra of emission, absorption and oscillation. *Journal of Experimental and Theoretical Physics*, **57**, (9), 850–863 (in Russian).

Potter, E., Herek, J., Pederson, S., Liu, Q., and Zewail, A. (1992). Femtosecond laser control of a chemical reaction. *Nature*, **355**, 66–68.

Pritchard, D. (1983). Cooling neutral atoms in a magnetic trap for precision spectroscopy. *Physical Review Letters*, **51**, 1336–1339.

Prodan, J., Phillips, W., and Metcalf, H. (1982). Laser production of a very slow monoenergetic atomic beam. *Physical Review Letters*, **49**, 1149–1152.

Purcell, E. M. (1946). Spontaneous emission probabilities at radiofrequencies. *Physical Review*, **69**, 681.

Puretzky, A. A., and Tyakht, V. V. (1989). Inverse electronic relaxation under IR multi-photon excitation of molecules. In *Laser spectroscopy of highly vibrationally excited molecules* (ed. V. S. Letokhov), pp. 329–378. Adam Hilger, Bristol.

Quack, M. (1978). Theory of unimolecular reactions induced by monochromatic infrared radiation. *Journal of Chemical Physics*, **69**, 1282–1307.

Quack, M. (1998). Multiphoton excitation. In *Encyclopedia of computational chemistry* (ed. P. V. Schleyer, N. L. Allinger, T. Clark, J. Gasteiger, P. A. Kollman, H. P. Schaefer III, and P. R. Schreiner), vol. 3, pp. 1775–1791. Wiley, Chichester.

Raab, E. L., Prentiss, M., Cable, A., Chu, S., and Prichard, D. L. (1987). Trapping of neutral sodium atoms with radiation pressure. *Physical Review Letters*, **23**, 2631–2634.

Rabi, I. I. (1937). Space quantization in a gyrating magnetic field. *Physical Review*, **51**, 652–654.

Rabitz, H., de Vivie-Riedle, R., Motzkus, M., and Kompa, K. (2000). Whither the future of controlling quantum phenomena? *Science*, **288**, 824–828.

Raether, H. (1988). *Surface plasmons*. Springer, Berlin.

Ramsey, N. F. (1950). A molecular beam resonance method with separated oscillating fields. *Physical Review*, **78**, 699–703.

Ramsey, N. F. (1956). *Molecular beam*. Clarendon, Oxford.

Ramsey, N. (1987). Experiments with separated oscillatory fields and hydrogen masers. *Reviews of Modern Physics*, **62**, 541–552.

Rasel, E. M., Oberthaler, M. K., Batelaan, H., Schmiedmayer, J., and Zeilinger, A. (1995). Atom wave interferometry with diffraction gratings of light. *Physical Review Letters*, **75**, 2633–2637.

Rautian, S. G., and Shalagin, A. M. (1970). Saturation effects for long-lived systems in spatially-limited fields. *Journal of Experimental and Theoretical Physics*, **58**(3), 962–974 (in Russian).

Ready, J. P. (1971). *Effects of high-power laser radiation*. Academic Press, New York.

Regal, C. A., Ticknor, C., Bohn, J. L., and Jin, D. S. (2003). Creation of ultracold molecules from a Fermi gas of atoms. *Nature*, **424**, 47–50.

Reilly, J. P., and Kompa, K. L. (1980). Laser induced multiphoton ionization mass spectrum of benzene. *Journal of Chemical Physics*, **73**, 5468–5476.

Rein, D. W. (1974). Some remarks on parity violation effects of intramolecular interactions. *Journal of Molecular Evolution*, **4**, 15–22.

Reisler, H., and Wittig, K. (1985). Multiphoton ionization of gaseous molecules. In *Photodissociation and photoionization* (ed. K. P. Lawley). Wiley, New York.

Renn, M. J., Montgomery, D., Vdovin, O., Anderson, D. Z., Wieman, C. E., and Cornell, E. A. (1995). Laser-guided atoms in hollow-core optical fibers. *Physical Review Letters*, **75**, 3253–3556.

Renn, M. J., Donley, E. A., Cornell, E. A., Wieman, C. E., and Anderson, D. Z. (1996). Evanescent-wave guiding of atoms in hollow optical fibers. *Physical Review A*, **53**, R648–R651.

Rentzepis, P. M. (1978). Picosecond chemical and biological events. *Science*, **202**, 174–182.

Rice, S. A., and Zhao, M. (2000). *Optimal control of molecular dynamics*. Wiley, New York.

Robert, A., Sirjean, O., Browaeys, A., Poupard, J., Nowak, S., Boiron, D., Westbrook, C. I., and Aspect, A. (2001). A Bose–Einstein condensate of metastable atoms. *Science*, **292**, 461–464.

Robieux, J. (2000). *High power laser interactions*. Intercept, London.

Robinson, P. I., and Holbrook, K. A. (1972). *Unimolecular reactions*. Wiley, London.

Rockwood, S., Reilly, J. P., Hohla, K., and Kompa, K. L. (1979). UV-laser induced molecular multiphoton ionization and fragmentation. *Optics Communications*, **28**, 175–178.

Rosker, M., Dantus, M., and Zewail, A. H. (1988). Femtosecond clocking of the chemical bond. *Science*, **241**, 1200–1202.

Rowe, M. A., Freedman, S. J., Fujikawa, B. K., Gwinner, G., Shang, S.-Q., and Vetter, P. A. (1999). Ground-state hyperfine measurement in laser-trapped radioactive ^{21}Na. *Physical Review A*, **59**, 1869–1873.

Ruska, E. (1980). *The early development of electron lenses and electron microscopy*. Hirzel, Stuttgart.

Salomon, C., Dalibard, J., Aspect, A., Metcalf, H., and Cohen-Tannoudji, C. (1987). Channeling atoms in a laser standing wave. *Physical Review Letters*, **59**, 1659–1662.

Sargent, M. III, Scully, M. O., Lamb, W. E. Jr. (1974). *Laser physics*. Addison-Wesley, Reading, MA.

Savage, C. M., Marksteiner, S., and Zoller, P. (1993). Atomic waveguides and cavities from hollow optical fibres. In *Fundamental of quantum optics* (ed. F. Ehlotzky), vol. 3. pp. 60–74. Springer, Berlin.

Schawlow, A. L., and Townes, C. H. (1958). Infrared and optical masers. *Physical Review*, **112**, 1940–1949.

Schlag, E. W., and Neusser, H. F. (1983). Multiphoton mass spectroscopy. *Accounts of Chemical Research*, **16**, 355–360.

Schlossberg, H. R., and Javan, A. (1966). Saturation behavior of a Doppler-broadened transition involving levels with closely spaced structure. *Physical Review*, **150**, 267–284.

Schmiedmayer, J., Chapman, M. S., Ekstrom, C. R., Hammond, S. T., Wehinger, S. T., and Pritchard, D. E. (1995). Index refraction of various gases for sodium matter waves. *Physical Review Letters*, **74**, 1043–1047.

Schreck, F., Khaykovich, K. L., Corwin, K. L., Ferrari, G., Bourdel, T., Cubizolles, J., and Salomon, C. (2001). Quasipure Bose–Einstein condensate immersed in a Fermi sea. *Physical Review Letters*, **87**, 080403.

Schultz, P. A., Sudbo, S. A., Krajnovich, D. L., Kwok, H. S., Shen, Y. R., and Lee, Y. T. (1979). Multiphoton dissociation of polyatomic molecules. *Annual Review of Physical Chemistry*, **30**, 379–409.

Schwefel, H.-P. (1995). *Evolution and optimal seeking*. Wiley, New York.

Seifert, W., Adams, C. S., Balykin, V. I., Heine, C., Ovchinnikov, Yu., and Mlynek, J. (1994). Reflection of metastable argon atoms from an evanescent wave. *Physical Review A*, **49**, 3814–3823.

Sewtz, M., Backe, H., Dretzke, A., Kube, G., Lauth, W., Schwamb, P., Eberhardt, K., Grüning, C., Thörle, P., Trautmann, N., Kunz, P., Lassen, J., Passler, G., Dong, C. Z., Fritzsche, S., and Haire, R. G. (2003). First observation of atomic levels for the element fermium ($Z=100$). *Physical Review Letters*, **90**, 163002.

Shank, C. V., and Ippen, E. P. (1974). Subpicosecond kilowatt pulses from mode-locked CW dye laser. *Applied Physics Letters*, **24**, 373–375.

Shapiro, M., and Brumer, P. (2001). On the origin of pulse shaping control of molecular dynamics. *Journal of Physical Chemistry A*, **105**, 2897–2902.

Shapiro, M., and Brumer, P. (2003). Coherent control of molecular dynamics. *Reports on Progress in Physics*, **66**, 859–942.

Shapiro, M., Frishman, E., and Brumer, P. (2000). Coherently controlled asymmetric synthesis with achiral light. *Physical Review Letters*, **84**, 1669–2000.

Shimizu, F. (2000). Atom holography. In *Advances in atomic, molecular and optical physics* (ed. B. Bederson and H. Walther), vol. 42, pp. 73–93. Academic Press, San Diego.

Shimizu, F., Shimizu, K., and Takuma, H. (1991). Four-beam laser trap of neutral atoms. *Optics Letters*, **16**, 339-341.

Shimizu, F., Shimizu, K., and Takuma, H. (1992). Double-slit interference with ultra-cold metastable neon atoms. *Physical Review A*, **46**, R17–R20.

Shnitman, A., Sofer, I., Golub, I., Yogev, A., Shapiro, M., Chen, Z., and Brumer, P. (1996). Experimental observation of laser control: electronic branching in the photodissociation of Na_2. *Physical Review Letters*, **76**, 2886–2889.

Shore, B. W. (1990). *The theory of coherent atomic excitation*. Vol. 1. *Simple atoms and fields*. Vol. 2. *Multilevel atoms and incoherence*. Wiley-Interscience, New York.

Silvera, I. F., and Walraven, J. T. M. (1980). Stabilization of atomic hydrogen at low temperature. *Physical Review Letters*, **44**, 164–168.

Snadden, M. J., McGuirk, J. M., Bouyer, P., Haritos, K. G., and Kasevich, M. A. (1998). Measurement of the Earth's gravity gradient with an atom interferometer-based gravity gradiometer. *Physical Review Letters*, **8**, 971–974.

Sobelman, I. I. (1979). *Atomic spectra and radiative transitions*. Springer, Berlin.

Söding, J., Grimm, R., and Ovchinnikov, Yu. B. (1995). Gravitational laser trap for atoms with evanescent-wave cooling . *Optics Communications*, **119**, 652–662.

Spence, D. E., Kean, P. N., and Sibbett, W. (1991). 60-fs pulse generation from a self-mode-locked Ti:sapphire laser. *Optics Letters*, **16**, 42–44.

Stebbings, R. F. (1976). High Rydberg atoms: newcomers to the atomic physics scene. *Science*, **193**, 537–542.

Stebbings, R. F., and Dunning, F. B. (ed.) (1983). *Rydberg states of atoms and molecules.* Cambridge University Press, Cambridge.

Strecker, K. E., Partridge, G. B., and Hulet, R. G. (2003). Conversion of an atomic gas to a long-lived molecular Bose gas. *Physical Review Letters*, **91**, 080406.

Stuchebrukhov, A. A., Kuzmin, M. V., Bagratashvili, V. N., and Letokhov, V. S. (1986). Threshold energy dependence of the intramolecular vibrational relaxation in polyatomic molecules. *Chemical Physics*, **107**, 429–443.

Stuke, M. (ed.) (1992). *Dye lasers: 25 years.* Springer, Berlin.

Stwalley, W. C. (1976). Stability of spin-aligned hydrogen at low temperatures and high magnetic fields: new field-dependent scattering resonances and predissociations. *Physical Review Letters*, **37**, 1628–1631.

Stwalley, W. C., and Nosanow, L. H. (1976). Possible "new" quantum systems. *Physical Review Letters*, **36**, 910–913.

Subbotin, M. V., Balykin, V. I., Laryushin, D. V., and Letokhov, V. S. (1997). Laser controlled atom waveguide as a source of ultracold atoms. *Optics Communications*, **139**, 107–116.

Sudbo, A., Schulz, P. A., Grant, E. R., Shen, Y. R., and Lee, Y. T. (1979). Simple bond rupture reactions in multiphoton dissociation of molecules. *Journal of Chemical Physics*, **70**, 912–929.

Sündström, W. (ed.) (1997). *Femtochemistry and femtobiology: ultrafast reaction dynamics at atomic-scale resolution*, Proceedings of Nobel Symposium. Imperial College Press, London.

Svoboda, K., and Block, S. M. (1994). Biological applications of optical forces. *Annual review of biophysical and biomolecular structure*, **23**, 247–285.

Szoke, A., and Javan, A. (1963). Isotope shift and saturation behaviour of the 1.15 μ transition of Ne. *Physical Review Letters*, **10**, 521–524.

Takamoto, M., Hong, F.-L., Higashi, R., and Katori, H. (2005). An optical lattice clock. *Nature*, **435**, 321–324.

Takasu, Y., Kenichi, M., Komori, K., Takano, T., Honda, K., Kumukura, M., Yabuzaki, T., and Takahashi, Y. (2003). Spin-singlet Bose–Einstein condensation of two-electron atoms. *Physical Review Letters*, **91**, 040404.

Takekoshi, T., Patterson, B. M., and Kurze, R. J. (1998). Observation of optically trapped cold cesium molecules. *Physical Review Letter*, **81**, 5105–5108.

Tanaka, K., Waki, H., Ido, Y., Akita, S., Yoshida, Y., and Yoshida, T. (1988). Protein and polymer analyses up to m/z 100,000 by laser ionization time-of-flight mass spectrometry. *Rapid Communications in Mass Spectrometry*, **2**(8), 151–153.

Tannor, D. J., and Rice, S. A. (1985). Control of selectivity of chemical reactions via control of wave packet evolution. *Journal of Chemical Physics*, **83**, 5013–5018.

Tannor, D., and Rice, S. A. (1988). Coherent pulse sequence control of product formation in chemical reactions. *Advances in Chemical Physics*, **70**, 441–523.

Tannor, D. J., Kosloff, T., and Rice, S. A. (1986). Coherent pulse sequence induced control of selectivity of reactions: exact quantum mechanical calculations. *Journal of Chemical Physics*, **85**, 5805–5820.

Teets, R., Feinberg, R., Hansch, T., and Schawlow, A. L. (1976). Simplification of spectra by polarization labeling. *Physical Review Letters*, **37**, 683–686.

Theis, M., Thalhammer, G., Winkler, K., Hellwig, M., Ruff, G., Grimm, R., and Denschlag, J. H. (2004). Tuning the scattering length with an optically induced Feshbach resonance. *Physical Review Letters*, **93**, 123001.

Thorsheim, H. R., Weiner, J., and Julienne, P. S. (1987). Laser-induced photoassociation of ultracold sodium atoms. *Physical Review Letters*, **58**, 2420–2423.

Tiesinga, E., Verhaar, B. J., and Stoof, H. T. C. (1993). Threshold and resonance phenomena in ultracold ground-state collisions. *Physical Review A*, **47**, 4114–4122.

Timp, G., Behringer, R. E., Tennant, D. M., Cunningham, J. E., Prentiss, M., and Berggren, K. K. (1992). Using light as a lens for submicron, neutral-atom lithography. *Physical Review Letters*, **69**, 1636–1639.

Tkachev, A. N., and Yakovlenko, S. I. (2003). On laser rare-isotope separation. *Quantum Electronics*, **33**, 581–592.

Turchette, Q. A., Hood, C. J., Lange, W., Mabuchi, H., and Kimble, J. H. (1995). Measurement of conditional phase shifts for quantum logic. *Physical Review Letters*, **75**, 4710–4713.

Udem, T., Holzwarth, R., and Hansch, T. W. (2002). Optical frequency metrology. *Nature*, **416**, 233–237.

Ungar, P. J., Weiss, D. S., Chu, S., and Riis, E. (1989) Optical molasses and multilevel atoms—theory. *Journal of the Optical Society of America B*, **6**, 2058–2071.

Varcoe, B. T. H., Brattke, S., Weidinger, M., and Walther, H. (2000). Preparation pure photon number states of the radiation field. *Nature*, **403**, 743–746.

Vasilenko, L. S., Chebotayev, V. P., and Shishayev, A. V. (1970). Two-photon absorption line shape for standing wave in gas. *Journal of Experimental and Theoretical Physics Letters*, **12**, 161–165.

Vogt, G., Krampert, G., Niklaus, P., Nuernberger, P., and Gerber, G. (2005). Optimal control of photoisomerization. *Physical Review Letters*, **94**, 068305.

von Helden, G., Holleman, I., van Roij, A. J. A., Knippels, G. M. H., van der Meer, A. F. G., and Meijer, G. (1998). Shedding new light on thermionic electron emission of fullerenes. *Physical Review Letters*, **81**, 1825–1828.

von Helden, G., van Heijnsbergen, D., and Meijer, G. (2003). Resonant ionization using IR light: a new tool to study the spectroscopy and dynamics of gas-phase molecules and clusters. *Journal of Physical Chemistry*, **107**, 1670–1688.

Vuletic, V., Chin, C., Kerman, A. I., and Chu, S. (1998). Degenerate Raman sideband cooling of trapped cesium atoms at very high atomic densities. *Physical Review Letters*, **81**, 5768–5771.

Walker, T. G., and Happer, W. (1997). Spin-exchange optical pumping of noble-gas nuclei. *Reviews of Modern Physics*, **69**, 3081–3093.

Wallis, H., and Ertmer, W. (1989). Broadband laser cooling on narrow transitions. *Journal of the Optical Society of America B*, **6**, 2211–2219.

Weinstock, H. (ed.) (1996). *SQUID sensors: fundamentals fabrication and applications*. Kluwer Academic, Dordrecht.

Weiss, D. S., Riis, E., Shevy, Y., Ungar, P. J., and Chu, S. (1989). Optical molasses and multilevel atoms: experiment. *Journal of the Optical Society of America B*, **6**, 2072–2083.

Westbrook, C. I., Watts, R. N., Tanner, C. E., Rolston, S. L., Phillips, W. D., Lett, P. D., and Gould, P. L. (1990). Localization of atoms in a three-dimensional standing wave. *Physical Review Letters*, **65**, 33–36.

Wieman, C., and Hansch, T. W. (1976). Doppler-free laser polarization spectroscopy. *Physical Review Letters*, **36**, 1170–1173.

Windhorn, L., Witte, T., Yeston, J., Proch, D., Motzkus, M., Kompa, K. L., and Fuss, W. (2002). Molecular dissociation by MID IR femtosecond laser pulses. *Chemical Physics Letters*, **357**, 85–90.

Windhorn, L., Yeston, J., Witte, T., Fuss, W., Motzkus, H., Proch, D., Kompa, K. L., and Moore, C. B. (2003). Getting ahead of IVR: a demonstration of mid-infrared induced molecular dissociation on sub-statistical time scale. (2003). *Journal of Chemical Physics*, **119**, 641–645.

Wineland, D., and Dehmelt, H. (1975). Proposed $10^{14}\Delta\nu < \nu$ laser fluorescence spectroscopy on Te$^+$ mono-ion oscillator. *Bulletin of the American Physical Society*, **20**, 637.

Wineland, D., and Itano, W. (1979). Laser cooling of atoms. *Physical Review A*, **20**, 1521–1540.

Witte, T., Hornung, T., Windhorn, L., Proch, D., de Vivie-Ridle, R., Motzkus, M., and Kompa, K. L. (2003). Controlling molecular ground-state dissociation by optimizing vibrational ladder climbing. *Journal of Chemical Physics*, **118**, 2021–2024.

Witte, T., Yeston, J. S., Motzkus, M., Heilueil, E. J., and Kompa, K. L. (2004). Femtosecond infrared coherent excitation of liquid phase vibrational distributions ($\nu > 5$). *Chemical Physics Letters*, **392**, 156–161.

Yablonovich, E. (1987). Inhibited spontaneous emission in solid-state physics and electronics. *Physical Review Letters*, **58**, 2059–2062.

Yang, K.-H., Stwalley, W. C., Heneghan, S. P., Bahns, J. T., Wang, K.-K., and Hess, T. R. (1986). Examination of effects of TEM$^*_{01}$ laser radiation in the trapping of neutral potassium atoms. *Physical Review A*, **34**, 2962–2967.

Ye, J., Vernooy, D.W., and Kimble, H. J. (1999). Trapping of single atoms in cavity QED. *Physical Review Letters*, **83**, 4987–4990.

Zandee, L., and Bernstein, R. B. (1979). Resonance-enhanced multiphoton ionization and fragmentation of molecular beam: NO, I$_2$, benzene and butadiene. *Journal of Chemical Physics*, **71**, 1359–1371.

Zel'dovich, Yu. B. (1973). Scattering and emission of quantum system in strong electromagnetic wave. *Uspekhi Fizicheskih Nauk*, **110**, 139-151 (in Russian).

Zel'dovich, B. Ya., Saakian, D. B., and Sobel'man, I. I. (1977). On the energy difference of left and right molecules due to parity violation at weak interaction of electrons with nucleus. *Journal of Experimental and Theoretical Physics Letters*, **25**, 95–98.

Zewail, A. H. (2000). Femtochemistry: atom-scale dynamics of the chemical bond. *Journal of Physical Chemistry A*, **104**, 5660–5694.

Zherikhin, A. N., Kompanets, O. N., Letokhov, V. S., Mishin, V. I., Fedoseyev, V. N., Alkhazov, G. D., Berlovich, E. E., Denisov, V. P., Dernyatin, A. G., and Ivanov, V. S. (1984). High-resolution laser photoionization spectroscopy of radioactive europium isotopes. *Journal of Experimental and Theoretical Physics*, **86**, 1249–1262.

Zhu, L., Kleinman, V., Li, X., Lu, S., Trentelman, K., and Gordon, R. J. (1995). Coherent laser control of the product distribution obtained in the photoexcitation of HI. *Science*, **270**, 77–80.

Zwierlein, M. W., Stan, C. A., Schunck, C. H. Raupach, S. M. F., Gupta, S., Hadzibabic, Z., and Ketterle, W. (2003). Observation of Bose–Einstein condensation of molecules. *Physical Review Letters*, **91**, 250401.

Index

Aberration
  chromatic 124, 125
  spherical 124, 126, 248
absorption multiple-photon 202
  saturation fluence 31
adiabatic approximation 217
  passage 162
Airy function 106
angular momentum 54, 55, 242
  orbital 54, 166, 241
  spin 54
  total 54
anharmonic interaction 213
  shift 209
anharmonicity 206, 221, 227
  intermode 204
  intramode 204, 205
anti-Stokes scattering 33
atom
  americium 172
  antihydrogen 110
  Cf 171
  chromium 123
  Cs 65
  Cu 180
  deuterium 219
  einsteinium 172
  fermium 172
  francium 172, 173
  Gd 163, 164, 167, 168
  He 90, 176
  Hf 62
  hydrogen 146, 219
  iridium 169
  kryptron 110
  lithium 145, 150, 156
  mercury 60
  neon 135
  osmium 219
  potassium 109, 151, 156
  Ra 173
  rhodium 170
  rubidium 65, 144, 154, 158
  ruthenium 169
  Sm 179
  sodium 20
  Sr 81, 90
  Tm 179, 180
  uranium 172, 175, 177, 180, 219
  Xe 66, 67
  Yb 146, 177
atom confinement 112
  hornfiber 100
  interferometry 130, 131, 134
  laser 109, 147
  lens 124
  localization 111
  mirror 120
  nanooptics 135
  optics 113, 114, 123
  traps 139
  waveguides 98, 100
atomic channeling 97
  clocks 64
  cooling 64, 81, 84, 87, 91
  focusing 121, 123, 125
  fountain 110
  holography 135
  mirror 114, 119, 120
  reflection 114
  wave packet 141
  wave-function splitter 132

attractive potential 99
Autler-Townes splitting 59
autoionization 160, 167, 189
   resonance 163, 164
   state 160, 163, 164, 167
AVLIS (see laser isotope separation) 175, 176, 219

Bennett hole 41
Bessel function 99
bond-selective chemistry 200
Born–Oppenheimer approximation 217
BEC (see Bose–Einstein condensation)
Bose–Einstein condensation 8, 138, 141–143, 146, 154, 156
   distribution 141
Bragg deflection 133
   interferometer 133
   scattering 133
broadening
   homogeneous 36, 38
   inhomogeneous 36
   natural 38
   power 31, 38
   radiative 38
   transit-time 38, 39, 50
   wall-collisional 38
Brownian motion 90, 240
Bunsen burner 206

cavity QED 111
channeling 96, 97
charge radius 177
chemoionization 182, 189
coherence 161
   lifetime 225
   time 224
coherent
   control 225, 226, 229–235, 237
   optimal control 225
   oscillations 25
   population trapping 47, 63, 64, 89
   quantum control 222
   reflection 119
   scattering 34
   states 229

collimation 80, 81
collisionless excitation 206
Compton scattering 247
conservative optical processes 74
control
   coherent 225, 226, 229–235, 237
   incoherent 224, 225
   noncoherent 8, 224, 225
   optimal 225, 230, 232
cooling
   longitudinal 78
   reflective 100
   temperature 88
Cooper pairing 150, 157
coupled transition 46
cross-section 17, 28, 30, 202, 215, 223
   excitation 235
Cs fountain 65

dark optical lattice 98
   resonance 64
de Broglie wavelength 113, 125, 129, 130, 139, 141, 248
deceleration length 79
decoherence time 235
deflection of atoms 97
degeneracy 18, 59, 161
degrees of freedom 188
density matrix 26
   of modes 5
   of states 217
   desorption 182, 190
diffraction 246, 247
   Fresnel 130
   of atoms 127, 128
diffusion coefficient 171
diffusional confinement 81
dipole
   gradient force 75
   far-off-resonance trap 94
   force 126
   interaction 23, 72, 73
   magnetic force 101
   matrix element 23
   moment 15, 27, 72
dispersion 47

dissociation 198, 206, 231
  energy 205
  limit 217
  multiple-photon 201, 208
dissociative ionization 233
Doppler 76
  broadening 21
  cooling 72, 82, 83, 104
  cooling limit 71, 72, 82
  regime 86
  effect 18, 35
  limit 83
  shift 19, 20, 35, 79
  first-order 111
  quadratic 21
  temperature 71, 104
  width 174, 175
Doppler-force spectroscopy 46, 70
double-resonance 58
  -slit interferometer 132
"doughnut" mode 94

Einstein coefficients 13, 16
electric field 23, 72
  ionization 165, 167
electromagnetic mirror 114
  separation 177
  trapping 110
electron 249
  beam 247, 248
  free 238
  microscopy 113, 249
  radius 244
  optics 113, 121, 238, 250
  spin 61
  vaporization 176
electronic predissociation 218
  relaxation 217
energy fluence 160, 203
  of interaction 24
  transfer 202
evanescent light wave 100, 107, 108, 114, 116, 249
  reflection 129
evaporation 196
excitation multistep 162

evaporative cooling 72, 91, 102, 143, 144
excited nuclei 62
extraction of ions 176

Fabry–Perot interferometer 131
Fano profile 163
femtosecond
  control 236
  IR pulses 222
  laser pulses 232
  lasers 224, 228
  pulse 221, 228
Fermi
  energy 140, 148
  quantum gases 156
  resonance 205, 214, 215
  degenerate gas 140, 148, 149
  degeneracy 140, 149
Feshbach resonance 153–155
field splitting 32
fine-structure constant 14, 245
first-order Doppler effect 111
focusing 114, 247, 248
Fokker–Planck equation 77, 82
forbidden transitions 83, 90
fragmentation 186, 206, 216, 228, 230, 231
Franck–Condon principle 58, 151
free electron 238
  laser 206
frequency synthesizer 111
Fresnel diffraction 130
friction coefficient 86
  force 75, 82

Gaussian beam 74, 93, 122, 123, 247
  curve 35
  distribution 227
gradient force 73, 116, 121, 122, 238, 239
gravito-magnetic trapping 92
  -optical trapping 106
grazing incidence 117, 129
guiding 114

Hamiltonian 22, 210, 226, 227, 233, 234, 244
high-frequency vibrations 219
hollow optical fiber 107
homogeneous broadening 36, 38
   linewidth 47
   spectral width 29, 41
hot cavity 172, 178
hyperfine interaction 57
   structure 57, 64, 83, 118, 180

incidence
   grazing 117, 129
   normal 129
incoherent
   control 224
   interaction 33, 162
   scattering 34
induced transparency 47
intramolecular vibrational redistribution 213
interaction
   coherent 28
   incoherent 29
interference
   of atoms 114
   constructive 229
   destructive 229
interferometer
   atomic 134
   Bragg 133
   Fabry–Perot 131
   Mach–Zehnder 131, 134
   Ramsey–Borde 133
   Talbot 130
   Talbot–Lau 133
intermodal anharmonicity 205
   resonances 205
intermolecular selectivity 198, 202, 218, 224
intramolecular vibrational redistribution (IVR) 202, 207, 212, 221, 226
   relaxation time 235
inverted Lamb dip 43, 44
Ioffe trap 101, 145

ionization 231
   dissociative 253
   electric field 165, 167
   isomer selective 160
   multiphoton 182, 187
   photoselective 158,
   potential 178
   thermal 178
   yield 178
IR Multi-Photon Excitation/ Dissociation (IR MPE/D) 201, 206, 208, 218, 220, 221
isobars 159, 175
isomer 62, 159
   selective excitation 198
   selective ionization 160
   separation 180
isomeric nuclear states 179
   nuclei 180
isomerization 206, 236
isotope 159, 175
   cosmogeneous 174
   radioactive 168, 177, 178
   rare 172, 174
   separation 218–220
   selective dissociation 198, 199
   excitation 200
   multistep ionization 176
   photodissociation 200
   shift 174, 177, 219
isotopical scrambling 177
   selectivity 174, 218
IVR effect 212, 213
   process 222
   rate 212, 215

Kapitza–Dirac effect 10, 244, 246, 247
Kirchoff's diffraction theory 126
Kramers–Kronig relation 47

Laguerre–Gaussian modes 242
Lamb–Dicke factor 97
   limit 69
   regime 96, 111
Lamb dip 42
Larmor precession 66

laser
  CO_2 210, 220
  chemistry 202, 203
  control 7, 198, 238, 244
  cooling 68, 110, 141
  Cu vapor 174
  femtosecond 224, 232
  focusing 120
  free electron 206
  induced photoassociation 151
  isotope separation 198, 218, 219
  mode-selective chemistry 202
  trapping 68, 92, 109, 238
  tweezers 240–244
  velocity-selective control 7
  ultracold-matter 4
learning algorithm 226, 230
level splitting 32
lifetime of the atom 12
light amplification 2
  pressure force 115
  shifts 60, 84, 93
LIS (laser isotope separation) 219, 221
linewidth 20
localization of atoms 97
longitudinal cooling 78
  relaxation 26, 59
Lorentzian function 15
  line 25
  profile 47

magnetic dipole 101
  moment 101, 177
  resonance 57, 66
  trap 101
  trapping 92, 100
magnetometer 64
magnetooptical trapping (MOT) 92, 103–105, 109–110
Majorana transitions 102, 103
MALDI (see matrix-assisted laser desorption/ionization) 195–197
mass separation 178–180, 183
  spectrometer 181, 184, 186, 230
  spectrum 187

matrix element 13
  -assisted laser desorption/ionization (MALDI) 195–197
matter wave 113
  interferometry 131, 132
  optics 114
Maxwell distribution 35
mean velocity 39
metastable state 90
microparticle 238
  sphere 238
Mie-sized particle 240
  enantiomers 52, 53, 235
mirror, atomic 106, 114, 117, 119
  electromagnetic 114
Molecular Laser Isotope Separaton (MLIS) 219
molecule, acetone 233, 234
  adenine 190, 191
  anthracene 189–191, 194, 217
  BCl_3 218
  benzaldehyde 185
  benzene 187
  CF_3Br 212–214
  $(CF_3)_3CI$ 216, 218
  CF_3H 219
  CF_3I 199, 200, 210
  CF_2HCl 219
  CH_2N_2 222
  $Cr(CO)_6$ 222
  diazomethane 222
  DNA 242, 243
  $Fe(CO)_5$ 221, 231
  freon-22 220
  fullerene 189
  H_2CO 182, 183
  $H{\equiv}C{-}CH_2CH_3$ 212
  left/right 235
  naphthalene 193, 194
  NH_3 200
  OsO_4 210, 217
  proteins 197
  RNA polymerases 243
  SF_6 201, 208, 209, 211
  tetrafluorethylene 220
  tripeptide 192

molecule, acetone (cont.)
   tryptophan 192
   $W(CO)_6$ 223
   ultracold 150, 152, 153
   ultraslow 51
Mollow triplet 97, 99
momentum diffusion 95
   distribution 90
monomolecular decay 188

nanofabrication 108, 123
nanooptics 135
nanoscale 125
   near-field 106, 108
neutron optics 113
NMR imaging 66, 67
nondemolition detection 112
normal mode 212
nuclear 181
   fission 171
   isomers 164, 175, 177, 178
   spin 54, 177
nuclei-acid base 190

occupation number 17, 34
octahedral splitting 27
off-diagonal matrix elements 27
   resonance dipole trap 94
optical dichroism 49
   force 238, 242, 243
   gradient 127
   lattice 70, 96, 98, 110–111
   magnetometry 65
   molasses 97
   nanofield 136
   orientation 54, 57, 61, 63
   potential 115
   pumping 56, 85, 86
   trap 242
   trapping 92, 108, 109, 239
   tweezers 240–243
   waveguide modes 98
optimal control 230
overexcitation 216

parity violation 51
Pauli blocking 149

Penning traps 181
phase relaxation 26–28
   space 87, 175
photoassociation 150, 152, 153
photodissociation 199, 201, 204, 208, 219, 229–231, 233
photoexcitation 203, 225
photofragmentation 182, 186
photoionization 159, 182, 183, 187, 189, 191, 229–231
   cavity 173
   detection 168
   separation 175
   spectra 179, 186
   yield 160, 179
photoisomerization 236
photon, angular momentum 241
   hole 136
   momentum 18, 88, 239
photonic crystals 6
photoselective ionization 158
Planck's distribution 16
plasmon 107, 119, 120
Poisson distribution 215
polarizability 165, 238, 239, 245
polarization cooling 72, 86
   saturation spectroscopy 49
   vector 23
pondermotive force 245, 247
   potential 247
population relaxation 26, 28, 30
potential energy 121, 245
   diagram 228
   curves 228
   surface 222, 226, 230, 232
   well 93, 99
power broadening 31, 38
principal quantum number 165, 166
probability amplitude 23–25, 32
pump-probe delay 228
Purcell effect 6

quadrupole magnetic trap 101, 102
   moment 177
   trap 101

quantum detect 174
  fluctuations 76
  interference 226, 229, 230, 233
  jumps 13, 14
  state-selective reflection 117
quasi-continuum 232
  -electrostatic trap 94
Rabi frequency 23, 28, 85, 88, 99, 111, 162, 215
  oscillation 24, 162
  radiation force 72, 74, 75
  intensity 28
  pressure 73
  scattering 33
radiative width 15, 17
Raman adiabatic passage 163
  cooling 87, 88
  excitation 152
  photoassociation 152
  scattering 33, 95
Ramsey method 133
rate equations 29
Rayleigh length 248
  scattering 238, 239
recoil, cooling limit 77, 83, 87
  effect 18
  energy 19, 77
reflection of atoms 114, 119, 249
  index 48, 114, 245, 246
  inelastic 147
relaxation time 202, 224
  transverse 26, 59
resonance broadening 30
  excitation 22, 161
  Fano 163
  feedback 2, 3
  Feshbach 153–155
    ionization 158, 159, 178, 189
  multiple-photon excitation 202
  paramagnetic 57
  photoionization 184, 185
  transitions 174
  velocity 40
  enhanced multiphoton excitation 187
    ionization (REMPI) 182, 183, 185, 194, 195

two-photon ionization (RETPI) 182, 187
rotating-wave approximation 24, 27
rotational relaxation 209
rubidium atomic clock 65
Rydberg atoms 165, 174
  constant 165
  states 165–167, 172–173, 175

saturation 162
  intensity 30
  parameter 30, 40
  power 29
  resonance 44, 52
  spectroscopy 45, 47, 52, 69
scattering cross-section 33
  force 238
  length 142, 154
Schrödinger equation 22, 26, 89, 127, 226, 227, 232, 244
selection rules 55
  factor 219
separation reactor 220
sideband cooling 68
single trapped ion 112
Sisyphus cooling 85
slow light 49
  molecules 51
spatially separated fields 64
specular reflection 116, 120, 129
spin 54
  exchange 61, 66
  flips 102
  forbidden transitions 83
  polarization 54, 56
  relaxation 59
spontaneous decay 38
  emission 5, 12, 15, 16
  force 75
  transition 1, 17
standing wave 75, 127, 134
Stark effect 60, 165, 166
  level shift 85, 87, 95
stationary states 23
statistical weight 161

stepwise photoionization 182, 183, 191
stimulated emission 16
  transition 28
stochastization 205, 210, 213, 214
  energy 205
  limit 207, 209, 210, 213
Stokes scattering 33
sub-Doppler cooling 77
  -recoil cooling 77
superposition 229
  principle 23
  state 64
supersonic jet 194, 195
surface plasmon 119

Talbot effect 130
temperature limit 76
Thomson effect 244
three-level atom 63, 64, 87, 89
time-of-flight (TOF) mass-spectrometer 106, 189, 196
transverse relaxation 26, 29
trap CO_2 laser 96
  dipole far-off resonance 94, 96, 146, 155
  Ioffe 101, 102
  Ioffe–Pritchard 145
  gravito-magnetic 92
  gravitooptical 107
  magnetic 92, 100, 101
  magnetooptical (MOT) 104, 105
  magnetostate 102
  parabolic 144
  Penning 181
  quadrupole 102
  quasi-electrostatic 94
travelling wave 73
triplet state 57, 61
tunneling 232
two-level atom 73, 87
two-photon absorption 210, 211
  Doppler-free spectroscopy 50, 87
  Raman transition 87
  resonance 8, 209

spectroscopy 50, 70
  transitions 210, 211
two-step dissociation 201
  excitation 146
  photoionization 158, 184, 185, 186
tweezers, optical 240–243

ultracold molecules 150, 152, 153
ultraslow molecules 51
unimolecular decay 205, 207, 216

van der Vaals potential 115
  velocity distribution 40, 79
velocity-selective coherent population trapping (VSCPT) 87, 89
  excitation 38
vibrational anharmonicity 221
  degrees of freedom 205, 216
  dynamics 227
  energy 204, 205, 213, 216
  distribution 225
  relaxation 202, 214
  stochastization 213
  excitation 202–204, 214, 215, 222
  heating 206
  levels 208
  modes 213
  overexcitation 204, 216, 217
  quasicontinuum 204, 214
  rotational levels 205
  spectroscopy 210
  states 215
  wave packet 227
vibrational mediated photodissociation 199, 200

wave packet 226, 228
Wigner time-frequency representation 227

Zeeman levels 58
  shift 79, 105
  splitting 57, 58
  sublevels 58, 63, 83–85, 103
  triplet 89